Springer Optimization and Its Applications

Volume 218

AF181030

Aims and Scope
Optimization has continued to expand in all directions at an astonishing rate. New algorithmic and theoretical techniques are continually developing and the diffusion into other disciplines is proceeding at a rapid pace, with a spot light on machine learning, artificial intelligence, and quantum computing. Our knowledge of all aspects of the field has grown even more profound. At the same time, one of the most striking trends in optimization is the constantly increasing emphasis on the interdisciplinary nature of the field. Optimization has been a basic tool in areas not limited to applied mathematics, engineering, medicine, economics, computer science, operations research, and other sciences.

The series **Springer Optimization and Its Applications (SOIA)** aims to publish state-of-the-art expository works (monographs, contributed volumes, textbooks, handbooks) that focus on theory, methods, and applications of optimization. Topics covered include, but are not limited to, nonlinear optimization, combinatorial optimization, continuous optimization, stochastic optimization, Bayesian optimization, optimal control, discrete optimization, multi-objective optimization, and more. New to the series portfolio include Works at the intersection of optimization and machine learning, artificial intelligence, and quantum computing.

Volumes from this series are indexed by Web of Science, zbMATH, Mathematical Reviews, and SCOPUS.

Maude J. Blondin • Iztok Fister Jr. •
Panos M. Pardalos
Editors

Artificial Intelligence, Optimization, and Data Sciences in Sports

Springer

Editors
Maude J. Blondin
Département Génie Électr. & Inform.
Universite de Sherbrooke
Sherbrooke, QC, Canada

Panos M. Pardalos
Industrial & Systems Engineering
University of Florida
Gainesville, FL, USA

Iztok Fister Jr.
Faculty of Electrical Engineering and
Computer Science
University of Maribor
Maribor, Slovenia

ISSN 1931-6828 ISSN 1931-6836 (electronic)
Springer Optimization and Its Applications
ISBN 978-3-031-76049-5 ISBN 978-3-031-76047-1 (eBook)
https://doi.org/10.1007/978-3-031-76047-1

This Springer imprint is published by the registered company Springer Nature Switzerland AG
The registered company address is: Gewerbestrasse 11, 6330 Cham, Switzerland

Acknowledgments

We are grateful for everyone who participated in this book, especially for the authors whose expertise and dedication made it possible. We also give special thanks to the reviewers. Their reviews enhanced the quality of this book. Finally, we want to thank the Springer editorial team, especially Elizabeth Loew, Kaylin Fussell, and Saveetha Balasundaram, for helping and supporting the publication of this contributed book.

Contents

Contributors

Alexander F. Almeida McDonough School of Business, Georgetown University, Washington, DC, USA

Adriano Alves Esporte Clube, Fortaleza, Brazil

Duarte Araújo CIPER, School of Human Kinetics of the Lisbon University, Cruz Quebrada, Portugal

Arnold Baca University of Vienna, Vienna, Austria

Maude J. Blondin Université de Sherbrooke, Sherbrooke, QC, Canada

Rory Bunker Nagoya University, Nagoya, Aichi, Japan

Sergiy Butenko TAMU-3131, College Station, TX, USA

Daniel Capanema Federal Center for Technological Education of Minas Gerais, Computing Department, Belo Horizonte, Brazil

Daniel Carrilho CIPER, School of Human Kinetics of the Lisbon University, Cruz Quebrada, Portugal

Vei S. Chan Faculty of Computing, University of Technology Malaysia, Johor Bahru, Malaysia

João Claudino Group of Research, Innovation and Technology Applied to Sport (GSporTech), Department of Physical Education, Center for Health Sciences, Federal University of Piauí, Teresina, Piauí, Brazil

Stefania Corsaro Department of Management and Quantitative Studies, University of Naples "Parthenope", Napoli, Italy

Kevin Dayaratna Department of Mathematics and Statistics, Georgetown University, Washington, DC, USA

Giuseppina Dello Ioio Department of Economics and Legal Studies, University of Naples "Parthenope", Napoli, Italy

Vincenzo Di Sauro Department of Economics and Legal Studies, University of Naples "Parthenope", Napoli, Italy

Carlo Dindorf Department of Sports Science, University of Kaiserslautern-Landau (RPTU), Kaiserslautern, Germany

Frédéric Domingue Département de génie électrique et génie informatique, Université du Québec à Trois-Rivières, Trois-Rivières, QC, Canada

Jonas Dully Department of Sports Science, University of Kaiserslautern-Landau (RPTU), Kaiserslautern, Germany

Bernhard Dumphart Institute of Health Sciences, St. Pölten University of Applied Sciences, St. Pölten, Austria

Sofia Ferreira CIPER, School of Human Kinetics of the Lisbon University, Cruz Quebrada, Portugal

Iztok Fister Jr. University of Maribor, Faculty of Electrical Engineering and Computer Science, Maribor, Slovenia

Michael Fröhlich Department of Sports Science, University of Kaiserslautern-Landau (RPTU), Kaiserslautern, Germany

Keisuke Fujii Nagoya University, Nagoya, Aichi, Japan
RIKEN Center for Advanced Intelligence Project, Osaka, Japan
JST Presto, Saitama, Japan

Akemi Gálvez Department of Applied Mathematics and Computational Sciences, University of Cantabria, Santander, Spain

Jaime Gil-Lafuente Departament d'Empresa, Universitat de Barcelona, Barcelona, Spain

Brian Horsak Center for Digital Health and Social Innovation, St. Pölten University of Applied Sciences, St. Pölten, Austria

Fabian Horst Institute of Sport Science, Johannes Gutenberg-University Mainz, Mainz, Germany

Andres Iglesias Department of Applied Mathematics and Computational Sciences, University of Cantabria, Santander, Spain

Andrew Johnson TAMU-3131, College Station, TX, USA

Domenico Marino Mediterranea University of Reggio Calabria, Reggio Calabria, Italy

Zelda Marino Department of Management and Quantitative Studies, University of Naples "Parthenope", Napoli, Italy

Steven J. Miller Department of Mathematics and Statistics, Williams College, Williamstown, MA, USA

Geoffrey Millour BeScored Institute, Valbonne, France

Erick Moreno-Centeno TAMU-3131, College Station, TX, USA

François Nougarou Département de génie électrique et génie informatique, Université du Québec à Trois-Rivières, Trois-Rivières, QC, Canada

Panos M. Pardalos University of Florida, Gainesville, FL, USA

Adriano Pereira Federal University of Minas Gerais (UFMG), Computer Science Department (DCC), Belo Horizonte, Brazil

Sara Pérez-Carabaza Department of Applied Mathematics and Computational Sciences, University of Cantabria, Santander, Spain

Alen Rajšp University of Maribor, Faculty of Electrical Engineering and Computer Science, Maribor, Slovenia

Djordje Slijepčević Institute for Creative Media and Technology, St. Pölten University of Applied Sciences, St. Pölten, Austria

Andres Torres Département de génie électrique et génie informatique, Université du Québec à Trois-Rivières, Trois-Rivières, QC, Canada

Andrew K. Yang Emmanuel College, University of Cambridge, Cambridge, UK

Justin Yates TAMU-3131, College Station, TX, USA

Maria Alejandra Yepez Département de génie électrique et génie informatique, Université du Québec à Trois-Rivières, Trois-Rivières, QC, Canada

Calvin Yeung Nagoya University, Nagoya, Aichi, Japan

Matthias Zeppelzauer Institute for Creative Media and Technology, St. Pölten University of Applied Sciences, St. Pölten, Austria

Artificial Intelligence, Optimization, and Data Sciences in Sports: Editorial

Maude J. Blondin, Iztok Fister Jr., and Panos M. Pardalos

Abstract The introductory chapter of this contributed book covers the primary motivation, achievements, and challenges. It also gives an overview of the opportunities of the modern information technology paradigms that have the potential to shape modern sports from the perspectives of sports training, sports data analysis, digital twins in sports, prediction of sports results, and the other essential aspects of knowledge discovery in data arising in various sports. Besides the recent overview of this research area, the authors also summarize the articles in this book.

Keywords Optimization in sports · Sports data analysis · Digital twins in sports · Prediction of sports results

1 Introduction

Did you know that hosting the Olympics can cost up to US$ 51 billion? [11] Thus, why does spending this amount of money worth it? Parts of the answer is that the sports industry yields important economic benefits by creating jobs, generating tourism [10], and developing infrastructure. A survey over 10 years on mega-sport events concluded that hosting a mega-event has a positive impact on tourism [7]. For instance, the Centre de Droit et d'Economie du Sport (CDES), of the University of

M. J. Blondin (✉)
Université de Sherbrooke, Sherbrooke, QC, Canada
e-mail: maude.blondin2@usherbrooke.ca

I. Fister Jr.
University of Maribor, Faculty of Electrical Engineering and Computer Science, Maribor, Slovenia
e-mail: iztok@iztok-jr-fister.eu

P. M. Pardalos
University of Florida, Gainesville, FL, USA
e-mail: pardalos@ufl.edu

M. J. Blondin et al. (eds.), *Artificial Intelligence, Optimization, and Data Sciences in Sports*, Springer Optimization and Its Applications 218,
https://doi.org/10.1007/978-3-031-76047-1_1

Limoges, estimated the economic benefits of the Olympic and Paralympic games of Paris 2024 between 6 and 12.2 BN USD [12]. Victorious sports teams and successful athletes get more media and sports fans attention, attracting more sponsorships, tourism during sports events, and investments.

Not only sports can be benificial for countries, but also on an individual level. In fact, practicing a sport can be a means to meet innate fundamental needs as defined in [2], i.e., the need for (i) competence, which refers to the need of people to accomplish activities that allow them to demonstrate their abilities, develop new skills, and overcome challenges. Feeling competent allows individuals to experience a sense of satisfaction, fulfillment, and intrinsic motivation. (ii) autonomy, which can be explained as individuals behaving according to their values, preferences, and interests rather than acting because of feeling pressured or controlled. Making decisions and pursuing activities that align with individuals' authentic selves provide opportunities to experience greater satisfaction, engagement, and vitality. (iii) relatedness, which is linked to the state of feeling valued, connected, and emotionally supported [5]. The sense of relatedness is a key element for social bonding, emotional resilience, and personal growth. These three needs are crucial for psychological functioning, healthy development, and overall well-being. Thus, the partial or complete fulfillment of these three individual needs combined with the benefits for countries may explain, in part, why sports play an important part in the industry.

In recent years, competitive sports have taken another performance level with the entrance of new devices, technologies, and scientific discoveries [3, 8, 1, 4]. Nowadays, elite athletes, as well as amateur athletes, must consider as many aspects of their training life as they can, such as techniques in the execution of their sports movements, sleeping, nutrition, training plan, mindset, and biological metrics, to put themselves competitive enough to have a chance to win a race or a game [5]. A favorable environment for athletes is composed of coaches, nutritionists, psychologists, and physiotherapists, to name a few. This team devices a plan for athletes based on scientific research, facts, and input from athletes. Before, input from athletes was primarily subjective, meaning it was based on how athletes felt. Today, objective inputs, such as data collected through wearable technology, are crucial in athlete preparation. The analysis of those data falls into the domain of sports analytics. Sports analysis goes beyond collecting real-time data and physical metrics on athletes and analyzing them to improve athlete performance. Indeed, sports analytics also provides methodologies to predict game outcomes, improve game strategy and real-time decision-making, enhance fan experience, and help in recruiting. On the other hand, some researchers also operate with the term Smart Sports Training (SST). In [9], SST is defined as sports training that utilizes wearables, sensors, Internet of Things (IoT) devices, and intelligent data analysis methods and tools to improve training performance and reduce workload while maintaining the same or better training performance. According to the definition, the data comes from various sensors capturing the athlete's state. At the same time, numerous intelligent data analysis methods are intended to discover knowledge from the data and thus help the athletes or their trainers with the extracted knowledge.

This edited book focuses on the synergy of data science, artificial intelligence, machine learning (ML), and optimization in sports. This book contains 12 standalone chapters and encompasses many disciplines, such as data mining, ML, and optimization, to better understand several aspects of sports, such as training, nutrition, and injury prevention, to improve performance. All papers were reviewed, and at least two reviewers reviewed each chapter. In particular, the book surveys the advances in ML and optimization related to sports to assess where the field stands and provide future research directions. Also, case studies for different applications and aspects of sports, such as training, recovery, and movement analysis, are presented.

The authors of the chapter "Machine Learning for Soccer Match Result Prediction" discuss ML for soccer match result predictions. This research is exciting due to the difficult nature of soccer and low-scoring sport with three outcomes. The authors look at available datasets, the types of models and features, and ways of evaluating model performance in this vibrant application domain. The main aim of this chapter is to give an extensive review of the current state of this application area and show future challenges and developments.

Chapter "Machine learning for Prediction of the index of effectiveness in cycling" deals with ML methods for predicting the index of effectiveness in cycling. The study highlights that the index of effectiveness can be predicted from variables easily measured by motion capture systems and that the proposed method is suitable for integration into bike fitting systems.

The authors of the chapter "Machine Learning in Biomechanics: Key Applications and Limitations in Walking, Running, and Sport Movements" have several recent applications of ML in sports, i.e., pose estimation, event detection, data exploration and clustering, and automated classification in gait (walking and running) and sports biomechanics. The potentials of ML methods used in biomechanical workflows are discussed. At the same time, the authors also give special attention to important data science methods, such as data availability and explainability, that fully utilize ML approaches in sports biomechanics.

The authors of the chapter "Artificial Intelligence, Machine Learning, and Data Analytics for Sports: General Overview and NBA Case Study" provide a historical view of the uses, tools, and applications of artificial intelligence and ML-based data analytics in sports while presenting one case study involving the NBA basketball league.

The authors of the paper "An Ecological Dynamics Approach to the Use of Artificial Intelligence and Machine Learning to Analyze Performance in Football" discuss the influence of AI/ML applications on football, mainly from a performance analysis perspective. The authors also present how the ecological dynamics theoretical approach can guide performance and provide a case study.

The authors of the chapter "A Supervised Learning Approach for evaluating Football Performances" tackled the ML approach and other statistical methods to model and predict the number of goals scored in football matches. Using these models, authors aim to improve accuracy and facilitate informed decision-making for betting professionals and fans.

The authors in their chapter "Bridging Route based Cycling Training with Digital Twins" proposed integrating a solution for generating adaptive cycling training routes with an artificial sports trainer (AST) [6] based on the digital twin concept. The proposed method allows athletes to use generated training routes as input in an AST Monitor device based on Raspberry Pi, which can be used while cycling to record the training activity and monitor the executed training.

The author of the chapter "Perspectives of Artificial Intelligence in Training and Exercise" discusses AI methods' potential in supporting the training process and planning. Feedback systems supporting the improvement of movement sequences, the assessment, and monitoring of movement execution are explored in the chapter, and real-world applications are presented.

The chapter "A Fuzzy Model for Optimizing the Football Rule Assuring Spectacle, Fair Play, objectivity, and Ethics" tackles the controversies that exist in elite professional matches, with fans, journalists, administrators, and coaches often finding fault with interpretations of the rules or perceived inconsistencies in how the game is regulated. A comprehensive algorithmic approach involving experts and fuzzy logic is proposed along with the crucial challenges for future research.

The authors of the chapter "Physical Efficiency in Soccer: Relevance, Correlations, and Impacts using Al Methods" also tackle soccer and AI to provide a more comprehensive elucidation of the concept of physical efficiency based on the literature review. The authors also go deeper into evaluating physical efficiency within the context of a professional soccer team.

The chapter "A PageRank-Based Method for College Football Recruiting Rankings" presents the PageRank method for ranking the recruiting classes for teams in the Football Bowl Subdivision based on data. The results show that the PageRank method is highly correlated with expert opinion.

Finally, the authors of the chapter "Applications of improvements to the Pythagorean won-lost expectation in optimizing rosters" touch the Bill James' Pythagorean formula that effectively estimates a baseball team's winning percentage using average runs scored and allowed, aiding in assessing player contributions to team wins. In their study, the authors extend the formula by modeling runs with independent Weibull distributions with varying shape parameters, improving prediction accuracy over the past 30 MLB seasons despite requiring more complex calculations.

References

1. Nader Chmait and Hans Westerbeek. Artificial intelligence and machine learning in sport research: An introduction for non-data scientists. *Frontiers in sports and active living*, 3:363, 2021.
2. Edward L Deci and Richard M Ryan. The" what" and" why" of goal pursuits: Human needs and the self-determination of behavior. *Psychological inquiry*, 11(4):227–268, 2000.
3. Alireza Farrokhi, Reza Farahbakhsh, Javad Rezazadeh, and Roberto Minerva. Application of internet of things and artificial intelligence for smart fitness: A survey. *Computer Networks*, 189:107859, 2021.

4. Iztok Fister and Iztok Fister Jr. Computational intelligence and data mining in sports, 2021.
5. Iztok Fister, Iztok Fister Jr, and Dušan Fister. *Computational intelligence in sports*, volume 22. Springer, 2019.
6. Iztok Fister, Andres Iglesias, Akemi Galvez, Suash Deb, and Dušan Fister. On deploying the artificial sport trainer into practice. In *2021 8th International Conference on Soft Computing & Machine Intelligence (ISCMI)*, pages 21–26. IEEE, 2021.
7. Johan Fourie and María Santana-Gallego. The impact of mega-sport events on tourist arrivals. *Tourism management*, 32(6):1364–1370, 2011.
8. Devansh Patel, Dhwanil Shah, and Manan Shah. The intertwine of brain and body: a quantitative analysis on how big data influences the system of sports. *Annals of Data Science*, 7(1):1–16, 2020.
9. Alen Rajšp and Iztok Fister Jr. A systematic literature review of intelligent data analysis methods for smart sport training. *Applied Sciences*, 10(9):3013, 2020.
10. Samo Rauter and Mojca Doupona Topič. Runners as sport tourists: The experience and travel behaviours of ljubljana marathon participants. *Collegium antropologicum*, 38(3):909–915, 2014.
11. The Guardian. Sochi 2014: the costliest olympics yet but where has all the money gone? *The Guardian*, October 9 2013. Retrieved May 30, 2024.
12. The International Olympic Committee. Economic benefits of hosting the olympic games. https://olympics.com/ioc/economic-benefits-hosting-olympic-games, 2024. Retrieved May 31, 2024.

Machine Learning for Soccer Match Result Prediction

Rory Bunker, Calvin Yeung, and Keisuke Fujii

Abstract Machine learning has become a common approach to predicting the outcomes of soccer matches, and the body of literature in this domain has grown substantially in the past decade and a half. This chapter discusses available datasets, the types of models and features, and ways of evaluating model performance in this application domain. The aim of this chapter is to give a broad overview of the current state and potential future developments in machine learning for soccer match result prediction, as a resource for those interested in conducting future studies in the area. Our main findings are that while gradient-boosted tree models such as CatBoost, applied to soccer-specific ratings such as pi-ratings, are currently the best-performing models on datasets containing only goals as the match features, there needs to be a more thorough comparison of the performance of deep learning models and random forest on a range of datasets with different types of features. Furthermore, new rating systems using both player- and team-level information and incorporating additional information from, e.g., spatiotemporal tracking and event data, could be investigated further. Finally, the interpretability of match result prediction models needs to be enhanced for them to be more useful for team management.

Keywords Machine learning · Sports analytics · Soccer match prediction · Gradient-boosted trees · Interpretable models

R. Bunker (✉) · C. Yeung · K. Fujii
Graduate School of Informatics, Nagoya University, Furo-cho, Nagoya, Japan
e-mail: rory.bunker@g.sp.m.is.nagoya-u.ac.jp; yeung.chikwong@g.sp.m.is.nagoya-u.ac.jp; fujii@i.nagoya-u.ac.jp

M. J. Blondin et al. (eds.), *Artificial Intelligence, Optimization, and Data Sciences in Sports*, Springer Optimization and Its Applications 218,
https://doi.org/10.1007/978-3-031-76047-1_2

1 Introduction

Predicting the results of professional soccer[1] matches is a challenging problem due to draws being a common outcome in the sport, as well as its low-scoring nature and often highly competitive leagues. Nonetheless, given the global popularity of the sport, both in terms of spectatorship and player numbers, it is a topic that is of interest to many groups including fans, bookmakers and bettors, as well as coaches, players, and performance analysts. While bettors and bookmakers require models that are highly accurate, coaches/management and sports performance analysts also require models that are interpretable, so that the most relevant match features (performance indicators [63]) for winning can be identified and improved upon in future matches. Once a predicted result for a specific match is obtained, an additional problem is to decide whether to actually bet on the match. While this is an important question [38], it is not the focus of the current chapter.

A large number of papers have been published that are related to machine learning (ML) for soccer match result prediction, particularly over the past decade [18]. Traditionally, however, statistical models were used to forecast soccer match results. Stefani [105] used least-squares regression to calculate team ratings, which could be updated on a weekly basis, to predict match results based on the difference in ratings between teams. Some early papers fit distributions to the number of goals scored by each team in a match. Maher [81] used an independent Poisson distribution to obtain the attacking and defensive ratings of teams. Dixon and Coles [37] modified this model to handle incomplete data and data from different divisions and to allow for temporal variations in the performance of teams. Goal distributions have also been fit using, e.g., the dependent Poisson, negative binomial, and extreme value distributions [51, 9, 10, 91]. Indeed, statistical models such as the bivariate Poisson still provide strong performance (e.g., [78] who used the engsoccerdata package [34]).

Constantinou [28] categorized soccer result prediction models into three groups, suggesting that the first and third types of studies tend to be published in statistical journals, while the second group tends to be published in journals focused on computer science and artificial intelligence.

1. **Statistical models**: including modeling goals scored, e.g., using the bivariate, independent, and dependent Poisson, negative binomial, and generalized extreme value distributions [81, 37, 51, 9, 10, 91], as well as ordered logistic regression models [4, 100].
2. **Machine learning and probabilistic graphical models**: including fuzzy or genetic algorithms [112, 87, 54, 95] and Bayesian methods (e.g., the studies of Joseph, Fenton, and Neil [68] and Constantinou, Fenton, and Neil [31]).

[1] Also known as "association football" or simply "football".

3. **Rating systems**: including Elo ratings [41, 65], pi-ratings [30], and more recently, Berrar ratings [8] and generalized attacking performance (GAP) ratings [118]. PageRank ratings [15, 59] also fit under this category.[2]

Nonetheless, this three-group model taxonomy is not entirely satisfactory. In recent years, researchers have incorporated models, techniques, or features from more than one of the above three groups to create hybrid approaches. For instance, the top-performing participants in the 2017 Soccer Prediction Challenge [7] applied machine learning models [8, 59, 28] to soccer-specific ratings. Statistical and machine learning models have also been combined to form hybrid approaches [40, 74, 53], with [53], for example, proposing a hybrid approach combining a random forest model with Poisson-based rankings. In addition, some models, e.g., logistic regression, fit under both the umbrella of statistics and that of machine learning. On the other hand, Bradley–Terry models [11], which also proved successful during the 2017 Soccer Prediction Challenge [116] and for ranking soccer teams based on current strength [78], could be considered to fit under both the statistical models and rating system groups. This is because, as will be explored later in this chapter in Sect. 4.1.1, Bradley–Terry forms the theoretical foundation for Elo ratings [33].

By covering available datasets—including an in-depth discussion of the Open International Database and the 2017 Soccer Prediction Challenge—as well as current and potential future models and features, as well as evaluation methods, this chapter aims to provide a broad overview of machine learning for soccer match result prediction and will hopefully act as a resource for those interested in carrying out future research in the domain.

The remainder of this chapter is organized as follows. In the next section, available datasets are outlined, as are the Open International Soccer Database and 2017 Soccer Prediction Challenge's in-competition and post-competition approaches and results. Subsequently, commonly used candidate models including conventional machine learning models, ensemble methods, and deep learning models are considered, as are model objectives and interpretability. Then, current and potential future model features are considered, e.g., match features, player and team statistics, and ratings (both general and soccer-specific). Feature selection methods are also briefly discussed. Model evaluation considerations—including baselines, target variable definition, as well as scoring rules and their properties, along with temporal splitting methods—are then described. Finally, conclusions and potential avenues for future research are discussed.

---

[2] How these different ratings are calculated is described in Sect. 4.1.

2 Data

In Sect. 2.1, a non-exhaustive list of the datasets that are available for engineering features and building models is described. The dataset used in the 2017 Soccer Prediction Challenge [7], the Open International Soccer Database [39], will then be discussed, which—despite not containing match features derived from in-play events during games (apart from goals scored)—now acts as somewhat of a benchmark dataset. The top-performing in-competition and post-competition studies that have used the Open International Soccer Database will be discussed in Sect. 2.2.

2.1 *Available Datasets*

An increasing amount of publicly available data from soccer matches is becoming publicly available online. Organizations such as StatsBomb (Bath, United Kingdom) make some event log data publicly available, although the data is generally held by professional teams who pay vendors such as StatsBomb or Stats Perform (Chicago, IL, USA)/Opta (London, United Kingdom) for access. A non-exhaustive list of datasets that can be used for engineering features and building soccer match result prediction ML models is presented in Table 1.

One challenge with many of these data sources is that they often cover different leagues and/or different seasons and may be missing some types of features. The Open International Soccer Database, for instance, does not contain match features apart from goals scored, nor does it contain betting odds. It may be necessary for researchers to source data from different datasets and merge them. The European Soccer Database on Kaggle has already performed such merging for a variety of feature types for a subset of their entire dataset (10,000 of the 25,000 matches). Spatiotemporal data is often only available to professional teams themselves, although sometimes small amounts of it are made publicly available.

2.2 *2017 Soccer Prediction Challenge*

For the 2017 Open International Soccer Prediction Challenge, the Open International Soccer Database [39], an open-source database containing over 216,000 matches from 52 soccer leagues and 35 countries was made available to participants and remains publicly available. As mentioned, this dataset now acts somewhat as a benchmark dataset in this domain despite not containing match features other than goals scored. The omission was likely a deliberate design choice by the challenge organizers, who likely integrated only data that are readily available for most soccer leagues worldwide—including lower-division leagues—so as to

Table 1 Datasets that can potentially be used for ML for soccer match result prediction. The attributes for each dataset can be confirmed by navigating to the corresponding URL

Dataset name	Description	URL(s)
European Soccer Database	Contains data on 25,000 matches and 10,000 players in 11 European leagues from 2008 to 2016. Player and team attributes are from the EA Sports FIFA video game. The database also contains team lineups and formations (x,y coordinates), as well as betting odds from up to 10 bookmakers. For over 10,000 matches, the database also contains events, e.g., goal types, possession, corners, crosses, fouls, and cards.	kaggle.com/datasets/hugomathien /soccer
Betting websites	Betting sites provide betting odds and sometimes additional features (but sometimes only short-time periods are available for free).	football-data.co.uk, betfair-datascientists.github.io, and oddsportal.com
engsoccerdata	An R package [34] that provides English and other European league data, along with US MPL and South African league data.	github.com/jalapic/engsoccerdata
Wyscout event data	Pappalardo et al. [90] and Wyscout provided spatiotemporal event data containing 1,941 matches from the top 5 European Soccer leagues, EURO 2016 and the 2018 World Cup.	doi.org/10.6084/m9.figshare.c. 4415000.v5
Statsbomb open data	Primarily event log data (with lineup and match metadata).	github.com/statsbomb/open-data
Open International Soccer Database	Contains the match results (season, league, date, home team, away team(Sea Lge Date HT AT HS AS GD WDL) of 216,743 games played between 19/03/2000 and 21/03/2017 from 52 leagues in 35 countries. This dataset was used in the 2017 Soccer Prediction Challenge.	osf.io/kqcye/
soccerdata	A scraper collection implemented in Python that scrapes soccer data from various websites including Club Elo, ESPN, FBref, FiveThirtyEight, Football-Data.co.uk, SoFIFA, and WhoScored.	github.com/probberechts/ soccerdata
World Football Elo Ratings	Contains the current Elo ratings of national soccer teams.	eloratings.net
Football Database	Contains the current Elo ratings of club teams from various leagues around the world.	footballdatabase.com
understat.com	Contains expected goals (xG) (however, the way in which these are computed is relatively opaque).	understat.com
football-data.co.uk	Contains, for various leagues, betting odds from multiple providers as well as match statistics for some seasons.	football-data.co.uk
FIFA Index	Contains player and team ratings from the EA Sports FIFA video game	fifaindex.com

maximize the size of the dataset. The papers related to the top-performing models from the 2017 Soccer Prediction Challenge were published in a special issue of the Machine Learning (Springer) Journal. Challenge participants trained models using the 216,000+ match training dataset to predict 206 future match results. The ranked probability score (RPS) [42, 29], which measures how good forecasts are compared to observed outcomes when the forecasts are expressed as probability distributions, was used as the evaluation metric in the competition. Specifically, the participants' models were evaluated based on the RPS averaged across the 206-match challenge test set.

2.2.1 2017 Soccer Prediction Challenge: Top Performers

We now cover three of the four papers that achieved the best performance in the competition.[3]

Hubáček, Sourek, and Zelezny [59] used relational- and feature-based methods to build models using the Open International Soccer Database. Pi-ratings and PageRank ratings were computed for each of the teams in each match. Both the regression and classification forms of XGBoost [26] were employed as the feature-based method, while boosted relational dependency networks (RDN-Boost) [85] were used as the relational method. The classification with XGBoost on the pi-ratings feature set performed best on both the validation set and the unseen challenge test set, achieving 0.5243 accuracy and an average RPS of 0.2063. Adding more model features, weighting aggregated data according to recency, and including expert guidance by using, e.g., active learning, were mentioned as possible avenues for further work.

Constantinou [28] created a model that combined dynamic ratings with a hybrid Bayesian network. The rating system used was a modified version of pi-ratings, which the author had proposed in previous work [30]. The computation of pi-ratings emphasized the match result (win, draw, or loss) to a greater degree than the goal margin, with the aim of dampening the influence of large goal differences. In contrast to the original pi-rating system, this version also incorporated a team form factor that aims to identify continued over- or underperformance. Four rating features (two for the home team and two for the away team) were used as the inputs to the hybrid Bayesian network. Notably, the proposed hybrid Bayesian network with modified pi-ratings was able to effectively predict a match between two teams, even when the prediction was based on historical match data that involved neither of the two teams. On the challenge test set, accuracy of 0.5146 and an average RPS of 0.2083 were obtained. Incorporating other key features or expert knowledge, e.g., player transfers, key player availability, international competition participation,

[3] The fourth top-performing paper [116] is not covered in this chapter because the methods used were more statistical in nature: e.g., a Bradley–Terry model, Poisson log-linear hierarchical model, and integrated nested Laplace approximation.

management, injuries, attack/defense ratings, and team motivation/psychology, was mentioned as a potential opportunity for future improvement.

Berrar et al. [8] applied two ML models to two different feature sets: recency features and rating features. Recency features were computed by averaging feature values over the previous nine matches and were based on four feature groups: attacking strength, defensive strength, home advantage, and opposition strength. A new soccer-specific rating system was proposed, subsequently referred to as "Berrar ratings."[4] XGBoost [26] and k-nearest neighbors (k-NN) were applied to each of the two feature sets, with both models performing better on the rating features compared to the recency feature set. XGBoost applied to the rating features provided the best performance with accuracy of 0.5194 and an average RPS of 0.2054; however, this result was obtained after the challenge had concluded.[5] The authors noted that soccer's low-scoring nature and narrow margins of victory make it challenging to predict based only on goals while also emphasizing the importance of effective feature engineering and incorporation of domain knowledge.[6] The authors also suggested including match features (e.g., yellow/red cards, fouls, possession, passing and running rates), player-level characteristics (e.g., salaries, ages, and physical conditions), and team-level characteristics (e.g., average height, attack running rate) could result in improved model performance.

Notably, the three top-performing studies in the 2017 Soccer Prediction Challenge applied machine learning models—namely, gradient-boosted tree models [8, 59] and Bayesian Networks [28]—to feature sets consisting of soccer-specific rating features.

2.2.2 Post-Challenge Studies That Have Used the Open International Soccer Database

Following the conclusion of the 2017 Soccer Prediction Challenge, several other researchers have utilized the Open International Soccer Database. The approaches and results of these studies in terms of RPS and accuracy are summarized, along with the results from the top 4 2017 Soccer Prediction Challenge studies, in Table 2.

Robberechts and Davis [100] used goals- and result-based offensive/defensive models and Elo ratings to predict FIFA World Cup results as well as matches in the Open International Soccer database. Elmiligi and Saad [40] also used the Open International Soccer Database and proposed a hybrid approach that combined machine learning and statistical models, achieving 0.4660 accuracy and an average RPS of 0.2176. The authors compared the performance when using all matches versus only part of the match history and when using individual league

[4] Please refer to Sect. 4.1.3 for a description of how Berrar ratings are calculated.

[5] k-NN applied to the rating features achieved 0.5049 accuracy and an average RPS of 0.2149 on the challenge test set, which was the best result achieved during the competition itself.

[6] Incorporation of domain knowledge is discussed in the appendix to this chapter.

Table 2 Ranked probability scores (RPSs) and accuracy results from the 2017 Soccer Prediction Challenge and subsequent studies that have used the Open International Soccer Database

Approach	RPS_{avg}	Accuracy	Paper
CatBoost+pi-ratings	0.1925	0.5582	Razali et al. [96]*
TabNet+pi-ratings	0.1956	0.5582	Razali et al. [97]*
Bookmaker odds	0.2020	0.5194	Bookmakers [100]*
Elo ordered logit model	0.2035	0.5146	Robberechts and Davis [100]*
XGBoost+Berrar ratings	0.2054	0.5194	Berrar et al. [8]*
XGBoost+pi-ratings	0.2063	0.5243	Hubáček et al. [59]
Double Poisson	0.2082	0.4888	Hubáček et al. [60]**
Hybrid Bayesian network+pi-ratings	0.2083	0.5146	Constantinou [28]
Bradley–Terry, Poisson log-linear hierarchical model, integrated nested Laplace approximation	0.2087	0.5388	Tsokos et al. [116]
Berrar ratings	0.2101	0.4854	Hubáček et al. [61]*
kNN+pi-ratings	0.2149	0.5049	Berrar et al. [8]

Studies that were not part of the 2017 Soccer Prediction Challenge are denoted by *.
The ** denotes that this study used a subset of the Open International Soccer Database, from the 2000/2001 to 2005/2006 seasons.

models versus a model for all leagues. Razali et al. [96] made use of the Open International Soccer Database, comparing the performance of gradient-boosting algorithms including XGBoost, LightGBM [71], and CatBoost [93] on goals- and result-based Elo ratings [65, 100], as well as pi-ratings. The authors found that applying CatBoost to pi-ratings yielded the best performance, with an average RPS of 0.1925 and accuracy of 0.5582, which outperformed the previous 2017 Soccer Prediction Challenge studies (and other studies that have used the Open International Soccer Database). In another recent study, Razali et al. [97] used a deep learning approach called TabNet [3], a deep neural network designed for tabular data, achieving an average RPS of 0.1956 and accuracy of 0.5582. In a study that only used rating systems and statistical models for result prediction, Hubáček, Šourek, and Železný [61] found that Berrar ratings provided the best performance (average RPS, 0.2101; accuracy, 0.4854), followed by bivariate Poisson, double Poisson, double Weibull, and pi-ratings, all of which obtained an average RPS of 0.2103. In earlier work, Hubáček, Šourek, and Železný [60] found that on a subset of the Open International Soccer Database (the 2000/2001 to 2005/2006 seasons), the double Poisson model provided the lowest average RPS of 0.2082 and accuracy of 0.4888 (pi-ratings, however, provided slightly higher accuracy).

Another soccer prediction challenge, using a similar dataset, was held in 2023; however, the results were not available at the time of writing. Please see the appendix of this chapter for a brief description of the 2023 Soccer Prediction Challenge.

3 Models

In this section, we first discuss the need for a clear model objective. Given this model objective, a guide to selecting a set of candidate models, as well as a list of commonly applied traditional machine learning models in this domain,[7] is provided. Then, we discuss model interpretability and its greater importance to some groups relative to others. We then discuss gradient-boosted tree models including XGBoost and CatBoost, which have provided some of the strongest performance in this domain of late. Following this, deep learning models, which have shown great promise in a number of application domains, are discussed in the context of soccer match result prediction.

3.1 Model Objective

Models can vary in their objectives, and it is important to establish what the objective of a model is at the outset of a match result prediction project or study. The objective of a model might be purely to achieve high predictive performance, e.g., to compete with expert predictions or in competitions. The model's objective may instead/also be used to place bets on match results. If the predictive model is used for betting, there also needs to be consideration of which matches should be bet on, e.g., using criteria such as the Kelly Index [114]. Finally, a model might be used for performance analysis purposes by coaches or analysts. Model interpretability is important for this group so that they are able to identify the most relevant features that are of importance to winning so that past performance can be analyzed and team strategy can be adjusted to increase the chance of winning future matches. Model interpretability is discussed further in Sect. 3.3.

3.2 Candidate Models

When deciding on a set of candidate ML models for soccer match result prediction, a useful starting point is to thoroughly review the literature and identify models that have performed well in predicting sports match results in general, but also those that have been effective in soccer in particular. The purpose of the model is again relevant when selecting the set of candidate models, as well as who the target audience is since, as mentioned, interpretability is of greater importance to certain groups. On the other hand, if the purpose of the model is purely to achieve the

[7] By traditional, we mean excluding gradient-boosted tree and deep learning models, which are relatively recent developments in this domain and are described later in this chapter.

highest performance, black-box models can be sufficient. While some studies have used a single type of model to predict match results, it is more common to conduct a comparative evaluation of the performance of a range of candidate ML models. Furthermore, as well as the 2017 Soccer Prediction Challenge, to confirm model generalizability, some researchers have applied their models to several leagues, e.g., the top 5 European leagues [35, 74, 127], the Greek/Dutch/English leagues [82], and the leagues of 12 different countries [23]. Before recent studies that have used gradient-boosted model trees and deep learning models, the most commonly applied classification ML models for soccer match result prediction included logistic regression, artificial neural networks (ANNs), Bayesian networks, decision trees, k-NN, naïve Bayes, random forest, and support vector machine (SVM).

3.3 Model Interpretability

Increasing the interpretability (understandability) of match result prediction models has become an area with increasing research activity recently. As mentioned, with greater interpretability, match result prediction models are of greater use to, e.g., coaches and performance analysts to identify the most relevant features that are within players' control and can be adapted to increase the chance of winning future matches. Models with greater interpretability can also be used by teams and/or management to adapt their tactical decisions, identify own and opposition team strengths and weaknesses, decide on appropriate formations, and make player selection and transfer decisions [124]. Moustakidis et al. [84] used explainable ML models using SHapley Additive exPlanations (SHAP) [80] to identify team performance indicator metrics that are of greatest relevance to predicting teams' average scoring performance per season. Another study that used the SHAP method is that of Ren and Susnjak [99]. Yeung, Bunker, and Fujii [124] proposed an interpretable ML model framework for soccer match result prediction that is conceptually similar to GAP ratings [118], but that predicts match statistics with—rather historical match statistics, which can be cumbersome to engineer—management decision- and player quality-related features. Player quality features were obtained from the EA Sports video game FIFA, a data source that has been used in other studies, e.g., [35, 92]. Random forest feature importance is another appealing approach to interpreting the most relevant model features and has been used in conjunction with a hybrid random forest model by Groll et al. [53]. Of course, some traditional ML models, e.g., decision trees and logistic regression, are also highly interpretable by design and may be useful if they can provide sufficient performance.

3.4 Ensemble Methods

Ensemble methods in ML can be broadly grouped into boosting methods, e.g., gradient-boosted tree models such as XGBoost and CatBoost, and bagging methods, e.g., random forest.

As previously mentioned, at least in the absence of match features other than goals scored, the 2017 Soccer Prediction Challenge and subsequent studies using the Open International Soccer Database have highlighted that gradient-boosted tree models, applied to soccer-specific ratings, are currently able to achieve some of the highest performance in this domain (Table 2).

However, on other datasets (some of which include match statistics), random forests have been found to be competitive with [5] and even exceed the performance of gradient-boosted tree models [107, 1, 43]. Despite ensemble methods appearing to provide generally better performance than statistical and other traditional ML models, a thorough comparative evaluation of the performance of ensemble methods with deep learning and (deep) neural network models is still required.

Extreme gradient boosting, or XGBoost [26], is one of the most popular gradient-boosted tree methods and has performed well in a variety of ML tasks. The foundation of XGBoost lies in gradient boosting. Gradient boosting, which can be used for both regression and classification, combines several weak learners (e.g., decision trees) into so-called strong learners (gradient-boosted trees). The gradient boosting process sequentially builds simple weak prediction models that each predict the residual error of the preceding model. The weak learners are added to the ensemble and their contributions are determined based on a gradient descent optimization problem, which minimizes an overall loss function representing the difference between the actual results and predictions of the strong learner.

CatBoost [93] has some similarities to XGBoost and is capable of handling both regression and classification tasks by aggregating multiple weak learners. One of the distinctive features of CatBoost is its ability to handle categorical features efficiently, alleviating the need for extensive data preprocessing. CatBoost employs a technique called "ordered target encoding," which involves encoding a categorical feature sequentially while excluding the needs of the target feature. This prevents information leakage (where the model accesses the target information of the current observation), ensuring an unbiased encoding process and preventing overfitting. The generation of match outcome probabilities (e.g., for win/draw/loss) that are well-calibrated is a distinct advantage of CatBoost, given its importance in match result forecasting [121].

Another boosted-tree model that has not been widely investigated, but that may have potential, for soccer match result prediction is the alternating decision tree (ADTree) model [46]. As opposed to XGBoost and CatBoost, ADTree uses AdaBoost [47] rather than gradient boosting and maintains an interpretable decision tree structure as the final model (see the appendix for more detail).

3.5 Deep Learning Models

Deep learning models have been effective in various domains including computer vision, trajectory analysis, and event prediction in sports. Nevertheless, there are still only a relatively small number of published articles related to deep learning for soccer match result prediction.

Danisik, Lacko, and Farkas [35] compared the performance of a long short-term memory (LSTM) model [56] with classification, numeric prediction, dense approaches, and also against average random guess, bookmaker prediction, and home team win baselines in predicting English Premier League matches. Rahman [94] used deep neural networks and ANNs to predict match results from the 2018 FIFA World Cup. Jain, Tiwari, and Sardar [67] used recurrent neural networks and LSTM networks for predicting English Premier League match results, engineering several features related to winning and losing streaks, points, and goal differences. Malamatinos, Vrochidou, and Papakostas [82] used k-NN, LogitBoost, SVM, random forest, and CatBoost—as well as convolutional neural networks and transfer learning with encoded tabular data converted to image models—to predict Greek Super League match results, with the best performing model also applied to the English Premier League and the Dutch Eredivisie. The best-performing model was found to be CatBoost, notably outperforming the deep learning convolutional neural network model. Given the time-series nature of soccer match result data, Joseph [69] considered time series-based approaches to predict English Premier League match results, including LSTM and Bayesian methods. As mentioned in Sect. 2.2.2, Razali et al. [97] recently achieved strong performance (Table 2) when applying TabNet [3]—a deep neural network model for tabular data—to the Open International Soccer Database.

Given the relatively small number of studies related to deep learning in soccer result prediction, there is the potential for greater future exploration into the potential of these models, e.g., to investigate whether deep learning models are able to outperform boosted tree and other ensemble models.

4 Features

This section describes some of the different types of features that can be used in soccer match result prediction models.

Features in sports match result data can generally be divided into different feature subsets. For instance, in basketball, Miljković et al. [83] categorized features into match-related and standing features. Tax and Joustra [113] compared the performance of a combined feature set consisting of betting odds and features derived from publicly available data to a feature set consisting only of betting odds. Hucaljuk and Rakipović [62] compared the performance of a separate expert-selected feature set against their own selected feature set. Feature selection algorithms can be applied to the entire feature set as well as subsets (e.g., feature selection algorithm-selected vs. human-selected features, betting odds included vs. betting odds excluded, rating features, match features, external features, etc.) of the original feature set, and the performance of the candidate models on each feature set can be compared.

In the remainder of this section, first, rating features are discussed, including general-purpose rating systems that are used across a range of sports, e.g., Elo ratings. Domain-specific (soccer-specific) rating systems including pi-ratings, Berrar ratings, and the recently proposed GAP ratings are then discussed, as are betting odds (which can also be considered a rating for which the calculation is opaque). Then, match, player, and team statistics are examined, as are external features that do not relate to events occurring within matches. Feature selection methods are also briefly discussed.

4.1 Ratings

The general idea of rating systems is to assign some initial rating for each team and update the ratings over time based on actual match results. Some soccer-specific ratings also consider the goal margin, match venue, and offensive/defensive strengths. There are several use cases for ratings, the first and most obvious of which is to rate/rank teams. In the context of match result prediction, rating systems can, of course, be used as a predictive model by simply predicting the team with the higher rating as the winner. However, using ratings as model features in machine learning models appears to provide better performance (Table 2), and for this reason, ratings are described here in the features section of this chapter rather than in the models section.

4.1.1 Elo Ratings

The Elo rating [41] system was originally used in chess but has since been applied in a wide range of sports including soccer. Extensions of Elo have been proposed, e.g., to account for goal margin [65] and home advantage [104].

In the Elo rating system, each team generally starts with a rating of 1500. The system can be decomposed into two steps: the estimation step (E-step) and the update step (U-step). The E-step involves estimating the probability that a team will win a particular match, while the U-step involves the update of the team's rating following a particular match result.

E-step The probability that team i beats team j is estimated by:

$$p_{i,j}(t) = \frac{1}{1 + 10^{\frac{-(R_i(t) - R_j(t))}{400}}} \tag{1}$$

Analogously, the probability that team j beats team i is:[8]

$$p_{j,i}(t) = \frac{1}{1 + 10^{\frac{-(R_j(t)-R_i(t))}{400}}} \tag{2}$$

Since these are probabilities, they add to one, i.e., $p_{i,j}(t) + p_{j,i}(t) = 1$.

U-step The Elo ratings are updated, e.g., for team i, according to:

$$R_i(t+1) = R_i(t) + K[A_i(t) - p_{i,j}(t)]$$

Team j's rating is updated in a similar manner:

$$R_j(t+1) = R_j(t) + K[A_j(t) - p_{j,i}(t)]$$

where $A_i(t)$ takes the value of 1, 0.5, or 0 for a win, draw, and loss, respectively.

K-factor The K term in the above U-step expressions is known as the K-factor and is often given a fixed value (e.g., 32 or 20). If K is too large, the Elo rating will vary excessively from match to match; however, if K is too small, the rating of a team will not change rapidly enough when it improves its performance [65].

There are different approaches to selecting the value of K: it can be a fixed value, a tunable parameter, or a function. Hvattum and Arntzen [65] proposed a goal-based Elo rating system that allows K to depend on the goal difference. In particular, $K = K_0(1+\delta)^{\lambda}$, where δ is the absolute goal margin, and K_0 and λ are parameters.

Other approaches to selecting the K-factor are described online, e.g., on eloratings.net and footballdatabase.com (see the appendix for further details).

Elo Ratings with Home Advantage Ryall and Bedford [104] extended Elo ratings to incorporate home advantage (albeit in Australian Rules Football, but it can equally be applied in soccer) by simply adding a home advantage term to the rating difference such that the estimated probability of team i beating team j from Eq. (1) becomes:

$$p_{i,j}(t) = \frac{1}{1 + 10^{\frac{-(R_i(t)-R_j(t))+H_{i,j}}{400}}} \tag{3}$$

where $H_{i,j}$ is the magnitude of home advantage that team i has over j. Note that in this setup, $H_{i,j}$ is a tunable parameter that does not vary by match, t.

Connection between Elo and Bradley–Terry As mentioned earlier, the Bradley–Terry paired comparisons model [11], which estimates the win probability in a

[8] The value of 400 was specified by Elo [41] to be twice the standard deviation of chess player ratings. The values of 10 and 400 merely serve to set a scale for the ratings and are generally used when the initial ratings are set to 1500 [65].

match using the strengths/abilities of each team, is the foundation of Elo ratings. In the Bradley–Terry model, the probability of team i beating team j is given by:

$$p(i \text{ beats } j) = \frac{s_i}{s_i + s_j}$$

where s_i and s_j denote the strength (ability) of teams i and j, respectively. The strength parameters are commonly estimated using maximum likelihood, e.g., based on observed win counts between the two teams. The Elo rating of a team can actually be expressed directly as a function of its Bradley–Terry strength [33]. In particular, the Elo rating of team i can be expressed as $R_i = 400log_{10}(s_i)$. Rearranging this expression enables one to calculate the Bradley–Terry strength of team i from its Elo rating: $s_i = 10^{\frac{R_i}{400}}$.

The Bradley–Terry model has been shown to be outperformed in soccer match result prediction by Poisson/Weibull-based models [61]. Furthermore, when using the ratings themselves for prediction, pi-ratings have been found to generally outperform Elo ratings.

4.1.2 Pi-Ratings

In the pi-rating system [30], each team is assigned two ratings corresponding to their home and away strengths, with their overall rating the average of their home and away ratings:

$$R_i = \frac{R_{i,H} + R_{i,A}}{2}$$

Then, for each match, the expected goal difference is calculated based on the home rating of the home team and the away rating of the away team. The actual match outcome is compared with the expected score, and if a team performs better than expected, its rating is increased based on the difference between actual and expected outcomes, as well as the learning rates (which are model parameters). The home and away ratings of both teams are updated after a match but using separate learning rates. Each team starts with an initial pi-rating of 0, and the pi-rating represents the rating of the average team relative to other teams. Given a match between teams α and β, the ratings are updated according to:

$$\hat{R}_{\alpha,H} = R_{\alpha H} + \lambda\psi_H(e)$$

$$\hat{R}_{\alpha,A} = R_{\alpha,A} + \gamma(\hat{R}_{\alpha,H} - R_{\alpha,H})$$

$$\hat{R}_{\beta,H} = R_{\beta,H} + \lambda\psi_A(e)$$

$$\hat{R}_{\alpha,H} = R_{\alpha,H} + \gamma(\hat{R}_{\alpha,A} - R_{\alpha,A})$$

where e is the error between the expected and actual goal difference. The expected goal difference for a team against the average opponent at ground G, which is either H (home) or A (away), is calculated as:

$$\hat{g}_{DG} = b^{\frac{|R_{iG}|}{c}} - 1$$

where $b = 10$ and $\hat{g}_{DG}$ is the expected goal difference for team i against the average opponent when playing at ground G (either H or A). Team ratings can potentially be negative, in which case the expected outcome is $-\hat{g}_{DG}$. The error, e, between the expected and actual goal difference is then given by:

$$e = |g_D - \hat{g}_D|$$

where $g_D = g_{DH}$ - g_{DA} and $\hat{g}_D = \hat{g}_{DH}$ - $\hat{g}_{DA}$ To dampen the influence of large goal margins when updating ratings, a reward/penalty function is specified to be a function of the goal difference error:

$$\psi(e) = c \times log_{10}(1 + e)$$

where $c = 3$. The probability distribution over potential match outcomes is computed using an ordered logit model.

4.1.3 Berrar Ratings

In the Berrar rating system [8], a logistic function is used to predict the expected number of goals scored using the teams' offensive and defensive strengths. In particular, the expected goals scored by the home and away teams are respectively given by:

$$\hat{G}_H = \frac{\alpha_H}{1 + exp(-\beta_H(o_H - d_A) - \gamma_H}$$

and

$$\hat{G}_A = \frac{\alpha_A}{1 + exp(-\beta_H(o_A - d_H) - \gamma_A}$$

where o_H and d_H and o_A and d_A denote the offensive and defensive ratings of the home and away teams, and α_H and α_A represent the maximum possible expected goals that can be scored in a match by the home and away teams. The β parameters determine how steep the logistic function is, and γ determines the threshold/bias.

The ratings for the home team are updated according to:

$$o_H = o_H + \omega_{o_H}(G_H - \hat{G}_H)$$

$$d_H = d_H + \omega_{d_h}(G_A - \hat{G}_A)$$

And similarly, for the away team:

$$o_A = o_A + \omega_{o_A}(G_A - \hat{G}_A)$$

$$d_A = d_A + \omega_{d_A}(G_H - \hat{G}_H)$$

4.1.4 GAP Ratings

Since soccer is a low-scoring sport and goals are rare, more recently, generalized attacking performance (GAP) ratings [118] were proposed, which build on pi-ratings but can also instead predict non-rare match statistics (e.g., shot attempts), not only goals scored. By predicting non-rare match statistics and excluding some of the early and latter rounds of seasons, GAP ratings were shown to be more informative (in terms of the Akaike Information Criterion) for soccer match result forecasting [120].

GAP ratings [118, 120] are inspired by pi-ratings [30] and are adjusted such that they can be applied to predict match statistics in advance. The GAP rating requires match statistics as input and can be used to predict goals, shots on and off target, and corners for both teams. The inputs, S_H and S_A, denote the match statistics for the home team (H) and away team (A), respectively. There are four ratings for each team i: H_i^a, H_i^d, A_i^a, A_i^d, which denote team i's home attack, home defense, away attack, and away defense ratings, respectively. Following a match between home team i and away team j, the GAP ratings for home team i are updated as follows:

$$H_i^a = \max\left(H_i^a + \lambda\phi_1\left(S_H - \frac{H_i^a + A_j^d}{2}\right), 0\right)$$

$$A_i^a = \max\left(A_i^a + \lambda(1-\phi_1)\left(S_H - \frac{H_i^a + A_j^d}{2}\right), 0\right)$$

$$H_i^d = \max\left(H_i^d + \lambda\phi_1\left(S_A - \frac{A_j^a + H_i^d}{2}\right), 0\right)$$

$$A_i^d = \max\left(A_i^d + \lambda(1-\phi_1)\left(S_A - \frac{A_j^a + H_i^a}{2}\right), 0\right)$$

The GAP ratings for away team j are updated as follows:

$$A_j^a = \max\left(A_j^a + \lambda\phi_2\left(S_A - \frac{A_j^a + H_i^d}{2}\right), 0\right)$$

$$H_j^a = \max\left(H_j^a + \lambda(1-\phi_2)\left(S_A - \frac{A_j^a + H_i^d}{2}\right), 0\right)$$

$$A_j^d = \max\left(A_j^d + \lambda\phi_2\left(S_H - \frac{H_i^a + A_j^d}{2}\right), 0\right)$$

$$H_j^d = \max\left(H_j^d + \lambda(1-\phi_2)\left(S_H - \frac{H_i^a + A_j^d}{2}\right), 0\right)$$

where $\lambda > 0$ is a parameter that weights the effect of the latest game on the rating, and $\phi_1 \in [0, 1]$ and $\phi_2 \in [0, 1]$ weights the effect of a home game on the away rating and an away game on the home rating, respectively.

Using goals as an example, H_i^a indicates the number of goals home team i will score and A_j^d indicates the number of goals the away team j will concede. The average of H_i^a and A_j^d represents the expected goals home team i should score and away team j should concede. Ratings are updated using the weights and the differences between actual and expected results. The GAP rating can be interpreted as the performance of a team compared to an average team in the league. Better teams will have higher attacking and lower defensive ratings. The estimates of S_H and S_A, which are denoted by $\widehat{S_H}$ and $\widehat{S_A}$, respectively, are given by:

$$\widehat{S_H} = \frac{H_i^a + A_j^d}{2}$$

$$\widehat{S_A} = \frac{A_j^a + H_i^d}{2}$$

To select the λ, ϕ_1, and ϕ_2 parameters, the least-squares method is applied. The cost function is given by:

$$Cost = \sum_{i=1}^{N} |S_H - \widehat{S_H}| + |S_A - \widehat{S_A}|$$

where N denotes the number of matches. All GAP ratings are initialized to zero. Parameter selection is repeated at the start of each season using all data. However, the first year of data is only used for parameter selection, and the first six games and final six games played by the home team in each season are ignored due to various factors that could increase the unpredictability of these matches. Also, the relegated teams' ratings are calculated as the average of the promoted teams' ratings, and similarly, the promoted teams' ratings are calculated as the average of the relegated teams' ratings.

4.1.5 Betting Odds

As mentioned, betting odds can be used as a baseline with which to model results, but they can also be considered a type of rating [122]. Therefore, like other ratings, odds can potentially be used as a model feature. Indeed, betting odds have been used as model features in many studies [106, 86, 113, 44], sometimes as the sole model feature. Since bookmakers establish betting odds, unlike the rating systems described above, the way in which odds are determined is opaque. However, it is reasonable to assume that bookmaker companies use models supplemented with observed betting volumes on each outcome to capture the "wisdom of crowds" [111, 17] in order to set appropriate odds for matches. Betting odds also differ from other ratings that generally only incorporate historical match result information, since odds also account for other factors such as market sentiment, player availability, and expert knowledge.

Deciding whether betting odds should be included as a model feature depends on the objective of the model. For instance, if a model is being used for betting purposes in an attempt to "beat the house," betting odds should not be included as a model feature [58]. As a baseline, betting odds have proven difficult to beat. For example, despite using a state-of-the-art gradient boosting model along with a sophisticated engineered set of features, Baboota and Kaur [5] were unable to outperform predictions based on bookmaker odds. Bookmaker predictions also outperformed the top-performing submissions from the 2017 Soccer Prediction Challenge (odds outperformed all competing models in [100], and this result is also shown in Table 2, along with the next-best result from that study, an Elo ordered logit model). On the other hand, if the objective is purely maximizing predictive performance, given the information inherent in betting markets, including betting odds model features may make sense.

Some studies have attempted to exploit bookmaker odds while using machine learning models, e.g., through arbitrage [74] and modern portfolio theory methods [58], including Talattinis et al. [112], who utilized the Sharpe ratio. Many betting sites provide betting odds data for future and historical matches (Table 1). When sourcing betting odds from such websites, to create a model feature, it is often useful to convert the raw decimal odds into probabilities by taking their reciprocal:

$$\text{Probability} = \frac{1}{\text{Decimal odds}} \tag{4}$$

The sum of the odds generally exceeds one due to the presence of a built-in bookmaker margin, which is sometimes referred to as the "over-round" [45, 125]. To ensure that the odds add to one and obtain the normalized odds-implied win probabilities, one simply divides the original odds-implied probabilities by the normalization factor, which is the sum of the probabilities obtained for each match outcome from Eq. (4). That is:

$$\text{Probability normalized} = P_{norm} = \frac{\text{Probability}}{\text{Normalization factor}}$$

For example, suppose that a match has only two possible outcomes and that the decimal bookmaker odds are such that team A is paying 1.34 to win and team B is paying 3.02. Taking the reciprocal as per Eq. (4), we obtain 0.746 and 0.331 for teams A and B, respectively. Then, 1.077 is the normalization factor ($0.746 + 0.331 = 1.077$). To obtain the odds-implied win probabilities, we have $P_{norm}(\text{team A win}) = 0.746/1.077 = 0.693$ and $P_{norm}(\text{team B win}) = 0.331/1.077 = 0.307$, so now $P_{norm}(\text{team A win}) + P_{norm}(\text{team B win}) = 0.693 + 0.307 = 1$.

Of course, betting on the match outcome is not the only option for bettors: it is possible to bet on a specific event occurring within a match, who will score the first goal, and so on. Betting on the over-under—whether a team scores over or under 2.5 goals in a match—is one such option. Wheatcroft [118] found that in the over/under 2.5 goals market, shots made and corners were better for probabilistic forecasting than actual goals.

4.2 *Match Features*

Simple Match Statistics and Performance Indicator Metrics As has been mentioned in previous sections, match features (also known as in-play, in-match, or in-game features) are features that relate to events that occur within matches. Match features are often performance indicator metrics [63], as they are referred to in the field of sports performance analysis, a subdiscipline of sports science. For instance, raw statistics such as passes, tackles, shot attempts, etc. can be aggregated into performance indicator metrics at the player, offensive/defensive unit, or team levels. The survey paper of Bunker and Susnjak [18] highlighted the potential benefits of greater future collaboration between researchers from sports performance analysis and machine learning for engineering relevant match features for machine learning models.

Of course, match features that are performance indicators are not known in their entirety until the match has concluded. Therefore, for match result forecasting, preprocessing is required, by aggregating (e.g., averaging) match features across a certain number of historical matches. There is no set number of matches with which to aggregate/average over since it will depend on the characteristics of the dataset. For example, Buursma [22] found that aggregating features over the past 20 matches provided the best performance, while Berrar et al. [8] found that using the past nine matches provided the best performance in building a set of what the authors referred to as "recency features" for the 2017 Soccer Prediction Challenge. Simple aggregation (e.g., averaging or summing) of features over a certain number of past matches does not, however, account for the fact that recent games are of greater relevance than matches that were played a long time ago. Exponential time weighting [37, 78, 60], which is used in the weighted likelihood function that is maximized in double Poisson models, could be used as an alternative to simple aggregation over historical matches.

Matches can be summarized in the form of simple statistics, some of which are performance indicators and some are not. Goals are the most commonly used match statistic since goals are also used to derive the match result and are therefore used in essentially every study on match result forecasting. Shots on target, shots off target, and corners are other commonly used match statistics. Other match statistics in soccer that can also be considered performance indicators include blocks, clearances, crosses, dribbles, interceptions, passes, possessions, saves, and tackles. Discipline-related match statistics include yellow and red cards, fouls, and offsides. As mentioned, match statistics prediction is an important part of the GAP rating method [118], which focuses on predicting non-rare events such as shot attempts (since actual goals are rare).

On-the-Ball Event Data Different vendors and data providers, e.g., StatsBomb and Stats Perform/Opta, have varying data formats and definitions that are used to describe and annotate events that occur within soccer matches. Decroos et al. [36] proposed a Soccer Player Action Description Language (SPADL) to standardize the event data format across vendors. The authors also proposed a method called valuing actions by estimating probabilities (VAEP), which uses event log data that has been converted into SPADL format, to value on-the-ball actions. VAEP can also be used to compute player ratings by aggregating each of their on-the-ball actions. Event data commonly includes the start and end times and locations of events, the player who performed the action and the team to which they belong, the action type (e.g., pass, cross, throw in, etc.), the body part used in the action, and whether the action was successful or unsuccessful. While event data focuses event on on-the-ball rather than off-ball events, supplementing on-the-ball event features with tracking data-derived features can address this.[9] Event data can provide more in-depth information than simple match statistics and has the potential to be further explored for engineering informative features for soccer match result prediction ML models.

Spatiotemporal Tracking Data Spatiotemporal tracking data is generally sourced from optical systems or wearable (GPS) tracking devices that record the locations of players and the ball. Another way in which spatiotemporal data can be obtained is from computer vision, machine learning, or deep learning methods applied to match video footage. As mentioned, it is valuable for event data and spatiotemporal data to be combined to be able to perform more holistic performance analyses; however, this is challenging—both in terms of cost but also technically—because it generally requires collaboration between researchers from disparate disciplines, i.e., computer/data scientists and sports scientists [49]. Furthermore, professional teams

[9] F. Goes (2021), The power of combining tracking and event data, https://www.scisports.com/the-power-of-combining-tracking-and-event-data/.

are generally unwilling to share data publicly, and obtaining it from professional data providers comes at significant cost. Rein and Memmert [98] also highlighted the value of greater collaboration between sports scientists and computer scientists in extracting tactical insights from spatiotemporal tracking data in soccer, which has traditionally been used for monitoring physiological demands. Toda et al. [115] proposed a method called valuing defense by estimating probabilities (VDEP), an adjustment of VAEP in which the defensive performance of a team is evaluated using on-the-ball event and tracking data. The generated metrics focus on actions that lead to penetration of the penalty area or a shot attempt (rather than a goal, which is rarer). The study showed that this resulted in higher predictability than VAEP. Deriving features from spatiotemporal tracking data also appears to have the potential for further research in match result prediction ML models in soccer. As mentioned, spatiotemporal tracking data has traditionally been used to assess physiological demands on athletes, and Tümer et al. [117] aimed to predict Turkish Super League rankings by applying three ML models—ANN, radial basis function, and linear regression—to physical and technical features (e.g., related to physical demands).

Expected Goals Expected goals (xG) refers to the expected number of goals a team should score in a game. For every shot a team takes in a match, the probability of a goal can be calculated using xG models, of which there are many types. Then, by aggregating all of these probabilities, one can obtain the xG for the match. To build an xG model, shot-related features are considered. For example, according to the Bundesliga,[10] the distance and angle to the goal, whether it is a pass or a header, and the type of pass are used. For the algorithm, any type of regression model, e.g., logistic regression or neural network, can be utilized. The target variable takes the value of 1 for a goal and 0 for no goal. Wheatcroft [120] showed that using goals could not provide statistically significant information. Replacing goals with expected goals (xG) as the target variable is a commonly suggested approach for further research because expected goals are not as rare.

4.3 *Player Statistics*

Player Characteristics and Ratings The incorporation of player-level attributes or ratings has been a popular line of enquiry in recent times. For example, Stübinger, Mangold, and Knoll [108] incorporated player and match attributes in an ensemble learning approach in a simulation on the top 5 European leagues. Chen [25] used FIFA video game player ratings to predict Spanish La Liga soccer results using three common ML models: ANN, random forest, and SVM. Another study that utilized

[10] Bundesliga (2019), xG stats explained: the science behind Sportec Solutions' Expected goals model, https://www.bundesliga.com/en/bundesliga/news/expected-goals-xg-model-what-is-it-and-why-is-it-useful-sportec-solutions-3177.

FIFA player ratings is that of Danisik, Lacko, and Farkas [35], whose deep learning approach using LSTM has been discussed previously in this chapter. Arntzen and Hvattum [4] compared the performance of an ordered logistic regression model and a competing risk model when applied to adapted plus-minus player ratings—which compare the performance of a team, e.g., in terms of goals, when a particular player is playing versus when they are not [64])—with when the two models were applied to team Elo ratings. The authors found no difference in performance between the ordered logit and competing risk models for match result prediction. Furthermore, they found that team ratings and player ratings had similar performance, but using a combined feature set with both player ratings and team ratings provided better performance than when utilizing only player ratings or only team ratings. Team and player ratings from EA Sports' FIFA are available via websites such as fifaindex.com (Table 1). Other video games such as Football Manager (Sports Interactive/Sega) also contain player ratings that are arguably better and could potentially be used in future studies.

Player Form Otting and Groll [88] applied hidden Markov models to investigate the existence of the "hot shoe" effect (known as the "hot hand" effect in basketball) and provided evidence that such an effect does exist. If such an effect does exist, it is reasonable to assume that players in the "hot" state are more likely to perform better than those in the "cold" state, which could provide more information for prediction.

4.4 Team Statistics

Team Ratings In Sect. 4.1, team ratings based on historical match results in terms of goals scored were discussed. However, there are other ways to obtain or compute team ratings. The first approach is to obtain team ratings from video games such as FIFA or Football Manager and directly use these as features in a predictive model [5, 124]. Aside from video games, another potential source of team ratings is the Union of European Football Associations (UEFA), which provides team ratings based on results in UEFA club competitions over the past 5 years [109]. Player ratings can be obtained from video games such as FIFA and then aggregated to engineer role- or team-level ratings [24]. Carpita, Ciavolino, and Pasca [24] used 33 player-related features from the FIFA video game to construct seven player-level performance indicators, which were then combined into performance indicators based on role (forward, midfielder, defender, and goalkeeper). A binomial logistic regression model was then applied to differences between the role-based performance indicators of opposing teams to analyze the degree to which these indicators affect the probability of winning. Alternatively, player ratings can be computed using VAEP or plus-minus ratings using event log data or historical match result data, respectively, and then aggregated into a team rating.

Some studies have incorporated or compared both player and team ratings. For instance, Pipatchatchawal and Phimoltares [92] incorporated both player and team

FIFA video game ratings, while Arntzen and Hvattum [4], as previously mentioned, compared the use of team Elo ratings and player plus-minus ratings.

Streak Streak is a similar concept to the hot shoe phenomenon; however, it generally refers to match result patterns rather than individual outcomes such as successful shots on goal. Baboota and Kaur [5] computed the team streak as of a team's jth game as:

$$Streak(j) = \sum_{i=j-k}^{j-1} Score_i / 3k$$

where $Score_i$ indicates the goals scored in game i, $Score_i \in \{0, 1, 3\}$ and k is a tunable hyperparameter indicating how many historical games are considered. To account for the fact that more recent matches are of greater relevance, the authors also engineered a time-weighted streak feature, which, for the jth game, is given by:

$$Weighted\ streak(j) = \sum_{i=j-k}^{j-1} 2 * \frac{i - (j - k - 1)Score_i}{3k(k + 1)}$$

The experimental results of the study showed that $k = 6$ provided the best performance for the various models.

Form The streak only considers a single team; thus, to consider the opponent teams, Baboota and Kaur [5] also engineered a form feature. The form of each team is set to an initial value of one and is updated after each game. The forms of teams A and B, for the jth game between these two teams, if team A wins, are given by:

$$Form_j^A = Form_{j-1}^A + \alpha Form_{j-1}^B$$

$$Form_j^B = Form_{j-1}^B - \alpha Form_{j-1}^A$$

If the match is a draw, the forms are given by:

$$Form_j^A = Form_{j-1}^A - \alpha(Form_{j-1}^A - Form_{j-1}^B)$$

$$Form_j^B = Form_{j-1}^B - \alpha(Form_{j-1}^B - Form_{j-1}^A)$$

The authors found that the hyperparameter $\alpha = 0.33$ provided the best performance.

Interplayer Chemistry Bransen and Van Haaren [12] suggested that the chemistry between players within the same team is more important than the past performance of a player in another team. Two offensive and defensive chemistry metrics were proposed based on VAEP, which considered the interaction between two players, to identify the squad with the best chemistry. Moreover, two CatBoost models were applied to player statistics features for a pair of players, with the two chemistry metrics as the respective target variables, in order to predict the chemistry of a specific player with players in other teams (which is potentially useful for scouting or transfers).

Passing Networks Some studies have used social or passing network analysis to predict match results in soccer. Using passing distribution data, Cho, Yoon, and Lee [27] used social network analysis to construct actual and predicted network indicators that reflect team performance, in conjunction with gradient boosting for match result prediction. Ievoli, Palazzo, and Ragozini [66] considered whether passing networks have a significant effect on match results, applying four ML models to passing network-derived indicators and an on-field feature set to predict 2016–2017 UEFA Champions League group stage results. The authors found that some network-derived variables were connected to the level of offensive actions and could improve the explanatory power of models beyond the models that included only on-field features.

4.5 External Features

External features are external to the match itself, i.e., these features are not derived from on-field events. These may include travel, player availability, match venue, match officials, weather, and so on. One study that incorporated weather as a model feature is that of Palinggi [89], who applied an SVM model to weather- and match-related features and achieved around 50% accuracy. Other potential external model features include the average age of a team's players and coach [52], the number of players who are international representatives [70], and club market values, transfer budgets, and operational costs [109]. Unlike match features, external features are generally already known prior to a match and can, therefore, often be used directly as model features, without the type of preprocessing that is required for match features through aggregation over historical matches. In terms of match venue having an effect, the home advantage phenomenon is a well-known effect in which the home team tends to outperform the away team, on average (the COVID-19 pandemic period, where some matches were played in front of empty stadiums, being a possible exception [77]).

Another potential source of external feature data is social media. Wunderlich and Memmert [123] applied sentiment analysis methods to tweets posted on Twitter during over 400 English Premier League matches, as well as those posted in the periods shortly before and after goals, to extract information (e.g., word clouds and other relevant features) to predict the total number of goals scored in a match and perform in-play forecasting. Their results suggested that in-play goal forecasting is very challenging, in-play information did not improve predictive accuracy, and its predictive value is small in comparison to pre-match information and that—perhaps due to overly high pre-match expectations from fans—fan sentiment on Twitter decreases during a match. Kinalioğlu and Kuş [72] proposed a hybrid clustering and classification method and applied it to data from 6,396 European league matches including—as well as team and player statistics—fan opinions from social media. Another study that used the Twitter posts of fans to predict soccer match outcomes is that of Kampakis and Adamides [70].

4.6 Feature Selection Methods

Feature selection techniques can be broadly grouped into three categories: filter, wrapper, and embedded methods [126].

The random forest algorithm [13] has embedded tree-based feature selection, in which feature importance is evaluated during the training process [79]. This embedded feature selection means that despite being a bagging method, random forest is a model with potentially high interpretability. In the sklearn package in Python, the SelectFromModel object can be used in conjunction with the RandomForestClassifier to automatically select features.

Filter feature selection methods generally rank features based on, e.g., the chi-squared, information gain (ratio), or correlation between each feature and the target variable. A disadvantage of using filter methods is that a cutoff needs to be arbitrarily selected to determine how many features to include in a model.

Other feature selection methods, e.g., CFS subset [55] and Relief/ReliefF subset feature selection methods [73, 75], consider the correlation or interaction among features as well as those between each feature and the target variable. Sequential forward selection is another possible approach to selecting features (e.g., [82]).

As previously mentioned, the performance of models applied to features selected by feature selection techniques can be compared with human expert-selected features (e.g., [62]).

5 Evaluation Methods

Selecting an appropriate evaluation metric is important when evaluating machine learning models. It is also necessary to clearly define the target variable that is to be predicted.

In sports with only two possible outcomes, assuming the league is competitive, class imbalance is generally not an issue since home team wins should only slightly outweigh away team wins due to the home advantage effect. In this case, classification accuracy can be an appropriate evaluation metric.

In soccer, which has three possible outcomes (with a draw much less common than the other two outcomes), accuracy has continued to be commonly used to evaluate classification ML models. However, more recently, scoring rules such as the ranked probability score (RPS) have become prevalent as evaluation metrics, in part due to its use in the 2017 Soccer Prediction Challenge.

5.1 Problem Setup and Target Variable Definition

Soccer match result prediction problems are commonly formulated as a three-class classification problem with a target variable defined with three discrete values (e.g., win/draw/loss),[11] as a numeric prediction problem predicting a goal margin target variable,[12] or by predicting the number of goals scored by each team and thus the match result. Since soccer is a low-scoring sport, numeric prediction of the goal difference is challenging compared to sports with match result margins that are larger in magnitude (e.g., basketball or rugby). In machine learning for soccer result prediction, classification is the most common approach, while statistical models commonly attempt to model and predict the number of goals scored by each team. However, given the 2023 Soccer Prediction Challenge (described further in the appendix) involved—as one of the tasks—predicting the number of goals scored by each team, more future studies may begin to consider this problem setup.

Other studies have sought to predict other target variables. For example, predicting the "over-under": whether more than 2.5 goals are scored in total in a match [89], or forecasting match statistics such as shot attempts or shots on target [118, 124]. Using a Bayesian approach for prediction, Robberechts, Van Haaren, and Davis [101] developed an in-game win probability model. Wunderlich and Memmert [123] considered both in-play forecasting and predicting the total goals scored in a match. Predicting whether both teams will score during a match has also been considered (e.g., by [32]). Others have attempted to predict the final team rankings in a league [117], or whether a team earns a competition point [43]. The per-season average scoring performance of teams has also been considered as a target variable to be predicted [84].

5.2 Baselines

Since datasets in this domain often cover different leagues and seasons and contain different variables from which model features can be derived, it is generally difficult

[11] In classification, some researchers, e.g., [72], excluded draws or merged draws into another class so as to create a binary problem, which naturally gives better performance compared to three-class classification models given the difficulty of predicting draws.

[12] Another approach proposed is to convert win/draw/loss to 1/0.5/0 and use a numeric prediction model. This was an approach followed by [35], who used LSTM regression and found that it outperformed the LSTM classifier.

to compare results across studies unless the studies being compared utilized a common dataset.

Baselines—including simple rule-based benchmarks, betting odds-derived predictions, and expert predictions—are another way in which researchers can evaluate model performance:

- **Betting odds:** Betting odds are somewhat unique in that they can be included as a model feature but can also act as a benchmark with which to evaluate model performance. It is straightforward to convert decimal odds into outcome probabilities by taking their reciprocal and normalizing such that the probabilities add up to one (Sect. 4.1.5). One can then predict the match outcome by simply predicting the outcome that has the highest probability. The odds of several bookmakers can also be combined to create a bookmaker consensus model [76]. Alternatively, provided the model is not being used for betting strategies to "beat the house," these probabilities can potentially be used as model features. The outcome with the highest probability will, of course, almost always be a non-draw outcome.
- **Simple rule-based benchmarks:** A rule that always predicts the majority class (also known as the zero rule algorithm) can be used as a baseline. The zero rule algorithm (ZeroR), which generally predicts a home team win because of the home team advantage phenomenon, is available in machine learning tools such as WEKA, as are other simple rule-based algorithms such as OneR [57], which can also be used as a benchmark. Random guesses are easily implementable and are also often used as a benchmark, e.g., [35].
- **Expert predictions:** Comparing model predictions with expert predictions, e.g., those of media or former players, is another approach to benchmarking; however, access to experts can be a challenge.[13] Butler, D., Butler, R., and Eakins [21] compared the accuracy of experts and laypeople over three English Premier League seasons and found that former professional soccer players in particular were superior in predicting match results.

5.3 Scoring Rules

Machine learning models often generate probabilities associated with each class label, and the performance of models—based on how closely these probabilities match the actual (true) class labels—can then be evaluated. Some machine learning models require that match instances are labeled with target variables encoded with a vector with three elements, e.g., in soccer, (1,0,0), (0,1,0), and (0,0,1) can denote win, draw, and loss outcomes, respectively. By representing labels as vectors in this

[13] Rather than getting experts to provide their match result predictions, other ways of incorporating expert opinion are into the model itself [68] or by engaging experts to select model features [62]. For more details on incorporating expert knowledge, please refer to the appendix.

manner, scoring rules such as the ranked probability or Brier scores can be utilized, in which a value of 0 indicates a perfect prediction and 1 represents a completely incorrect prediction (thus, lower scores are preferred).

5.3.1 Accuracy

Classification accuracy is a widely used evaluation metric and has also been used in many studies in ML for soccer match result prediction. Accuracy is computed by taking the total number of instances (matches) correctly classified by the model (where the predicted value equals the true value) and dividing this by the total number of instances, i.e.:

$$Accuracy = \frac{C}{N}$$

where C denotes the number of instances correctly classified by the model and N is the total number of instances.

5.3.2 Brier Score (BS)

The original Brier score [14] can be applied for multi-class prediction. The lower the BS, the smaller the prediction error. The original Brier score is given by:

$$\text{BS original} = \frac{1}{N}\sum_{n=1}^{N}\sum_{i=1}^{R}(\widehat{y_{in}} - y_{in})^2$$

The most common version of the Brier score is for a binary prediction, in which case it is equivalent to the mean square error (MSE) and is given by:

$$BS = \frac{1}{N}\sum_{n=1}^{N}(\widehat{y_n} - y_n)^2$$

The BS was deemed inappropriate in the 2017 Soccer Prediction Challenge since it only measures the difference between the predicted and actual scores but does not account for the ordinal nature of the win/draw/loss target variable.

5.3.3 Ranked Probability Score (RPS)

The ranked probability score (RPS) [42] does account for the ordinal nature of the three match outcomes in soccer [29] and was thus deemed more appropriate for model evaluation in the 2017 Soccer Prediction Challenge [39]. The RPS is given by:

$$RPS = \frac{1}{r-1}\sum_{i=1}^{r-1}\left(\sum_{j=1}^{i}(p_j - a_j)\right)^2 \tag{5}$$

where r denotes the number of potential match outcomes (e.g., $r = 3$ if there are three possible outcomes: home win, draw, and away win). The RPS value always lies within the interval [0, 1], with an RPS closer to 0 representing a better prediction. The RPS can be averaged across multiple instances as follows:

$$\text{RPS}_{\text{avg}} = \frac{1}{N}\sum_{i=1}^{N}\text{RPS}_i \tag{6}$$

where i denotes the ith instance in the dataset, RPS_i is the RPS value corresponding to that instance, and N is the total number of instances.

Participants in the 2017 Soccer Prediction Challenge aimed to minimize the average RPS across the $N = 206$ matches in the challenge prediction set (containing matches that participants aimed to predict):

$$\text{RPS}_{\text{avg}} = \frac{1}{206}\sum_{i=1}^{N=206}\text{RPS}_i \tag{7}$$

5.3.4 Ignorance Score (IGN)/Log Loss

The ignorance score (IGN) was proposed by Good [50], with its foundation lying in information theory. In short, the IGN penalizes predictions with larger logarithmic errors and is given by:

$$IGN = \frac{1}{N}\sum_{n=1}^{N} -(ylog_2(p) + (1-y)log_2(1-p))$$

where $y \in \{0, 1\}$ and $p = P(y = 1)$. Its value falls in the range $IGN \in [0, \infty)$, with a lower score representing better model performance. If the log base is changed to base e (multiplied by a constant scalar), the IGN is equivalent to log loss.

Based on its properties, IGN has been suggested as being more appropriate than RPS for evaluating the probabilistic forecasts of soccer match result prediction models [119].

5.3.5 Root Mean Squared Error (RMSE)

The root mean squared error (RMSE) evaluates numeric prediction models, e.g., that predict the goals scored or goals margin. The RMSE is calculated by first computing the error, the predicted value minus the actual value, and then taking the square root across all instances:

$$\text{RMSE} = \sqrt{\frac{1}{N}\sum_{i=1}^{N}(y_i - \hat{y}_i)^2}$$

where N denotes the number of instances, y_i is the actual observed value of the ith instance, and $\hat{y}_i$ is the predicted value ith instance.

5.3.6 Scoring Rule Selection and Properties

For numeric prediction problems, RMSE is the most commonly applied scoring rule, while for outcome prediction, the properties of the scoring rule need to be considered. The following three properties are the most commonly described:

- **Proper**: A scoring rule is considered proper when—given inputs $S(P, x)$ with a predicted distribution P, an actual distribution Q, an outcome $x \sim Q$, and scoring rule S—the rule returns a real value $\mathbb{R}$, which we aim to maximize. The expected score is given by $E[S(P, x)] = S(P, Q)$.
  A scoring rule is *strictly proper* if $S(Q, Q) \geq S(P, Q)$ if and only if $Q = P$, which means that the true distribution is given the lowest (most favorable) score and any other distribution receives a higher (less favorable) score.
  On the other hand, a scoring rule is *proper* if $S(Q, Q) \geq S(P, Q)$ for all distributions Q and P. A more theoretical definition can be found in [48].
- **Locality**: A scoring rule is considered local if it only considers the prediction of the observed (target) class. On the other hand, if the scoring rule considers unobserved classes, it is classified as nonlocal [119].
- **Sensitivity to distance**: In the context of three-class (home win, draw, away win) prediction, a scoring rule is considered sensitive to distance if it takes into account the ordinal nature of the outcomes of the match. For example, in soccer, when a team is leading by one goal, it only takes one goal from the opponent to turn the match into a draw and two goals for the opponent to secure a win. Consequently, the outcome space is ordered based on the goal differences.

Of the abovementioned scoring rules, the Brier score, RPS, and IGN/log loss are strictly proper, only IGN/log loss is local, and only RPS is sensitive to distance.

For the outcome prediction problem, it is essential to use a strictly proper scoring rule to incentivize the model to truthfully report its predictions [48, 16]. The objective is to develop a model that accurately reflects the true distribution of match outcomes.

Regarding the scoring rule for three-class prediction in soccer, Constantinou and Fenton [29] proposed using the RPS because of its sensitivity to distance, i.e., it takes into account the ordering in the outcomes (e.g., a home win is "closer" to a draw than it is to an away win). However, Wheatcroft [119] challenged this perspective, arguing that the sensitivity to distance does not add anything in terms of the usual aims of using scoring rules and that football outcomes may be drawn from an unknown distribution. Therefore, instead of evaluating predictions based on exact outcomes, it was suggested that it is more appropriate to consider the sample distribution of outcomes. In this context, the ignorance (log loss) score is a better choice due to its local property, which is more efficient compared to other nonlocal scoring rules.

While the sensitivity to distance property is still a topic of debate, the scoring rule that is selected should align with the objective of the prediction task.

5.4 *Temporal Splitting Methods*

The temporal order of matches should be preserved when evaluating results so that upcoming games are predicted based on past games only. One challenge in ML for sports match result prediction is that—unlike many applications of supervised machine learning techniques—the instances representing sports matches are temporally ordered. Furthermore, there is a hierarchy in terms of the way matches are structured, i.e., there are multiple matches in a particular round and multiple rounds that comprise a particular season. As a result, conventional cross-validation, which shuffles instances randomly, cannot be used since future matches may end up being incorrectly used to predict past matches [20, 18, 113]. Performance that appears extremely strong relative to other studies for an equivalent two- or three-class problem (e.g., a study reporting over 80% or 90% accuracy for a three-class problem when most other studies report between 45% and 55%) may be evidence that this mistake has been made in a study.[14] There is an option to preserve the order of the instances in the machine learning software WEKA (Fig. 1), and code in programming languages such as Python and R can be written to ensure that the temporal order of matches is preserved when training and testing models.

Deciding how to split match data into training and validation sets depends on the amount of data that the researcher has on hand, e.g., whether there is one season of match data or multiple seasons. Seasons further in the past become less relevant for predicting matches in the current or future seasons because of, e.g., changes in team rosters and team strengths. However, if player-level features are included, player changes from season to season can be accounted for. The time series nature of soccer match result data should thus be accounted for when researchers evaluate their

[14] For example, [102] used 10-fold cross-validation in WEKA and reported obtaining accuracy of 99.56% with a decision tree model.

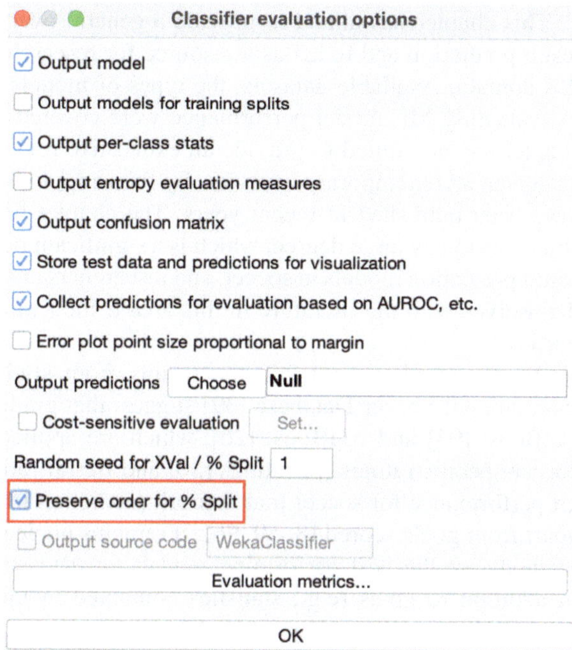

Fig. 1 The "Preserve order for % Split" option can be set in WEKA to ensure that the temporal order of instances in a dataset is maintained so that future matches are not incorrectly used to predict historical matches

models, e.g., using cross-validation for time series. Furthermore, any hierarchies that are present in terms of a particular league being made up of seasons that are comprised of rounds, which, in turn, consist of matches, should be considered. If the competition is structured such that teams sometimes play more than one match per round, this should also be taken into account, as should if teams are promoted and relegated at the end of seasons.

Important

Any match result prediction project must ensure that only matches in the past are being used to predict current or future matches. Therefore, traditional cross-validation, which shuffles data instances randomly, should not be used.

6 Conclusions and Future Directions

Soccer match result prediction is inherently difficult due to soccer being a low-scoring sport with three outcomes and generally highly competitive leagues at the professional level. There is also an element of luck in any sport, including soccer [2], which is, of course, a major part of what attracts fans.

This chapter has aimed to provide a general overview of ML for soccer match result prediction and to act as a resource for researchers conducting future work in this domain. Available datasets, the types of models and features, and approaches to evaluating ML model performance were covered in this chapter, and while the chapter has not aimed to provide an exhaustive review of all studies in the area, it has given a contemporary overview by focusing, to a greater extent, on studies that have been published in recent years. The chapter has also not considered betting strategies to any large degree, which is a significant potential use case for ML match result prediction models in soccer, and researchers are encouraged to also familiarize themselves with the literature in this area if their models will be deployed for this purpose.

As discussed in Sect. 2.2, the results from studies that have used the Open International Soccer Database [39] suggest that gradient-boosted tree models, e.g., CatBoost [93] and XGBoost [26], which are applied to feature sets consisting of soccer-specific ratings (pi-ratings [30] and Berrar ratings [8]), provide state-of-the-art performance for soccer match result prediction in the absence of match features apart from goals scored [8, 59, 96]. It remains unclear, however, whether ensemble methods are the best-performing models on datasets that contain match statistics in addition to goals (e.g., statistics contained in the European Soccer Database, football-data.co.uk) and/or external features and also whether deep learning models can outperform ensemble models. On other datasets containing match statistics and other types of features, random forests have been competitive with [5] and even surpassed the performance gradient-boosted tree models [107, 1, 43]. Thus, comparing the performance of bagging methods (e.g., random forest) with boosting methods (e.g., CatBoost) and deep learning models on different datasets with varying features is an interesting avenue for further work. Furthermore, using the recently proposed generalized attack performance (GAP) [118] ratings as model features is yet to be investigated. Rating systems themselves can also be used as predictive models by simply predicting the team with the higher rating as the winner; however, using the ratings as model features has been shown to provide better performance. Despite many sophisticated models having been developed, including hybrid approaches that combine statistical and machine learning models and rating systems [40, 74, 53], predictions derived from bookmaker odds still provide strong baseline performance [100, 5]. There are different ways in which expert knowledge can be incorporated (see the appendix) into the model itself: through feature selection, or by using the match predictions of experts as a baseline (as mentioned above, former players appear to be better at predicting results [21]). Player or team ratings can be obtained from video games, constructed from performance indicator metrics, or by aggregating player ratings—computed using plus-minus ratings or VAEP—into a team-level rating.

The interpretability of models is of greater importance to certain groups such as sports coaches and performance analysts, and techniques including random forest feature importance, SHAP, and models new to the domain such as alternating decision trees (see the appendix) may be of use in identifying and interpreting performance indicator model features that are most relevant for winning. In the context of deep learning models, which are generally black-box, mimic learning [110] may be worth exploring to aid in their interpretability for match result prediction.

Predicting a different target variable, e.g., expected goals/xG, shot attempts, shots on target, etc., to predict the match result may be superior for prediction than using a three-class win/draw/loss outcome target variable. The best evaluation metric for probabilistic forecasting in this domain remains a subject of debate [30, 119]. Researchers should ensure that their model evaluation adequately accounts for the time series nature of soccer match result data and its hierarchical nature (i.e., match, round, season).

While all existing rating systems account for historical results—as well as accounting for the venue (home/away) and creating separate offensive and defensive ratings in the case of the soccer-specific rating systems—developing new rating systems that incorporate additional factors is an avenue for further research. For instance, rating systems could incorporate information derived from, e.g., event log data, spatiotemporal tracking data, player ratings, player/team attributes and performance indicators, passing networks, interplayer chemistry, and even information from social media. Streak and form features could potentially be engineered at both the player (hot shoe) and team levels (form and streaks in terms of match outcomes [5]) and their value as model features compared.

Acknowledgments This chapter was partly supported by JSPS KAKENHI (grant number 20H04075) and JST Presto (grant number JPMJPR20CA).

Appendix

Incorporating Expert Knowledge

The incorporation of domain expert knowledge has been a key area of enquiry for some time. Expert knowledge can be incorporated into the modelling process itself: e.g., Joseph, Fenton, and Neil [68] did so in constructing their Bayesian network. Alternatively, expert knowledge can be incorporated by obtaining expert predictions as a benchmark with which to compare model prediction performance (see also Sect. 5.2). Another way in which expert domain knowledge of the sport can be incorporated is by asking experts to select features. Hucaljuk and Rakipović [62] considered 96 matches from the European Champions League and compared the performance of ML models that included features selected by feature selection algorithms with models that included features selected by experts. The feature selection algorithms produced a basic feature set of 20 features in their initial feature set, which they compared with a feature set consisting of this initial set plus the expert-selected features. With their particular models and on their specific dataset, however, the expert-selected features were not found to yield any improvement in model performance. More recently, Beal et al. [6] developed a benchmark dataset and results from natural language processing and ML models for soccer match result prediction, and the performance of these models was compared with statistical models. The authors provided a prediction accuracy baseline that utilizes match statistics and match preview articles written by sports journalists in The Guardian (London, United Kingdom) over six English Premier League seasons.

Elo K-Factor Selection Techniques Described Online

The World Football Elo Ratings Website (eloratings.net) [103] provides Elo ratings of national teams and uses a higher value of K for matches that are of greater importance. For example, $K = 60$ for World Cup finals matches and $K = 20$ for international friendlies. The K-factor is also adjusted based on the goal margin. Specifically, the K-factor is increased by 1/2 for a win by 2 goals, by 3/4 for a win by 3 goals, and by $\frac{3}{4} + \frac{(M-3)}{8}$ if the team wins by 4 or more goals, where M denotes the goal margin.

The Football Database website (footballdatabase.com) provides Elo ratings for club teams from various leagues around the world, also applying different values of K based on the league or importance of the match.[15] The K-factor is multiplied by a value, G, which accounts for the goal margin. In particular, $G = 1$ if the match was drawn, $G = 1.5$ if the goal margin was 1 or 2 goals, and $G = \frac{11+M}{8}$ if the goal margin was 3 or more, where M again denotes the goals margin.

2023 Soccer Prediction Challenge

A subsequent Soccer Prediction Challenge, using a similar dataset to the 2017 challenge, was held in 2023.[16] However, unlike the 2017 competition, the 2023 version required two tasks: first, predicting match results in terms of the exact goals scored by each team, and second, as per the 2017 challenge, predicting match results by computing the probabilities for a win, draw, and loss. The exact score prediction models were evaluated using the root-mean-squared-error metric. Similar to the 2017 challenge, studies will be submitted to the Machine Learning (Springer) journal; however, these studies were not available at the time the current chapter was written.

---

[15] See footballdatabase.com/methodology.php for further details.

[16] https://sites.google.com/view/2023soccerpredictionchallenge.

Alternating Decision Trees (ADTrees)

An interesting boosted-tree model that is relatively unexplored in sports result prediction is the alternating decision tree (ADTree) model [46]. As opposed to XGBoost and CatBoost, both of which use gradient boosting and gradient descent optimization, ADTree uses the well-known AdaBoost [47] algorithm. AdaBoost is used to grow the ADTree, with each iteration of AdaBoost adding a branch to the ADTree. The general idea of AdaBoost is to iteratively place more weight on instances that were previously incorrectly classified in a previous iteration. The ADTree algorithm combines the accuracy-increasing benefits of boosting while producing an interpretable decision tree structure as the final ADTree model. Given the interpretability of the ADTree, an enhanced version of ADTree that can compete with the likes of CatBoost would be a great asset to the sports match result prediction domain.[17]

References

1. Alfredo, Y.F., Isa, S.M.: Football match prediction with tree based model classification. International Journal of Intelligent Systems and Applications **11**(7), 20–28 (2019)
2. Aoki, R.Y., Assuncao, R.M., Vaz de Melo, P.O.: Luck is hard to beat: The difficulty of sports prediction. In: Proceedings of the 23rd ACM SIGKDD International Conference on Knowledge Discovery and Data Mining, pp. 1367–1376 (2017)
3. Arik, S.Ö., Pfister, T.: Tabnet: Attentive interpretable tabular learning. In: Proceedings of the AAAI Conference on Artificial Intelligence, vol. 35, pp. 6679–6687 (2021)
4. Arntzen, H., Hvattum, L.M.: Predicting match outcomes in association football using team ratings and player ratings. Statistical Modelling **21**(5), 449–470 (2021)
5. Baboota, R., Kaur, H.: Predictive analysis and modelling football results using machine learning approach for english premier league. International Journal of Forecasting **35**(2), 741–755 (2019)
6. Beal, R., Middleton, S.E., Norman, T.J., Ramchurn, S.D.: Combining machine learning and human experts to predict match outcomes in football: A baseline model. In: Proceedings of the AAAI Conference on Artificial Intelligence, vol. 35, pp. 15447–15451 (2021)
7. Berrar, D., Lopes, P., Davis, J., Dubitzky, W.: Guest editorial: special issue on machine learning for soccer. Machine Learning **108**, 1–7 (2019)
8. Berrar, D., Lopes, P., Dubitzky, W.: Incorporating domain knowledge in machine learning for soccer outcome prediction. Machine Learning **108**, 97–126 (2019)
9. Bittner, E., Nußbaumer, A., Janke, W., Weigel, M.: Self-affirmation model for football goal distributions. Europhysics Letters **78**(5), 58002 (2007)

[17] Bunker, Yeung, Susnjak, Espie, and Fujii [19] recently found that ADTrees performed well in predicting professional ATP tennis match results when applied to the difference in average betting odds across 11 different bookmaker companies (the companies included in the betting odds data on tennis-data.co.uk). Betting odds, as mentioned before, can be considered a rating [122], so this is another example of boosted-tree models, applied to rating features, being effective for sports match result prediction.

10. Bittner, E., Nußbaumer, A., Janke, W., Weigel, M.: Football fever: goal distributions and non-gaussian statistics. The European Physical Journal B **67**, 459–471 (2009)
11. Bradley, R.A., Terry, M.E.: Rank analysis of incomplete block designs: I. the method of paired comparisons. Biometrika **39**(3/4), 324–345 (1952)
12. Bransen, L., Van Haaren, J.: Player chemistry: Striving for a perfectly balanced soccer team. arXiv preprint arXiv:2003.01712 (2020)
13. Breiman, L.: Random forests. Machine Learning **45**, 5–32 (2001)
14. Brier, G.W.: Verification of forecasts expressed in terms of probability. Monthly Weather Review **78**(1), 1–3 (1950)
15. Brin, S., Page, L.: The anatomy of a large-scale hypertextual web search engine. Computer Networks and ISDN systems **30**(1–7), 107–117 (1998)
16. Bröcker, J., Smith, L.A.: Scoring probabilistic forecasts: The importance of being proper. Weather and Forecasting **22**(2), 382–388 (2007)
17. Brown, A., Reade, J.J.: The wisdom of amateur crowds: Evidence from an online community of sports tipsters. European Journal of Operational Research **272**(3), 1073–1081 (2019)
18. Bunker, R., Susnjak, T.: The application of machine learning techniques for predicting match results in team sport: A review. Journal of Artificial Intelligence Research **73**, 1285–1322 (2022)
19. Bunker, R., Yeung, C., Susnjak, T., Espie, C., Fujii, K.: A comparative evaluation of elo ratings-and machine learning-based methods for tennis match result prediction. Proceedings of the Institution of Mechanical Engineers, Part P: Journal of Sports Engineering and Technology p. 17543371231212235 (2023)
20. Bunker, R.P., Thabtah, F.: A machine learning framework for sport result prediction. Applied Computing and Informatics **15**(1), 27–33 (2019)
21. Butler, D., Butler, R., Eakins, J.: Expert performance and crowd wisdom: Evidence from english premier league predictions. European Journal of Operational Research **288**(1), 170–182 (2021)
22. Buursma, D.: Predicting sports events from past results towards effective betting on football matches. In: Conference Paper, presented at 14th Twente Student Conference on IT, Twente, Holland, vol. 21. sn (2011)
23. Carloni, L., De Angelis, A., Sansonetti, G., Micarelli, A.: A machine learning approach to football match result prediction. In: HCI International 2021-Posters: 23rd HCI International Conference, HCII 2021, Virtual Event, July 24–29, 2021, Proceedings, Part II 23, pp. 473–480. Springer (2021)
24. Carpita, M., Ciavolino, E., Pasca, P.: Exploring and modelling team performances of the kaggle european soccer database. Statistical Modelling **19**(1), 74–101 (2019)
25. Chen, H.: Neural network algorithm in predicting football match outcome based on player ability index. Advances in Physical Education **9**(4), 215–222 (2019)
26. Chen, T., Guestrin, C.: Xgboost: A scalable tree boosting system. In: Proceedings of the 22nd ACM SIGKDD International conference on Knowledge Discovery and Data Mining, pp. 785–794 (2016)
27. Cho, Y., Yoon, J., Lee, S.: Using social network analysis and gradient boosting to develop a soccer win–lose prediction model. Engineering Applications of Artificial Intelligence **72**, 228–240 (2018)
28. Constantinou, A.C.: Dolores: a model that predicts football match outcomes from all over the world. Machine Learning **108**(1), 49–75 (2019)
29. Constantinou, A.C., Fenton, N.E.: Solving the problem of inadequate scoring rules for assessing probabilistic football forecast models. Journal of Quantitative Analysis in Sports **8**(1) (2012)
30. Constantinou, A.C., Fenton, N.E.: Determining the level of ability of football teams by dynamic ratings based on the relative discrepancies in scores between adversaries. Journal of Quantitative Analysis in Sports **9**(1), 37–50 (2013)
31. Constantinou, A.C., Fenton, N.E., Neil, M.: Profiting from an inefficient association football gambling market: Prediction, risk and uncertainty using bayesian networks. Knowledge-Based Systems **50**, 60–86 (2013)

32. da Costa, I.B., Marinho, L.B., Pires, C.E.S.: Forecasting football results and exploiting betting markets: The case of "both teams to score". International Journal of Forecasting **38**(3), 895–909 (2022)
33. Coulom, R.: Computing "elo ratings" of move patterns in the game of go. ICGA Journal **30**(4), 198–208 (2007)
34. Curley, J.P.: engsoccerdata: English soccer data 1871–2016. R package version 0.1.5 DOI
35. Danisik, N., Lacko, P., Farkas, M.: Football match prediction using players attributes. In: 2018 World Symposium on Digital Intelligence for Systems and Machines (DISA), pp. 201–206. IEEE (2018)
36. Decroos, T., Bransen, L., Van Haaren, J., Davis, J.: Actions speak louder than goals: Valuing player actions in soccer. In: Proceedings of the 25th ACM SIGKDD International Conference on Knowledge Discovery & Data Mining, pp. 1851–1861 (2019)
37. Dixon, M.J., Coles, S.G.: Modelling association football scores and inefficiencies in the football betting market. Journal of the Royal Statistical Society: Series C (Applied Statistics) **46**(2), 265–280 (1997)
38. Dmochowski, J.P.: A statistical theory of optimal decision-making in sports betting. PLOS One **18**(6), e0287601 (2023)
39. Dubitzky, W., Lopes, P., Davis, J., Berrar, D.: The open international soccer database for machine learning. Machine Learning **108**, 9–28 (2019)
40. Elmiligi, H., Saad, S.: Predicting the outcome of soccer matches using machine learning and statistical analysis. In: 2022 IEEE 12th Annual Computing and Communication Workshop and Conference (CCWC), pp. 1–8. IEEE (2022)
41. Elo, A.: The rating of chessplayers, past and present (1978)
42. Epstein, E.S.: A scoring system for probability forecasts of ranked categories. Journal of Applied Meteorology (1962–1982) **8**(6), 985–987 (1969)
43. Eryarsoy, E., Delen, D.: Predicting the outcome of a football game: A comparative analysis of single and ensemble analytics methods. In: Proceedings of the 52nd Hawaii International Conference on System Sciences | 2019 (2019)
44. Esme, E., Kiran, M.S.: Prediction of football match outcomes based on bookmaker odds by using k-nearest neighbor algorithm. International Journal of Machine Learning and Computing **8**(1), 26–32 (2018)
45. Forrest, D., Goddard, J., Simmons, R.: Odds-setters as forecasters: The case of english football. International Journal of Forecasting **21**(3), 551–564 (2005)
46. Freund, Y., Mason, L.: The alternating decision tree learning algorithm. In: ICML, vol. 99, pp. 124–133 (1999)
47. Freund, Y., Schapire, R.E.: A decision-theoretic generalization of on-line learning and an application to boosting. Journal of Computer and System Sciences **55**(1), 119–139 (1997)
48. Gneiting, T., Raftery, A.E.: Strictly proper scoring rules, prediction, and estimation. Journal of the American Statistical Association **102**(477), 359–378 (2007)
49. Goes, F., Meerhoff, L., Bueno, M., Rodrigues, D., Moura, F., Brink, M., Elferink-Gemser, M., Knobbe, A., Cunha, S., Torres, R., et al.: Unlocking the potential of big data to support tactical performance analysis in professional soccer: A systematic review. European Journal of Sport Science **21**(4), 481–496 (2021)
50. Good, I.J.: Rational Decisions, pp. 365–377. Springer New York, New York, NY (1992). DOI 10.1007/978-1-4612-0919-5_24. URL https://doi.org/10.1007/978-1-4612-0919-5_24
51. Greenhough, J., Birch, P., Chapman, S.C., Rowlands, G.: Football goal distributions and extremal statistics. Physica A: Statistical Mechanics and its Applications **316**(1–4), 615–624 (2002)
52. Groll, A., Ley, C., Schauberger, G., Van Eetvelde, H.: Prediction of the fifa world cup 2018-a random forest approach with an emphasis on estimated team ability parameters. arXiv preprint arXiv:1806.03208 (2018)
53. Groll, A., Ley, C., Schauberger, G., Van Eetvelde, H.: A hybrid random forest to predict soccer matches in international tournaments. Journal of Quantitative Analysis in Sports **15**(4), 271–287 (2019)

54. Guan, S., Wang, X.: Optimization analysis of football match prediction model based on neural network. Neural Computing and Applications pp. 1–17 (2022)
55. Hall, M.A.: Correlation-based feature selection for machine learning. Ph.D. thesis, The University of Waikato (1999)
56. Hochreiter, S., Schmidhuber, J.: Long short-term memory. Neural Computation **9**(8), 1735–1780 (1997)
57. Holte, R.: Very simple classification rules perform well on most commonly used datasets. Machine Learning **11**, 63–91 (1993)
58. Hubáček, O., Šourek, G., Železný, F.: Exploiting sports-betting market using machine learning. International Journal of Forecasting **35**(2), 783–796 (2019)
59. Hubáček, O., Šourek, G., Železný, F.: Learning to predict soccer results from relational data with gradient boosted trees. Machine Learning **108**, 29–47 (2019)
60. Hubácek, O., Sourek, G., Zelezny, F.: Score-based soccer match outcome modeling–an experimental review. MathSport International (2019)
61. Hubáček, O., Šourek, G., Železný, F.: Forty years of score-based soccer match outcome prediction: an experimental review. IMA Journal of Management Mathematics **33**(1), 1–18 (2022)
62. Hucaljuk, J., Rakipović, A.: Predicting football scores using machine learning techniques. In: 2011 Proceedings of the 34th International Convention MIPRO, pp. 1623–1627. IEEE (2011)
63. Hughes, M.D., Bartlett, R.M.: The use of performance indicators in performance analysis. Journal of Sports Sciences **20**(10), 739–754 (2002)
64. Hvattum, L.M.: A comprehensive review of plus-minus ratings for evaluating individual players in team sports (2019)
65. Hvattum, L.M., Arntzen, H.: Using elo ratings for match result prediction in association football. International Journal of Forecasting **26**(3), 460–470 (2010)
66. Ievoli, R., Palazzo, L., Ragozini, G.: On the use of passing network indicators to predict football outcomes. Knowledge-Based Systems **222**, 106997 (2021)
67. Jain, S., Tiwari, E., Sardar, P.: Soccer result prediction using deep learning and neural networks. In: Intelligent Data Communication Technologies and Internet of Things: Proceedings of ICICI 2020, pp. 697–707. Springer (2021)
68. Joseph, A., Fenton, N.E., Neil, M.: Predicting football results using bayesian nets and other machine learning techniques. Knowledge-Based Systems **19**(7), 544–553 (2006)
69. Joseph, L.D.: Time series approaches to predict soccer match outcome. Ph.D. thesis, Dublin, National College of Ireland (2022)
70. Kampakis, S., Adamides, A.: Using twitter to predict football outcomes. arXiv preprint arXiv:1411.1243 (2014)
71. Ke, G., Meng, Q., Finley, T., Wang, T., Chen, W., Ma, W., Ye, Q., Liu, T.Y.: Lightgbm: A highly efficient gradient boosting decision tree. Advances in Neural Information Processing Systems **30** (2017)
72. Kinalioğlu, İ.H., Kuş, C.: Prediction of football match results by using artificial intelligence-based methods and proposal of hybrid methods. International Journal of Nonlinear Analysis and Applications **14**(1), 2939–2969 (2023)
73. Kira, K., Rendell, L.A.: A practical approach to feature selection. In: D.H. Sleeman, P. Edwards (eds.) Ninth International Workshop on Machine Learning, pp. 249–256. Morgan Kaufmann (1992)
74. Knoll, J., Stübinger, J.: Machine-learning-based statistical arbitrage football betting. KI-Künstliche Intelligenz **34**(1), 69–80 (2020)
75. Kononenko, I.: Estimating attributes: Analysis and extensions of relief. In: F. Bergadano, L.D. Raedt (eds.) European Conference on Machine Learning, pp. 171–182. Springer (1994)
76. Leitner, C., Zeileis, A., Hornik, K.: Is federer stronger in a tournament without nadal? an evaluation of odds and seedings for wimbledon 2009. Austrian Journal of Statistics **38**(4), 277–286 (2009)
77. Leitner, M.C., Daumann, F., Follert, F., Richlan, F.: The cauldron has cooled down: a systematic literature review on home advantage in football during the covid-19 pandemic from a socio-economic and psychological perspective. Management Review Quarterly **73**(2), 605–633 (2023)

78. Ley, C., Wiele, T.V.d., Eetvelde, H.V.: Ranking soccer teams on the basis of their current strength: A comparison of maximum likelihood approaches. Statistical Modelling **19**(1), 55–73 (2019)
79. Li, J., Zhang, H., Zhao, J., Guo, X., Rihan, W., Deng, G.: Embedded feature selection and machine learning methods for flash flood susceptibility-mapping in the mainstream songhua river basin, china. Remote Sensing **14**(21), 5523 (2022)
80. Lundberg, S.M., Lee, S.I.: A unified approach to interpreting model predictions. Advances in Neural Information Processing Systems **30** (2017)
81. Maher, M.J.: Modelling association football scores. Statistica Neerlandica **36**(3), 109–118 (1982)
82. Malamatinos, M.C., Vrochidou, E., Papakostas, G.A.: On predicting soccer outcomes in the greek league using machine learning. Computers **11**(9), 133 (2022)
83. Miljković, D., Gajić, L., Kovačević, A., Konjović, Z.: The use of data mining for basketball matches outcomes prediction. In: IEEE 8th International Symposium on Intelligent Systems and Informatics, pp. 309–312. IEEE (2010)
84. Moustakidis, S., Plakias, S., Kokkotis, C., Tsatalas, T., Tsaopoulos, D.: Predicting football team performance with explainable ai: Leveraging shap to identify key team-level performance metrics. Future Internet **15**(5), 174 (2023)
85. Natarajan, S., Khot, T., Kersting, K., Gutmann, B., Shavlik, J.: Gradient-based boosting for statistical relational learning: The relational dependency network case. Machine Learning **86,** 25–56 (2012)
86. Odachowski, K., Grekow, J.: Using bookmaker odds to predict the final result of football matches. In: Knowledge Engineering, Machine Learning and Lattice Computing with Applications: 16th International Conference, KES 2012, San Sebastian, Spain, September 10–12, 2012, Revised Selected Papers 16, pp. 196–205. Springer (2013)
87. Omomule, T., Ibinuolapo, A., Ajayi, O.: Fuzzy-based model for predicting football match results. Int. J. Sci. Res. in Computer Science and Engineering Vol **8**(1) (2020)
88. Ötting, M., Andreas, G.: A regularized hidden markov model for analyzing the 'hot shoe' in football. Statistical Modelling **22**(6), 546–565 (2022)
89. Palinggi, D.A.: Predicting soccer outcome with machine learning based on weather condition. Ph.D. thesis (2019)
90. Pappalardo, L., Cintia, P., Rossi, A., Massucco, E., Ferragina, P., Pedreschi, D., Giannotti, F.: A public data set of spatio-temporal match events in soccer competitions. Scientific Data **6**(1), 236 (2019)
91. Petretta, M., Schiavon, L., Diquigiovanni, J.: On the dependence in football match outcomes: traditional model assumptions and an alternative proposal. arXiv preprint arXiv:2103.07272 (2021)
92. Pipatchatchawal, C., Phimoltares, S.: Predicting football match result using fusion-based classification models. In: 2021 18th International Joint Conference on Computer Science and Software Engineering (JCSSE), pp. 1–6. IEEE (2021)
93. Prokhorenkova, L., Gusev, G., Vorobev, A., Dorogush, A.V., Gulin, A.: Catboost: unbiased boosting with categorical features. Advances in Neural Information Processing Systems **31** (2018)
94. Rahman, M.A.: A deep learning framework for football match prediction. SN Applied Sciences **2**(2), 165 (2020)
95. Randrianasolo, A.S.: Predicting euro games using an ensemble technique involving genetic algorithms and machine learning. In: 2023 IEEE 13th Annual Computing and Communication Workshop and Conference (CCWC), pp. 0470–0475. IEEE (2023)
96. Razali, M.N., Mustapha, A., Mostafa, S.A., Gunasekaran, S.S.: Football matches outcomes prediction based on gradient boosting algorithms and football rating system. Human Factors in Software and Systems Engineering **61**, 57 (2022)
97. Razali, N., Mustapha, A., Arbaiy, N., Lin, P.C.: Deep learning for football outcomes prediction based on football rating system. In: AIP Conference Proceedings, vol. 2644. AIP Publishing (2022)

98. Rein, R., Memmert, D.: Big data and tactical analysis in elite soccer: future challenges and opportunities for sports science. SpringerPlus **5**(1), 1–13 (2016)
99. Ren, Y., Susnjak, T.: Predicting football match outcomes with explainable machine learning and the kelly index. arXiv preprint arXiv:2211.15734 (2022)
100. Robberechts, P., Davis, J.: Forecasting the fifa world cup–combining result-and goal-based team ability parameters. In: Machine Learning and Data Mining for Sports Analytics: 5th International Workshop, MLSA 2018, Co-located with ECML/PKDD 2018, Dublin, Ireland, September 10, 2018, Proceedings 5, pp. 16–30. Springer (2019)
101. Robberechts, P., Van Haaren, J., Davis, J.: A bayesian approach to in-game win probability in soccer. In: Proceedings of the 27th ACM SIGKDD Conference on Knowledge Discovery & Data Mining, pp. 3512–3521 (2021)
102. Rosli, C.M.F.C.M., Saringat, M.Z., Razali, N., Mustapha, A.: A comparative study of data mining techniques on football match prediction. In: Journal of Physics: Conference Series, vol. 1020, p. 012003. IOP Publishing (2018)
103. Runyan, B.: The world football elo rating system. website (1997)
104. Ryall, R., Bedford, A.: An optimized ratings-based model for forecasting australian rules football. International Journal of Forecasting **26**(3), 511–517 (2010)
105. Stefani, R.T.: Football and basketball predictions using least squares. IEEE Transactions on systems, man, and cybernetics **7**(2), 117–21 (1977)
106. Štrumbelj, E., Šikonja, M.R.: Online bookmakers' odds as forecasts: The case of european soccer leagues. International Journal of Forecasting **26**(3), 482–488 (2010)
107. Stübinger, J., Knoll, J.: Beat the bookmaker–winning football bets with machine learning (best application paper). In: Artificial Intelligence XXXV: 38th SGAI International Conference on Artificial Intelligence, AI 2018, Cambridge, UK, December 11–13, 2018, Proceedings 38, pp. 219–233. Springer (2018)
108. Stübinger, J., Mangold, B., Knoll, J.: Machine learning in football betting: Prediction of match results based on player characteristics. Applied Sciences **10**(1), 46 (2019)
109. Sujatha, K., Godhavari, T., Bhavani, N.P.: Football match statistics prediction using artificial neural networks. International Journal of Mathematical and Computational Methods **3** (2018)
110. Sun, X., Davis, J., Schulte, O., Liu, G.: Cracking the black box: Distilling deep sports analytics. In: Proceedings of the 26th ACM SIGKDD International Conference on Knowledge Discovery & Data Mining, pp. 3154–3162 (2020)
111. Surowiecki, J.: The wisdom of crowds. Anchor (2005)
112. Talattinis, K., Kyriakides, G., Kapantai, E., Stephanides, G.: Forecasting soccer outcome using cost-sensitive models oriented to investment opportunities. International Journal of Computer Science in Sport **18**(1) (2019)
113. Tax, N., Joustra, Y.: Predicting the dutch football competition using public data: A machine learning approach. Transactions on Knowledge and Data Engineering **10**(10), 1–13 (2015)
114. Thorp, E.O.: Portfolio choice and the kelly criterion. In: Stochastic Optimization Models in Finance, pp. 599–619. Elsevier (1975)
115. Toda, K., Teranishi, M., Kushiro, K., Fujii, K.: Evaluation of soccer team defense based on prediction models of ball recovery and being attacked: A pilot study. PLOS One **17**(1), e0263051 (2022)
116. Tsokos, A., Narayanan, S., Kosmidis, I., Baio, G., Cucuringu, M., Whitaker, G., Király, F.: Modeling outcomes of soccer matches. Machine Learning **108**, 77–95 (2019)
117. Tümer, A.E., Akyıldız, Z., Güler, A.H., Saka, E.K., Ievoli, R., Palazzo, L., Clemente, F.M.: Prediction of soccer clubs' league rankings by machine learning methods: The case of turkish super league. Proceedings of the Institution of Mechanical Engineers, Part P: Journal of Sports Engineering and Technology p. 17543371221140492 (2022)
118. Wheatcroft, E.: A profitable model for predicting the over/under market in football. International Journal of Forecasting **36**(3), 916–932 (2020)
119. Wheatcroft, E.: Evaluating probabilistic forecasts of football matches: the case against the ranked probability score. Journal of Quantitative Analysis in Sports **17**(4), 273–287 (2021)

120. Wheatcroft, E.: Forecasting football matches by predicting match statistics. Journal of Sports Analytics **7**(2), 77–97 (2021)
121. Wheatcroft, E., Sienkiewicz, E.: A probabilistic model for predicting shot success in football. arXiv preprint arXiv:2101.02104 (2021)
122. Wunderlich, F., Memmert, D.: The betting odds rating system: Using soccer forecasts to forecast soccer. PLOS One **13**(6), e0198668 (2018)
123. Wunderlich, F., Memmert, D.: A big data analysis of twitter data during premier league matches: do tweets contain information valuable for in-play forecasting of goals in football? Social Network Analysis and Mining **12**, 1–15 (2022)
124. Yeung, C.C., Bunker, R., Fujii, K.: A framework of interpretable match results prediction in football with fifa ratings and team formation. PLOS One **18**(4), e0284318 (2023)
125. Zeileis, A., Leitner, C., Hornik, K.: Probabilistic forecasts for the 2018 fifa world cup based on the bookmaker consensus model. Tech. rep., Working Papers in Economics and Statistics (2018)
126. Zheng, A., Casari, A.: Feature engineering for machine learning: principles and techniques for data scientists. O'Reilly Media, Inc. (2018)
127. Zulkifli, S., Mustapha, A.B., Ismail, S., Razali, N.: Comparative analysis of statistical and machine learning methods for classification of match outcomes in association football. In: Proceedings of the 7th International Conference on the Applications of Science and Mathematics 2021: Sciemathic 2021, pp. 351–365. Springer (2022)

Machine Learning for the Prediction of the Index of Effectiveness in Cycling

A. Torres, M. A. Yepez, G. Millour, F. Nougarou, and F. Domingue

Abstract Biomechanical research permits the description, analysis, and assessment of the mechanics and biophysics of the musculoskeletal system during the execution of a movement. It is a science that is related to different fields such as medicine, sports, ergonomics, and engineering. It involves carefully analyzing each movement position, velocity, and acceleration to understand how the body moves and functions. Biomechanics has benefited from technological advances in recent years resulting in a large amount of data for kinematic and kinetic analysis. In this way, it has contributed by developing innovative, effective, and practical solutions to improve athlete performance, optimize training plans, improve movement patterns, prevent injuries, and develop better equipment. Nowadays, machine learning algorithms have transformed different aspects of the world, from modifying the process of performing tasks to health care and sports, increasing the performance of athletes without neglecting their health and comfort. Furthermore, machine learning is a solution to solve complex problems in biomechanics. Nevertheless, some of these techniques are often black boxes, in which the contribution of each predictor is unknown, which may be a disadvantage.

In the context of cycling, there are many biomechanical studies applied to improve the cyclist's performance and understand his movement technique. Pedaling technique is one of the most important and complex aspects of a cyclist because it affects his efficiency, power, endurance, and comfort. For this purpose, it is necessary to generate an effective force on the cranks through a combination of the cyclist's muscle recruitment and joint kinematics. Currently, for measuring that, a few motion capture systems and instrumental devices have been developed. They make it possible to identify kinematics and kinetics variables, as well as power and cadence, to know if the pedaling force production is optimal.

A. Torres (✉) · M. A. Yepez · F. Nougarou · F. Domingue
Université du Québec à Trois-Rivières, Trois-Rivières, QC, Canada
e-mail: Andres.Torres2@uqtr.ca

G. Millour
BeScored Institute, Valbonne, France

M. J. Blondin et al. (eds.), *Artificial Intelligence, Optimization, and Data Sciences in Sports*, Springer Optimization and Its Applications 218,
https://doi.org/10.1007/978-3-031-76047-1_3

Considering these multiple variables are obtained from different measurement systems, it is difficult to analyze and compare analytically the results. Machine learning algorithms have been considered as alternatives for modeling in biomechanics, and there is no doubt that it has the potential to revolutionize the sports industry and push the boundaries of what is possible. The prediction of variables in cycling through machine learning allows a better understanding of the phenomena involved in the generation of pedaling force and power. Therefore, in this chapter, we present the methodology for the creation of an interpretable model for predicting pedaling efficiency from lower-limb kinematics.

For this purpose, we have analyzed the strengths and limitations of some machine learning methodologies to propose the most optimal and interpretable model according to the knowledge of biomechanics. Therefore, the chapter is organized with the first part introducing the basic concepts of biomechanics. Then some machine learning concepts are presented with emphasis on multiple linear regression model. Subsequently, biomechanical modeling and some applications in cycling are presented. It also presents the methodology for the development of machine learning models. In addition, an example of the application of a methodology applied to cycling is illustrated. In the example, a multiple linear regression model for the prediction of the index of effectiveness is clearly explained with emphasis on the selection of predictors to obtain an interpretable model. Finally, an analysis of the model obtained with the conclusions of the example shown is provided, leaving open the possibility of applying other machine learning methodologies, not only to the case of cycling but also to other sports, thanks to the versatility of this type of methodologies.

Keywords Cycling biomechanics · Machine learning · Modeling · Movement analysis · Pedaling kinetics · Index of effectiveness

1 Introduction

The biomechanics of human movement is defined as the interdiscipline that describes, analyzes, and assesses human movement [82]. Human movement is a phenomenon that involves multiple movements of different parts of the body. Biomechanics uses the principles of classical mechanics and applies them to the musculoskeletal system to analyze the execution of a movement. In classical mechanics, to analyze a movement it is necessary to know the displacements, linear velocities, linear accelerations, angles, angular velocities, and angular accelerations of the body in motion. This type of analysis is known as kinematics [82]. On the other hand, the analysis of the internal or external forces that cause movement is known as kinetics [82]. External forces generally come from contact with the earth's surface or from a device, usually the floor, which is touching a body segment. In addition, forces can be originated from the muscular contraction of the muscles surrounding a joint, and therefore it is necessary to know the signal activation. To know this activation signal, controlled by the nervous system, the technique known as surface electromyography was developed [13]. It provides information when the muscles are activated, a phenomenon known as muscle recruitment [82]. From the information obtained from kinematics, kinetics, and surface electromyography, the task is to apply a method to analyze the movement, whether it is the simple movement of a joint or the analysis of a sporting gesture. These methods can vary depending on the technology available and in the case of sport will depend on multiple factors that can range from environmental to physical. Computational advances and measurement technology have facilitated these analyses, allowing new options, such as machine learning, which in recent years has emerged as an alternative method for use in biomechanics. Machine learning techniques can be used for injury prediction and prevention [79] or estimation of some variable of interest, e.g., knee joint forces [70]. In addition, it is possible to predict the performance of a cyclist by means of a multiple linear regression model [73]. Therefore, in this chapter, we will present a case of the application of machine learning for sports analysis, specifically in cycling, which can be extrapolated to other sports or analysis in the science of human movement.

2 Fundamental Concepts

2.1 Kinematics in Biomechanics

To describe the kinematics, three-dimensional (3D) motion capture systems are used. These systems are based on cameras or inertial measurement units (IMUs) that integrate accelerometers, gyroscopes, and magnetometers. Using commercial or custom software, the data is processed to determine the joint angles, joint angular velocities, and joint angular accelerations according to a biomechanical model. Despite the advances with IMUs, there is still debate about their accuracy, and therefore camera acquisition systems are considered the gold standard. For this purpose, these motion capture systems use biomechanical protocols to place markers in anatomical sites, which have been validated and proposed as standard protocols by institutions such as the International Society of Biomechanics [83]. There are different protocols for marker positioning: Cleveland Clinic Marker Set, IOR model [3], and the Newington-Helen Hayes gait model, i.e., 42 anatomical markers [17], which is the most recognized and widely implemented. In the case of cycling, the number of anatomical markers is modified compared to the Helen Hayes marker set [35], to facilitate the work of the bike fitter in terms of time and to avoid marker occlusion problems. As a result, the number of markers is simplified to define the lower limbs, upper limbs, and some points of the column.

Regarding the anatomical markers, there are two methods to allow the cameras to detect the anatomical points. First, passive markers, which are covered by a reflective film or reflective paint, the camera has an infrared filter to receive only the light reflection of the marker. This methodology is used in human gait analyses and sports in general. Nevertheless, the room and the individual must be free of any element that reflects light. This is a difficulty in the case of cycling since both the bike and the user have several reflective points, i.e., uniform, shoes, and certain parts of the bicycle, causing errors in the measurement as the system confuses these points as anatomical markers. For this reason, the use of active markers, which emit infrared light while blocking the camera's infrared filter, has become popular in cycling, detecting points of interest with high fidelity. The difficulty lies in the wiring or in the considerable size and weight of the marker and battery pack. However, it is the most widely used system in cycling also because of the abovementioned use of simplified marker positioning protocols.

2.2 *Kinetics in Biomechanics*

To know the forces in a biomechanical system, the main technology implemented is the strain gauges, which are attached to a plate or to any system where the athlete's body part is interacting. Depending on the direction in which the strain gauges are placed, it is possible to know the forces, i.e., F_x, F_y, F_z, in the three Cartesian axes, i.e., *X*, *Y*, *Z*. This technology is widely used in biomechanics research, and its use has been recognized by the International Society of Biomechanics, which has defined standards for its application [83]. In the case of cycling, the strain gauges are integrated in the pedal. In this way, the external forces that reach the foot are obtained, specifically in the metatarsal heads where the cleats of the shoe are approximately located. Knowing these forces and using trigonometric formulas, tangential or effective force F_{T}, radial force F_{R}, mediolateral force F_{ML}, and total force F_{tot} are calculated as shown in Fig. 1.

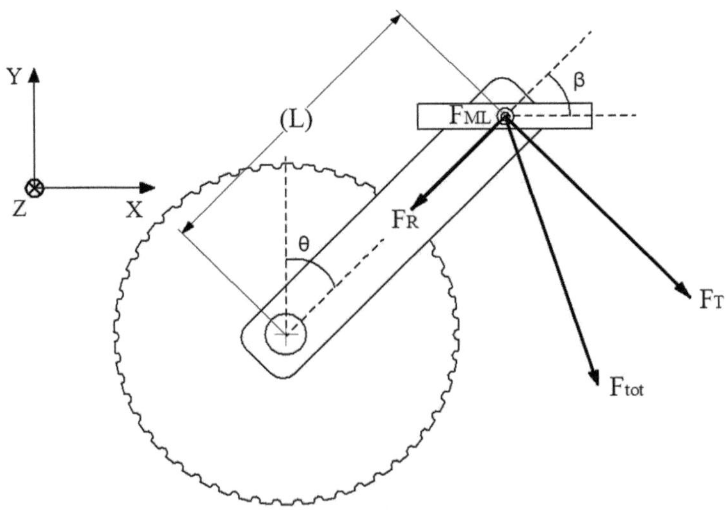

Fig. 1 Decomposition of forces during pedaling

Regarding the analysis of the forces, several metrics are proposed in the literature to analyze the pedaling technique based on kinetic variables. Among these metrics, the index of effectiveness (IE), defined as the ratio of the tangential force also called effective force to the total force applied on the pedals, is considered the gold standard [6]. Elite cyclists would have a better IE than non-cyclists (47.4% ± 4.5% vs 39.2% ± 6.2%) at 90 rpm and 60% of maximal aerobic power [51]. The IE is related to the gross efficiency, which demonstrates the close relationship between pedaling technique and performance [41, 86]. A higher pulling force would cause the better pedaling effectiveness in elite cyclists during the pedal upstroke and a greater ankle range of motion, which highlights the link between pedaling technique and joint kinematics [23, 51]. Regarding instrumented pedals, there are several challenges due to the speed at which the whole system rotates and the need for wireless transmission of all signals coming from strain gauges. This has caused the marginal development to be restricted to each research laboratory developing its own pedals. To the best of our knowledge, unlike motion capture systems, very few companies commercialize the instrumented pedals. Some of the companies that have manufactured these pedals are Radlabor (Radlabor GmbH, Frankfurt, Germany), Forped (Forped, Cycling Science, Ljubljana, Slovenia), and some systems adapted from industrial load cells such as INSTRON (Instron, Norwood, MA, US). This motivates the exploration of alternative methods to know these forces, considering their importance in understanding the biomechanical phenomena that occur in pedaling.

2.3 *Surface Electromyography*

Surface electromyography is one of the most important techniques for analyzing human movement, especially in sports. Human movement is a complex system that combines multiple muscular actions and must be coordinated. Therefore, it is necessary to evaluate to understand how a movement is performed. For this purpose, noninvasive sensors are placed on the muscle surface, which detect the myoelectric signal, and then through bioinstrumentation techniques, it is possible to see the signal and then analyze it. The Surface Electromyography for the Non-Invasive Assessment of Muscles (SENIAM) organization [47] issues recommendations to standardize the correct positioning of the electrode on the muscle surface as well as the skin preparation that ensures a clean signal free of artifacts.

2.4 *Machine Learning Models*

In machine learning modeling, there are two types of models depending on the type of data available: supervised learning and unsupervised learning [19].

Supervised learning predicts a dependent variable with labelled data. On the contrary, unsupervised learning identifies characteristics from unlabelled data. In this chapter, supervised learning is of interest because in most cases in sports analysis the data from a kinematic or kinetic study are fully labelled. Today, in sports, many approaches can be used to create predictive models in the context of supervised learning, starting with linear regression, multiple linear regression, ridge regression, logistic regression, least absolute shrinkage and selection operator (LASSO) regression, polynomial regression, Bayesian linear regression, and others [33]. Also, there are also more complex techniques that offer greater robustness and accuracy to make predictions, such as artificial neural networks (ANN), support vector machines (SVM), decision trees, random forest (RF), k-nearest neighbors (K-NN) [33].

Notably, it is of interest to apply a machine learning methodology that finds an interpretable model, which means that it behaves like a transparent box that allows knowing the factors involved in the estimation of a dependent variable. The concept of interpretability, according to Miller [49], is the degree that an observer understands how the model achieves the prediction. The term interpretable is also known as explainable AI, which is the most general term [53]. Another quality sought by the model is that it seeks to be parsimonious, in other words, the desired level of prediction with the least number of predictors [71]. Although the result is important, it is of interest to know how the result is reached, especially in the biomechanics area of human movement, where multiple movement phenomena occur at the same instant. Even though there is an extensive literature on these methods, some points are important to emphasize based on the theoretical concepts of these techniques to select the most appropriate method. Nowadays, machine learning techniques are implemented in sports to create mathematical models to predict data [63].

Particularly in this chapter, the prediction of a cycling metric is presented. In biomechanics, to analyze a movement of a limb, there are kinematic, kinetic, and physical variables, such as mass and geometric positions, referred to as independent variables or predictors. All groups of these variables allow applying machine learning methods to estimate or predict a dependent variable. The selection of the correct machine learning technique is an important task that depends on many factors. In general, any machine learning technique could be applied to predict a variable, but the most important task is to evaluate and optimize the model to find the best accuracy and generalize the model. In the present chapter, the application of a multiple regression model to predict the IE will be presented, thanks to its versatility in modeling any phenomenon. In addition, a multiple regression model is characterized by its simplicity, its easy application for coding in any software, and because it is an interpretable model [14]. Table 1 presents a summary of the interpretability, accuracy, and computational requirements of some machine learning methodologies.

Table 1 Summary of the characteristics of the machine learning methodologies

Method	Interpretability	Accuracy	Computational requirements
Neural networks	Low	High	High
Support vector machines	Low	Very high	High
Decision trees	High	Low	Medium
Random forest	Low	Medium	Medium
Multiple linear regression	Very high	High	Low
LASSO regression	High	Good	Medium
Ridge regression	Low	Good	Medium
Elastic net	Low	Very good	Medium
Partial last square regression	Low	Good	Medium

2.5 *Multiple Linear Regression Model*

Multiple linear regression is a machine learning methodology that provides an interpretable model. The basic theory of a multiple linear regression consists of predicting a variable of interest by means of a few predictors or independent variables. Each predictor is accompanied by a number called regression coefficients or regression weights. Thus, we obtain an equation that shows the relationship between the dependent variable and the predictors multiplied with the respective weights. The ratio of the regression weights is the ratio of the predictive importance of the predictors. In other words, a multiple linear regression shows the mathematical relationships of each predictor on the dependent variable. It is considered a technique to model the complex real world in a simple and practical way [62] and is one of the most transparent and easily understandable methods for humans [48]. Nevertheless, in order to build a correct model, it is necessary (1) to have a good amount of data, (2) to apply a variable selection method to avoid overfitting, and (3) for the model to comply with four assumptions. The assumptions of multiple linear regression model are linearity, no multicollinearity, homoscedasticity, and normally distributed variables [59]. The assumption of normally distributed variables can be evaluated by means of graphs and histograms such as the Q-Q graph. To prevent non-normality of the variables, a treatment of outliers must be implemented. The assumption of a linear relationship between the predictors and the dependent variable is assessed visually by plotting the predicted data against the residuals and expecting to find a horizontal trend line on the 0-axis of the residuals. In some cases, it is possible to make a transformation of independent variables with a nonlinear tendency to linear tendencies. There are mathematical methods that allow exponential and logarithmic transformations to linear tendencies, but their application should be evaluated. The assumption of homoscedasticity, i.e., residuals are equally distributed throughout the ranges of the predictors, is evaluated using a scale-location plot seeking a symmetrical distribution of data. Assumption of no multicollinearity, usually it must be analyzed by the correlation matrix. To analyze the correlation matrix, the correlation coefficient (r) between pairs of independent variables is calculated, which will be between -1 and 1. If the coefficient r is near to -1 or 1 there is a strong relationship; if it is close to 0, there is a weak relationship. It is recommended that the correlation coefficient be less than 0.8 [69]. This correlation could lead to collinearity problems, and therefore, from a statistical point of view, it would be advisable to eliminate these variables if there is a strong relationship. Nevertheless, this can lead to the loss of information, and therefore it is more convenient to support the decision with mathematical algorithms developed for this purpose. These analyses are not conclusive; they must be complemented with other metrics such as the variance inflation factor (VIF) value once the regression is built. The VIF value is expected to be less than 5 [16]. However, VIF ranges between 5 and 10 are considered acceptable for non-collinearity. This has been discussed by authors such as Schroeder et al. [67] and García García et al. [24].

Since we are interested in understanding the phenomena of human movement in the field of sport, a complete scientific understanding of how a task is generated is of interest. Therefore, the multiple linear regression method is considered as a suitable starting point in the determination of variables that explain the phenomena of force production, which in our case is cycling. However, a difficulty arises since, as mentioned above, the task of generating a movement involves multiple biomechanical phenomena acting at the same instant. More concretely, the movement of a joint involves multiple muscles acting in a complex synergy. Depending on the movement, the muscles work as agonists, i.e., producing movements through their own contraction, and antagonists, i.e., interfering with the physiological action of another, thus generating angular velocities and accelerations at the same instant and in different directions. Also, some muscles act only in one joint, i.e., mono-articular muscles, and two joints, i.e., bi-articular muscles, creating multiple relationships between the kinematic variables, which, in other words, could result in multicollinearity that is by nature present in the phenomenon of modeling the human lower limb. Therefore, in the application of machine learning methodology, a crucial step is the analysis of the biomechanical phenomena that occur to make a correct selection of predictors. This will ensure model accuracy while minimizing the impact of multicollinearity.

2.6 *Algorithms for Predictor Selection*

A critical point for implementing a machine learning method is the selection of predictors or independent variables, called feature selection, which means finding the best independent variables that allow to estimate the dependent variable with high accuracy and the best independent variables that work best together in the model. The consequences of an improper selection of variables can lead to phenomena such as multicollinearity, low precision of the predictor's weights, and computational instability in model estimation [75]. The selection of predictors is a mathematical method that allows for the selection of variables that must later be verified. Several methods exist in the literature, such as best subset selection, backward stepwise selection, forward stepwise selection, and hybrid subset selection [30]. Another way to reduce multicollinearity is by regularization methods or by reducing the size of the coefficient estimates (shrinkage) by making them tend to zero. The LASSO regression, ridge regression, and elastic net regression [2] are some of the most used methods. The selection of the method is related to the degree of multicollinearity detected, the number of independent variables, and the amount of data available. The following is a brief explanation of each feature selection method and the evaluation method of a machine learning model.

2.6.1 Best Subset Selection

Best subset selection is a mathematical method developed by Gupta and Yang [28]. The objective is to find the best subset of independent variables of a random size that contains the best of all possible subsets. The procedure consists of calculating a statistical information criterion for each independent variable for all possible models, $2^k - 1$, where k is the number of independent variables [11]. If the independent variable does not meet the given conditions, the estimated coefficient is equal to 0 [12]. Best subset selection is considered one of the most computationally demanding procedures if the number of independent variables is considerable [11], but it manages to find the best model of all possible combinations given a criterion. Also, best subset selection manages to find a set of independent variables with a high level of significance, i.e., p-value < 0.05, and therefore, the best subset selection method is considered one of the best performing in variable selection methodologies [80].

2.6.2 Backward Stepwise Selection

Backward stepwise selection is an algorithm that starts with a complete model that considers all the available independent variables and then eliminates them one at a time. The procedure eliminates the independent variable that has the least contribution in decreasing the prediction error. The process ends when all variables are significant [85]. Some authors suggest that this method is very inefficient in large database models [36].

2.6.3 Forward Stepwise Selection

In the contrary, forward stepwise selection procedure consists of starting with an empty model and gradually adding variables with high correlation with the dependent variable. The algorithm adjusts the weights of the independent variables each time a new one is entered and terminates when the last variable has an insignificant coefficient [85]. This method is considered a version of the best subset selection method. Even though an extensive discussion in the literature about its validity is ongoing [30], many authors question that sometimes in the process of eliminating variables, they cannot be re-included as the subset selection method does.

2.6.4 Least Absolute Shrinkage and Selection Operators (LASSO)

The LASSO regression and elastic net regression can be used when severe collinearity exists, but, in general, LASSO regression is used in the case of large databases [2]. The LASSO method helps to determine the variables by means of a contraction or "shrinks" of the independent variables towards zero. Thus, the method limits the model complexity since the variables with coefficients equal to zero are eliminated from the model [61]. LASSO has earned the prestige of being one of the best methods for the selection of variables. However, this method has been criticized, as problems of overfitting and an overly optimistic bias have been detected [2]. Another underlying problem is that there are doubts in the interpretability of the model. Indeed, the objective is more focused on the accuracy of the model and not on the contribution of each independent variable [61].

2.6.5 Ridge Regression

Ridge regression is another method used in the case of multicollinearity. This method was introduced in the 1970s by Hoerl and Kennard [32] to solve problems in the areas of engineering. It uses regression techniques modifying an ordinary least squares (OLS) regression and determining the parameters of the linear regression model by means of a standard least squares linear solution. This solution is the best linear unbiased estimation that minimizes the variance by introducing a small amount of bias toward the dependent variable [38]. The problem is that the resulting model is not interpretable because it keeps all predictors in the model [87].

Best subset selection, stepwise forward selection, and LASSO are popular methods for the selection of parameters in a linear model [30]. In general, any variable selection method can be applied, and their application is discussed in statistical research [22, 72]. Despite this, we recommend evaluating several methods to select variables that allow the most accurate results. Parallel to the above, a metric must be selected to measure the explanatory power of each independent variable. For this purpose, the information criteria generally used are Akaike's information criterion (AIC), Mallows's Cp (Cp), adjusted R^2, and Bayesian information criterion (BIC) [42]. AIC and BIC are the criteria widely discussed in the statistical literature. Both offer a good penalty criterion; however, some authors suggest the BIC criterion because it penalizes more than the AIC criterion [40].

2.7 Model Evaluation

The final step in model selection is the evaluation of the performance of the models. This stage consists of estimating the prediction error of the model and adjusting the prediction parameters if necessary [5]. From the data, which is at disposal, it should be divided into training, validation, and test data. To evaluate the model, different strategies of training, validation, and testing data selection will be applied to test the model by estimating the prediction error, i.e., relative standard error (RSE), and the model fit with the different data selections using the adjusted R^2 coefficient. If the variation of R^2 and the RSE between folds is very high, it indicates that the model does not fit the data correctly. The problem may be linked to the need for more data or to the inadequate selection of predictors. Several strategies for subsampling data exist, such as single holdout random subsampling, k-fold random subsampling, k-fold cross-validation, leave-one-out cross-validation, and jackknife [5]. In general, k-fold cross-validation with $k = 10$ is the most recommended strategy for validation in machine learning.

2.8 Biomechanics Modeling

From the information obtained by the kinematics and kinetics systems, it is possible to create musculoskeletal models in three different ways: (1) forward models, (2) inverse dynamic models [64], and (3) modeling with the use of machine learning methods [55]. Forward models consider the muscle actions that are made through the muscles, which generate forces and torques in the joints that will cause the movement, i.e., acceleration and decelerations, of the body segment; thus, the model can determine the final acting force. An example is the work of Bogey et al. [9], who present a method for modeling ankle torque and ankle muscle forces from electromyography activity. In the opposite case, the inverse dynamic method analyzes the forces that reach the joint by means of Newtonian mechanics equations. It seeks to estimate the torques in the articulations and finally determine the forces that are produced at the muscular level, up to the last articulation where the movement originates. This approach has been one of the most widely used over the years, from human gait analysis to sports analysis. Some approaches seek to determine the internal loads that reach the joints, such as the knee, which is a joint where most sports injuries occur [27], or calculate joint moments when performing a movement [15]. The problem with these two methods is the redundancy of variables, i.e., there are many variables and few equations, mainly because several muscles control a single joint, resulting in an underdetermined system. To solve this problem, optimization techniques have been proposed, with a good mathematical support, but there is controversy in its application due to the anatomical and functional simplifications that must be made. In addition, another way to solve this problem is to use computational methods by means of software such as Anybody Modeling

System [54] and OpenSim [34] that allows modeling the kinematics, muscular activity, and kinetics of motion. Computational modeling methods have been defined to relate muscle activity to the force produced by the muscle [9]. Despite the advances achieved with these computational methods, their implementation is complex and requires a tedious learning curve. Another difficulty is the necessity to do certain assumptions about the mass, dimensions, and center of mass of the body segments. Therefore, machine learning modeling applied to biomechanics is an alternative that can solve some of the problems described above. In this context, some authors have opted to use machine learning techniques, such as multiple linear regression, to estimate the grip strength of the upper limb [37]. Neural networks also have countless applications in biomechanics, and authors such as Stetter et al. [70] implemented a neural network to determine the internal knee joint forces in different sport movements. Another good example of machine learning application is in injury prevention. Xia et al. [84] show the implementation of a recurrent neural network (RNNs) called long short-term memory (LSTM) networks that can predict an Achilles tendon injury.

2.9 Cycling Biomechanics

Cycling biomechanics analyzes the limb kinematic, muscle recruitment, and the internal or external forces acting on the joints of the cyclist, in particular the lower limb, using motion capture, bioinstrumentation, and force sensors adapted to cycling [21]. The motion capture systems focused on cycling can measure joint angles, joint angular velocity, and joint angular accelerations of each lower limb joint, i.e., hip, knee, and ankle, during pedaling. In addition, although the pelvis is limited in movement, the saddle and the upper limb tend to be positioned with the brake levers; the analysis of the static angles of the upper limb and the spine is critical to determining correct posture on the bicycle. Despite all the information that can be obtained during motion analysis, cycling kinematics on the cyclist's fit is generally limited to lower-limb joint angles during flexion and extension, which is commonly used to determine the optimal saddle height. Also in cycling kinematics, the horizontal knee position relative to the pedal axis at a 90° crank angle from the top dead center, also called knee over pedal spindle (KOPS), is usually evaluated to indicate the correct saddle setback for proper pedaling efficiency [45]. The correct KOPS is related to the reduction of knee injuries [81]. Nevertheless, the joint angular velocities of the hip, knee, and ankle are valuable information that should be evaluated because there is a close relationship between the pedaling cadence, the joint angular velocity, and the speed of muscle shortening [43]. The angular velocity analysis allows understanding the muscle recruitment of the lower limb and how the muscle coordination achieves an optimal force and power delivery on the pedals [44].

Although it is possible to obtain information of IE and have all the articular kinematic information, it is complex to relate all this large amount of data to identify strategies to improve the pedaling technique. Therefore, it is of interest to present in this chapter an alternative to not only guide the athlete but also to propose interpretable machine learning model to predict metrics of interest in the environment of cycling. There are some examples of these applications, such as the determination of the saddle height based on anthropometry, bicycle geometry, and user-defined joint kinematics [26]. Another example is the prediction of cycling performance based on cadence analysis [73]. The authors Gatti et al. [25] found a machine learning model to predict knee loads based on joint kinematics. In the study, the authors implemented several models based on multiple linear regression.

Regardless of the activity to be evaluated, the application of a machine learning methodology allows in any sport to determine a variable of interest based on kinematic and kinetic data obtained with the available technology.

3 Methodology of Multiple Linear Regression Model

First, the predictors or independent variables and dependent variables should be defined, according to the subject of study. Then, a sample size study should be implemented by means of a statistical power analysis. A statistical power analysis is calculated to provide the minimum sample size needed for an experiment, given a required significance level, statistical power, and effect size [77]. Afterwards, a data capture protocol must be designed to obtain the data with which the models will be built. The experimental protocol must first guarantee a correct selection of the participants by means of certain inclusion criteria to be met by the research team. An ethical authorization must be signed beforehand to ensure safe handling of the data and minimize the risks inherent to experiments. It is also important to apply an informed consent and certain questionnaires on the athlete's state of health, which must be approved by the ethics committee that governs the research. A protocol should include with sufficient data richness to have a wide variety of features to ensure that the machine learning model learns from these features, so that the model is generalizable as much as possible. For each technology used in biomechanics, a preprocessing of the raw signal must be performed, including resampling, filtering, and synchronization of the signals. Then, it is recommended to perform an analysis and treatment of the data. Data analysis aims to detect abnormal data, also called outliers, and missing values or not a number data (NaN). Several methodologies to perform the treatment of these data exist, starting from statistical methods such as interquartile range analysis, z-score analysis, and Bayesian statistics analysis or simply deleting the data [1]. Then, as a preliminary step to the construction of a multiple linear regression model, data normalization must be performed, to place all variables in the same range, which is recommended between 0 and 1. This is useful when there are different units of measurement on the independent variables and ensures that all independent variables have a similar contribution. There are multiple methods to normalize variables such as min-max normalization, z-score normalization, and mean-centered normalization, among others. The next step is

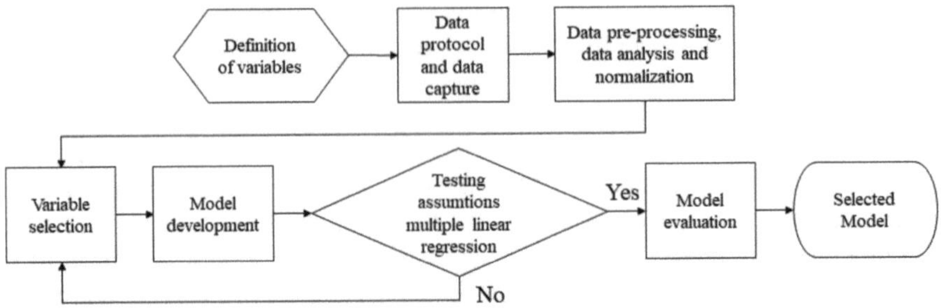

Fig. 2 Methodology used to develop a multiple linear regression model

the selection of independent variables by one of the methods explained earlier in the chapter to prevent multicollinearity. Once the model is obtained, the assumptions of a multiple linear regression must be verified. Finally, the cross-validation procedure must be performed to select the best model. Figure 2 shows the methodology used in this chapter to develop a multiple linear regression model.

3.1 Applying Machine Learning Methodology

As mentioned previously, as an illustrative example, we propose in this chapter the application of machine learning regression methodology adapted to the case of cycling. The aim is to predict the IE using a multiple linear regression from lower limbs kinematics. Nevertheless, the different aspects of this methodology can be extrapolated to other modeling problems in sports or any human movement. The idea is to use the kinematic variables of the cyclist's lower limbs at different cadences and power conditions, thus defining possible predictors. In this way, this amount of data will allow creating a multiple regression model that could be generalized to estimate the IE defined as a dependent variable. As explained, in a multivariate regression analysis, the objective is to find the predictors or independent variables with their respective weights and model error [78]. In addition, independent variables that have no relationship with the dependent variable are discarded. In the case proposed in cycling, it is very useful to determine exactly the biomechanical factors, i.e., joint angles, joint angular velocities, and joint angular accelerations, of corporal segments that are related to pedaling efficiency. All data analyses and multiple regression models were constructed using RStudio, i.e., R Core Team, Vienna, Austria, 2021; https://www.R-project.org/.

3.1.1 Data Protocol and Data Capture

In the example shown, 17 healthy cyclists volunteered to participate in this study (2 women and 15 men, 36 ± 12 years old, 1.74 ± 0.06 m, 72.4 ± 8.6 kg). All participants were recreational road cyclists, according to the classification by Priego Quesada et al. [60]. The participants did not report pain or pathology that could affect their pedaling technique. After being equipped, the participants warmed up for 5 min, selecting a comfortable cadence and easy power output. A Borg scale [10] was presented to the cyclists, who had to adopt a power corresponding to 1/10 on the Borg scale. The cyclists performed three 1-min tests at an easy (2/10 on the Borg scale), moderate (3/10 on the Borg scale), and difficult (5/10 on the Borg scale) perceived exertion with their preferred pedaling cadence (based on their comfort perception), interspersed with 1 min of recovery at a very easy intensity (1/10 on the Borg scale) and free-pedaling cadence. The test session was reproduced twice with their preferred pedaling cadence of +10 and −10 rpm (the three powers chosen during the first session were maintained for the second and third sessions). Thus, a total of 153 data were obtained. Therefore, we obtained data at different intensities and pedaling cadences to include in the models in order to cover a wide range of IE since this index is closely related to power output and cadence [6]. The study was carried out at a perceived exertion ≤5/10 to limit the influence of fatigue on the kinetic and kinematic variables [66]. Finally, the participants rested with a 5-min cooldown at a free cadence, as shown in Fig. 3. Kinetic and kinematic measurements were recorded during all 1-min tests.

The data capture of joint kinematics was assessed using the Cycling 3DMA motion capture system (STT Systems, San Sebastian, Spain). The STT 3DMA system allows the selection of several motion capture protocols for cycling, e.g., right side and left side, each of them with eight markers. In addition, a full-body bike fitting protocol composed of 20 markers is also possible. This system is composed of eight OptiTrack cameras (NaturalPoint, Inc., Corvallis, United States) positioned around the rider with a sampling frequency of 50 Hz. In this example, the bike fitting left-side protocol was used. Eight active markers were placed on the left side of the cyclist at the foot, i.e., in alignment with the pedal axle (lateral malleolus and calcaneus markers), and at 90° between the foot (lateral malleolus, greater trochanter, femoral lateral epicondyle, greater tubercle of the humerus, humerus lateral epicondyle, and distal radioulnar markers). Three additional markers were placed on the floor to define the reference plan: two in front of the bike, separated by 0.5 m, and one at the rear of the bike about 2 m in a straight line with the front right marker. Figure 4 shows the capture environment with the software used.

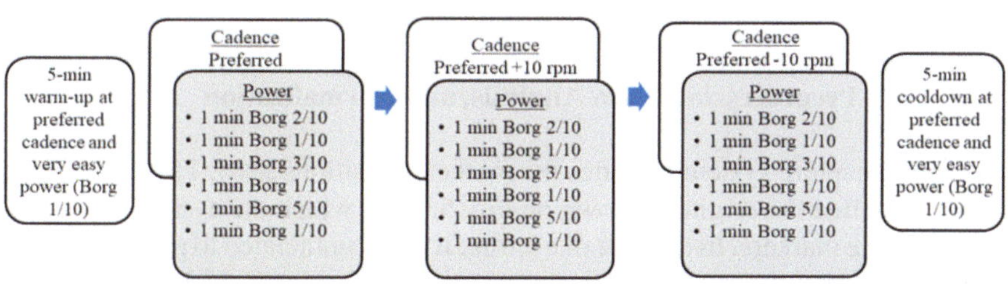

Fig. 3 Experimental protocol

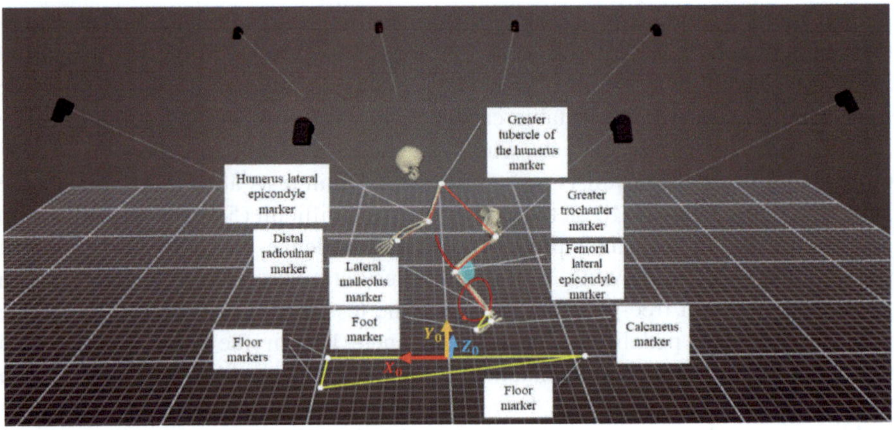

Fig. 4 Motion capture environment. Taken from STT 3DMA motion capture [52]

The raw data of 3D capture systems can provide the angle information in the sagittal, transverse, and frontal planes. For the present study, only the sagittal plane, i.e., 2D measures, was considered to simplify the lower limb model. The sagittal plane provides the planar angles that are easier to interpret and understand by bike fitters. In addition, for purposes of adjusting the rider to the bike, the lateral side is the most popularly used. Therefore, we use this reference in the development of the model. In addition, only the trajectories of the lower limb markers were considered, because the lower limb is strongly linked to the production of force in cycling [18]. The Forped instrumented pedals were installed on the bicycle to measure pedal forces with a sampling frequency of 450 Hz. Before the experiment, the instrumented pedals were calibrated by the Forped company. Then, their validity was tested with standard loads (Troemner Metric Weight, Troemner, Thorofare, USA), and before each use, they were calibrated according to the manufacturer's instructions. The tests were performed on the participant's personal road bike, which was mounted on an electromagnetic ergometer, i.e., Saris, H3 Direct Drive Smart Trainer Bundle, Fitchburg, WI, United States, to maintain a constant crank power output regardless of the pedaling cadence. To know pedaling cadence and the power delivered to the pedals, the Garmin cycling microcomputer was used (Garmin International, Kansas City, MO, USA) with a sampling frequency of 1 Hz.

3.1.2 Data Preprocessing, Data Analysis, and Normalization

The motion capture system provides the marker's positions $X[n]$, $Y[n]$, $Z[n]$, which X_0, Y_0, Z_0 called the origin as shown on Fig. 5, from which the system delivers the position of the markers. In the case of cycling, it is recommended to place the bicycle in such a way that the origin of the capture area matches with the bicycle's bottom bracket. Therefore, in the example presented, the motion capture software provides

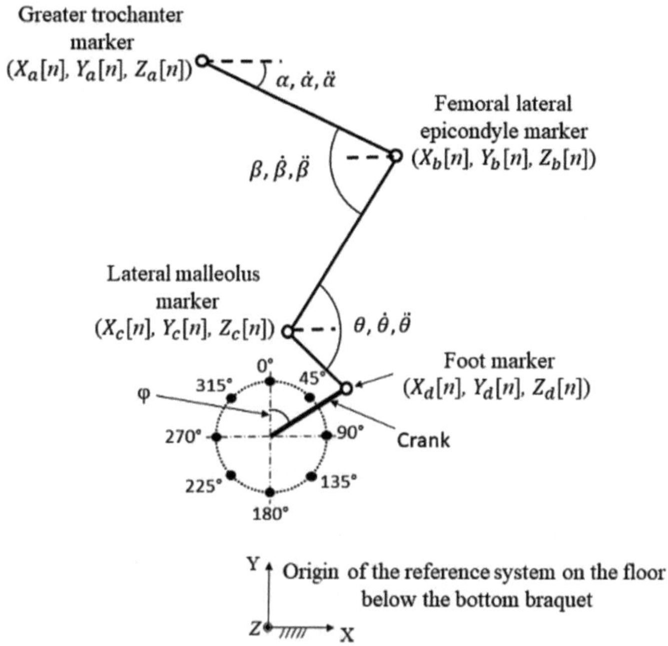

Fig. 5 Nomenclature of the experimental model developed

the raw data coordinates ($X[n]$, $Y[n]$, $Z[n]$) of all the markers $n = 1, 2, \ldots, N$, where N is the number of samples. Given the 50 Hz sampling frequency and 60 s of capture for each test, we obtained $N = 3000$ samples for each marker. All kinematic calculations were performed in the sagittal plane; therefore, only the X and Y coordinates were considered. MATLAB code (MATLAB 2021, MathWorks, Inc., Natick, Massachusetts, USA). The MathWorks Inc. (2021). Optimization Toolbox version: 9.4 (R2021b), Natick, Massachusetts: The MathWorks Inc. was developed to determine the joint angles, i.e., α, β, and θ; joint angular velocities, i.e., $\dot{\alpha}, \dot{\beta}, \dot{\theta}$; and joint angular accelerations, i.e., $\ddot{\alpha}$, $\ddot{\beta}$, and $\ddot{\theta}$, of the femur, knee, and ankle and the position of the crank during the capture, as shown in Fig. 5.

Figure 5 illustrates the marker locations of the greater trochanter marker a, femoral lateral epicondyle marker b, lateral malleolus marker c, and foot marker d. The angle φ is the crank angle.

The kinematic and kinetics data were filtered with a Butterworth filter of order 2, with a 10 Hz cut-off frequency. The α will be defined as femur angle relative to the horizontal. The calculations are presented in Appendix 1 (Eqs. 1–9). All kinematic variables were computed relative to the crank angle, from 0° to 360°. Subsequently, to define the independent variables, only the values every 45° were used, i.e., 72 kinematic data points per participant (8 points × 3 joints × 3 kinematic values). Moreover, additional variables, i.e., pedaling cadence, power output, mass of the individual, and KOPS, were included with the kinematic variables, resulting in 76 independent variables to be examined. Furthermore, the instrumented pedals provided kinetic data for each power output and cadence selection to determine the IE as a dependent variable. The equations for calculating the IE are shown in Appendix 2 (Eqs. 10 and 11). Using the RStudio function "pwr.f2.test," the minimum sample size with a power of 0.8, significance level of 0.05, effect size of 0.8, and 76 variables is 107. The data analysis was performed, and no NaN or outlier data were found. Then, before constructing the models, a min-max normalization was made.

3.1.3 Variable Selection and Model Development

To develop the different models, it is first hypothesized that all 76 independent variables described above are candidates to form the multiple linear regression model. In general, from a biomechanical point of view, the force production in a body segment combines different phenomena manifested by the angle between the body segments; the angular velocity of the segments, i.e., the indicator of a muscle contraction [68]; and the angular acceleration that is related to the torque produced at the joint [31]. In addition, since the IE is a function of force transmission, the power transition, pedaling cadence, and KOPS are variables to be taken into account in the model [20]. In the case of cycling, several critical points around the crank rotation exist. These points have been studied because around these points, there are critical points of force transmission, i.e., 0° top dead center (TDC), 180° bottom dead center (BDC), 90°, and 270°. In particular, around 90° and 270° the tangential forces can reach maximum positive and maximum negative values respectively that influence the IE.

For explanatory purposes, the first model developed is a multiple linear regression taking into account all the variables, i.e., 76 variables. Surprisingly, the model presents a low error, i.e., RSE = 2.39, and a good fit, i.e., adjusted $R^2 = 0.90$, as shown in Table 2. Nevertheless, 57 predictors are non-significant, i.e., p-value > 0.05. Likewise, when evaluating collinearity, the model presents multicollinearity, which would most likely lead to an overfitting and non-generalizable model. Using an RStudio function, the optimal number of independent variables was defined in such a way that the highest R^2 was reached. Therefore, it was determined that the best number of variables is 11 according to the BIC criterion, which has the lowest residual sum of squares and the highest R^2, as shown in Fig. 6. The procedure also proves that despite the increase to more than 11 variables, the value of R^2 does not significantly increase, and the residual sum of squares remains stable.

Below, several variable selection procedures with different penalization metrics were tested to find the best model, identifying the corresponding variables. Also from Fig. 6, the models obtained with the stepwise forward and stepwise backward methods were evaluated by setting the number of variables to 11. In addition, the sample size was adjusted with this number of predictors, a power of 0.8, significance level of 0.05, effect size of 0.93 founded a value of 29 sample size. The results of each procedure are shown in Tables 2, 3, 4, and 5. The RStudio function used for

Table 2 Results of the best subset selection procedure

Value/name	Multiple linear regression	Best subset selection
Information criteria	N/A	BIC
# Variables selected	76	11
RSE	2.39	2.81
Adjusted R^2	0.90	0.86
Most significant variables (***) p-value = 0	POWER	HA45 KA225 AA45 AA270 HV90 HV270 KV180 KV270 AV270 MASS POWER
Most significant variables (**) p-value < 0.001	AAC135 KAC135 HAC135	/
Most significant variables (*) p-value < 0.01	HV135 MASS AAC45 KAC45 HAC45 AV135 KV135 KV90 KV45 AV135	/
Other variables (.) p-value < 0.05	HA270 HV90 AV90 HAC225 KAC0 KAC225 AAC225	/
Variables with no significance	HA0 HA45 HA90 HA135 HA180 HA225 HA315 KA0 KA45 KA90 KA135 KA180 KA225 AA135 KA270 KA315 AA0 AA45 AA90 AA180 AA225 AA270 AA315 HV0 HV45 HV180 HV225 HV270 HV315 KV0 KV180 KV225 KV270 KV315 AV0 AV45 AV180 AV225 AV270 AV315 CAD HAC0 HAC90 HAC180 HAC225 HAC270 HAC315 KAC90 KAC180 KAC270 KAC315 AAC0 AAC90 AAC180 AAC270 AAC315 KOPS	/

variable selection was “regsubset.” The RStudio function used for the development of the multiple linear regressions was “lm.”

In the following tables, to identify the independent variable, the first letter indicates the joint, i.e., H = hip, K = knee, and A = ankle; the second letter indicates the type of kinematic data, i.e., A = angle, V = angular velocity, and AC = angular acceleration; and the number after the letters indicates the crank position between 0° and 360°. Also, MASS is the cyclist’s mass, CAD is pedaling cadence, and POWER is the power delivered to the cranks.

Based on the information in the tables, the objective is to find the best model based on the evaluations proposed by the literature and the biomechanical analysis of the results. Firstly, there is no specific recommendation in the literature on how many predictors or independent variables should be selected. The main objective

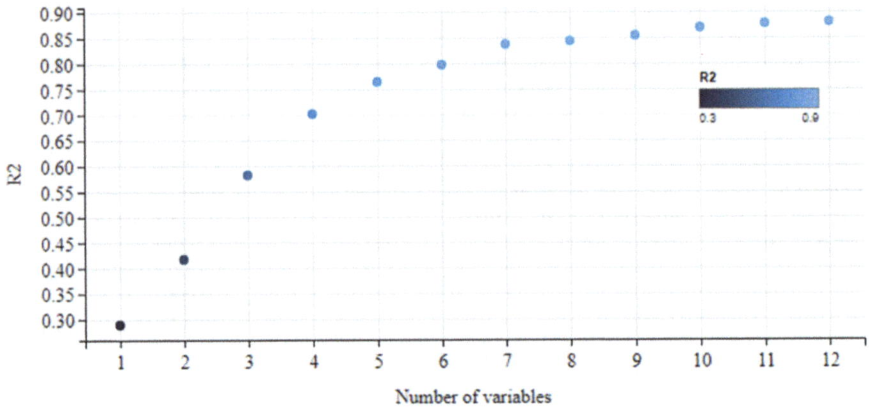

Fig. 6 Values of R^2 obtained in relation to the number of selected variables

is to find the best parsimonious model. Cycling is a special case where a good number of predictors were defined. In general, with all the methods tested, the average RSE was 2.69 ± 0.395, which is a low value, and theaverageadjustedR^2 was 0.85 ± 0.0313, which is an equally appropriate value. Given the multicollinearity that was found in the 76-variable model, it is suggested to look for the model with the fewest number of independent variables and that these are the most significant, i.e., p-value $= 0$. The model with less independent variables is the 11-variable model built with the best subset selection method, evaluated with BIC criteria. The other variable selection methods that were fitted to 11 variables, i.e., stepwise forward selection and stepwise backward selection, also give a similar adjusted R^2, but an increase in the RSE and a group of independent variables with less significance were observed. The rest of the models contain more variables that could have multicollinearity problems, and therefore, they will not be analyzed. Therefore, analyzing the three models of 11 independent variables, another aspect is to evaluate the significance of each variable, i.e., p-value < 0.05. The model obtained with the best subset selection method, according to BIC criteria, achieves all the independent variables with a maximum level of significance, i.e., p-value $= 0$. The other models, composed of 11 independent variables, are not within the exceptional range of significance of the variables, and therefore, they would be discarded as models. In addition to the above, this model has the lowest RSE $= 2.81$ and the highest adjusted $R^2 = 0.86$, and therefore, it can be said that it is the best model of all. The above is corroborated by Tamura et al. [75] and Hastie et al. [30], stating that the process of eliminating variables with a high penalty criterion guarantees the elimination of multicollinearity. All independent variables have a p-value < 0.001, which indicates a high relationship with the dependent variable of the model. As for the ridge regression model, the accuracy value is slightly lower than the others, so it will not be considered. Finally, the LASSO regression provides a good value of accuracy and a low error, which could be a good alternative for further study. Nevertheless,

Table 3 Results of stepwise forward selection

Value/name	Stepwise forward selection	Stepwise forward selection	Stepwise forward selection	Stepwise forward selection	Stepwise forward selection
Information criteria	adjR^2	Cp	BIC	AIC	N/A
# Variables selected	58	36	23	73	11
RSE	2.29	2.36	2.63	2.39	3.24
Adjusted R^2	0.91	0.90	0.88	0.90	0.82
Most significant variables (***) *p*-value = 0	POWER AV270 KV270 HV270	KAC90 AV315 KV270 KV45 HV270 HV90 POWER	AV0 KV45 HV180 HV90 HV0 KA315 KA45 POWER CAD	POWER	HA90 KA135 KA315 AA180 AV0 POWER MASS
Most significant variables (**) *p*-value < 0.001	HAC135 AV135 HV135	KAC0 HAC0 AV270 AV90 HV0 AA270 KA315 KA135 HA135 HA45 CAD MASS	KAC90 AV90 HA45	AAC135 KAC135 HAC135	HAC0
Most significant variables (*) *p*-value < 0.01	AAC135 KAC135 KAC90 AV90 KV180 KV135 KV90 KV45 HA270 HA225	HAC180 HAC90 AA45 KA225 KA45 HA0	AAC90 HAC180 HAC90 HV315	AAC45 KAC45 AV135 KV135 KV90 KV45 HV135MASS	HV180
Other variables (.) *p*-value < 0.05	MASS KAC180 AV180 HV180 AA225 AA135	KV315 KV180 KA0	/	AAC225 KAC225 KAC0 HAC225 HAC45 AV90 HV90 HA270	AAC90

Table 4 Results of stepwise backward selection

Value/name	Stepwise backward selection	Stepwise backward selection	Stepwise backward selection	Stepwise backward selection	Stepwise backward selection
Information criteria	adjR^2	Cp	BIC	AIC	N/A
# Variables selected	50	26	21	73	11
RSE	2.2	2.4	2.51	2.20	3.24
Adjusted R^2	0.91	0.90	0.89	0.91	0.89
Most significant variables (***) p-value = 0	POWER AAC270 AAC225 AAC135 AAC45 AAC0 KAC270 KAC225 KAC135 KAC90 KAC45 KAC0 HAC270 HAC225 HAC135 HAC45 AV90 AV45 KV180 KV135 KV90 KV45 HV135 HV90 HA270	AAC225 AAC180 AAC135 AAC45 KAC225 KAC90 HAC270 HAC225 HAC180 HAC135 AV90 AV45 AV0 KV180 KV90 KV45 AA225 HA270 MASS POWER	HA270 AA225 HV90 KV45 KV90 KV180 AV0 AV45 AV90 HAC135 HAC180 HAC225 KAC90 KAC225 AAC45 AAC135 AAC180 AAC225 POWER MASS	HA45 HA270 HV90 HV135 KV45 KV90 KV135 KV180 AV45 AV90 HAC45 HAC135 HAC225 HAC270 KAC0 KAC45 KAC90 KAC135 KAC225 KAC270 AAC0 AAC45 AAC135 AAC225 AAC270 POWER	HA270 AA225 KV180 AV90 HAC225 KAC225 AAC225 POWER MASS
Most significant variables (**) p-value < 0.001	MASS AAC180 HAC0 AV135 KV315 AA135 AA0 KA270 HA225 HA45	AAC270 KAC270 HV90	/	MASS AAC180 HAC0 AV135 KV315 AA135 AA0 KA270	AAC135 AAC180
Most significant variables (*) p-value < 0.01	AV0 HV180 AA225 KA135 HA315	HA225 HA45	HAC270	HA225 HA315 KA135 AA45 AA225 HV180 HV315 AV0	/
Other variables (.) p-value < 0.05	AA45 HV315	KAC135	/	/	/

Table 5 Results of ridge regression and LASSO regression

Value/name	Ridge regression	LASSO regression
Information criteria	N/A	N/A
# Variables selected	73	45
RSE	3.87	3.04
Adjusted R^2	0.74	0.84
Most significant variables	HA0 HA45 HA90 HA135 HA180 HA225 HA270 HA315 KA0 KA45 KA90 KA135 KA180 KA225 KA270 KA315 AA0 AA45 AA90 AA135 AA180 AA225 AA270 AA315 HV0 HV45 HV90 HV135 HV180 HV225 HV270 HV315 KV0 KV45 KV90 KV135 KV180 KV225 KV270 KV315 AV0 AV45 AV90 AV135 AV180 AV225 AV270 AV315 CAD HAC0 HAC45 HAC90 HAC135 HAC180 HAC225 HAC270 KAC0 KAC45 KAC90 KAC135 KAC180 KAC225 KAC270 AAC0 AAC45 AAC90 AAC135 AAC180 AAC225 AAC270 POWER KOPS MASS	HA0 HA45 HA90 HA270 KA0 KA180 KA225 KA270 KA315 AA0 AA90 AA135 AA225 AA270 HV0 HV90 HV135 HV180 HV225 HV270 HV315 KV45 KV135 KV180 KV225 KV270 KV315 AV45 AV135 AV180 AV270 AV315 CAD HAC90 HAC135 KAC0 KAC45 KAC90 KAC225 AAC0 AAC90 AAC135 AAC225 POWER MASS

since the model needs 45 predictors, it does not achieve a parsimonious model and is therefore discarded.

3.1.4 Testing the Assumptions of Multiple Linear Regression

Multicollinearity was evaluated using the correlation matrix and the evaluation of the VIF value of each independent variable. RStudio allows multiple ways to represent the correlation matrix. One of them is presented in Fig. 7, in which two color codes are presented, blue in the range of 0 to 1 and red in the range of 0 to −1. The size of the circle and the intensity of the color indicate the strongest relationship, i.e., an r close to 1 or −1. The RStudio function used for the development of the correlation matrix was "corrplot."

Table 6 shows the calculation of the VIF for each independent variable of the selected model.

Evaluating the correlation matrix, two variables were found with correlation values of $r = 0.86$, i.e., KV270 and HV270, namely, the angular velocity values at the hip and knee in the 270° position. Then, considering the VIF values, it was found that, as expected, they were in a range outside the recommended value of

Fig. 7 Correlation matrix of the 11-variable model

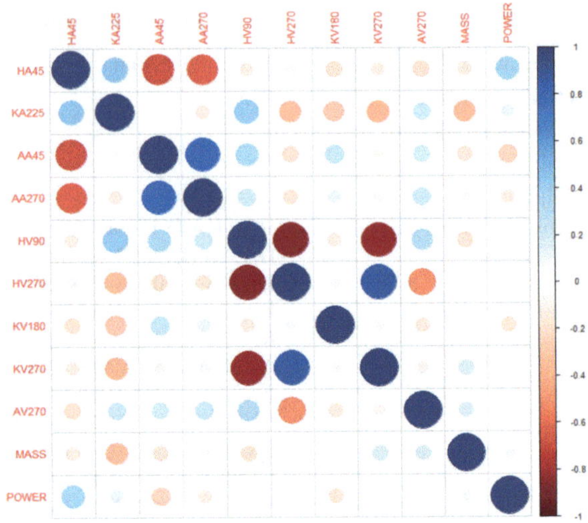

Table 6 VIF values of the 11-variable model

Variable	VIF
HA45	5.50
KA225	3.23
AA45	7.78
AA270	6.46
HV90	7.64
HV270	20.22
KV180	1.97
KV270	25.86
AV270	7.84
MASS	1.70
POWER	1.36

VIF > 10. Thus, it can be argued that there is a moderate correlation between these two variables. The first recommendation that the literature suggests is to eliminate one of these two variables. The difficulty is in justifying the elimination of one or the other variable in the context of biomechanics, knowing that these variables are the manifestation of biomechanical phenomena of muscle contraction, which is extremely important in cycling. In the first step, the effect of eliminating one or another variable was observed, and it was found that the model loses fit, i.e., R^2 decreases, and some selected independent variables lose their significance. In cycling, in the process of moving the legs from BDC to top TDC, at the point around 270°, there is a synchronization of the hip flexors and the knee flexors, making this work crucial in pedaling phenomenon. Both joints are connected by bi-articular muscles and work in a coordinated way, i.e., agonist-antagonist, with the opposite muscles to quickly raise the leg. It is a pulling action that is crucial

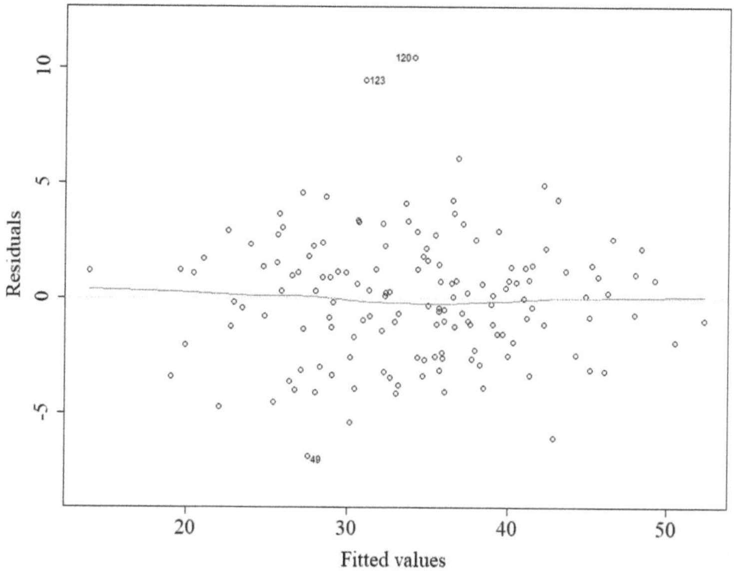

Fig. 8 Linear relationship graph

for pedaling efficiency since it generates angular velocities that allow diminishing negative forces, directly affecting the IE. Because of the above and observing a moderate level of correlation in the analysis of the correlation matrix, it was decided to include these two variables in the model, given their importance. Other variables that have a high correlation coefficient are HV270 and HV90, i.e., $r = 0.85$, namely, the hip angular velocity at 270° and 90° position. Here again HV270 appears, but as justified above, it will not be eliminated. In addition, as for HV90, its VIF value is less than 8, so it will not be eliminated either. The rest of the relationships between independent variables are $r < 0.8$ and VIF < 8, so the variables are maintained. Subsequently, with the validation of the model via cross-validation, emphasis will be placed on this decision. In this way, even though the statistical parameters are a guide to make decisions in the optimization of a model, researchers will be able to adjust the model according to their experience and knowledge.

In addition, the model respects the assumptions for conducting a multiple linear regression, i.e., linearity, normality, and homoscedasticity, which was evaluated using specific tests and graphs. The assumption of a linear relationship was assessed by plotting the predicted data against the residuals, finding a horizontal trend line on the 0-axis of the residuals, as shown in Fig. 8. Normality was assessed with the Q-Q graph, where there is a line from the lower left to the upper right, indicating that the normality assumption is satisfied, as shown in Fig. 9. Homoscedasticity was assessed using the scale-location plot, which showed that the residuals are distributed approximately according to the red line, as shown in Fig. 10.

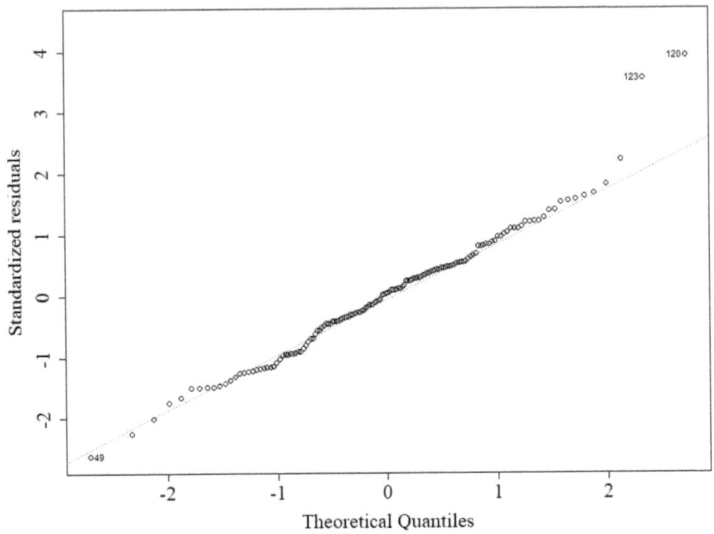

Fig. 9 Q-Q graph

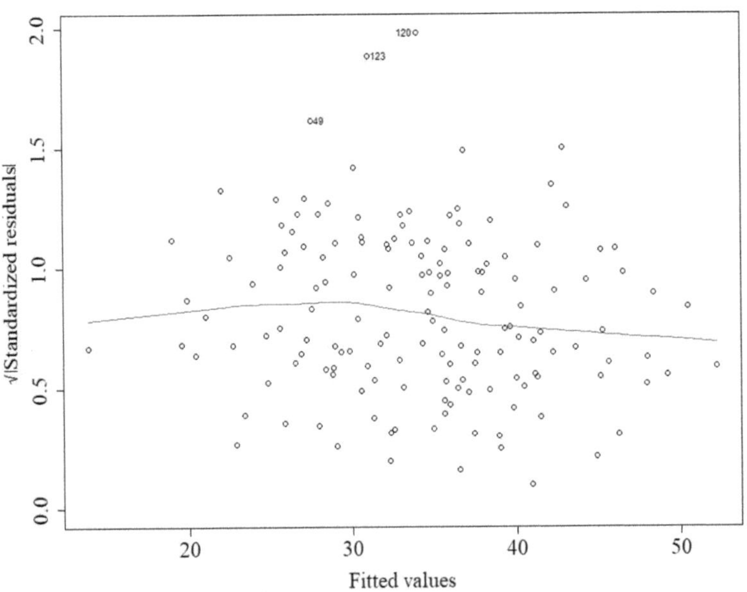

Fig. 10 Scale-location plot

3.1.5 Model Evaluation

The adjusted coefficient of determination, i.e., adjusted R^2, was used to evaluate how well the data fit the regression model. To determine the differences between the predicted and measured values, RSE was calculated. The average knee angles during lower-limb extension, i.e., close to the bottom dead center, were 137 ± 4°. The average horizontal knee position was −9 ± 24 mm. A RSE of 2.81 and an adjusted R^2 of 0.86 characterized this model. The Bland-Altman plot illustrates the discrepancies between the measured and modelled IE. It was constructed by calculating the biases and limits of agreement between the measured and predicted variables [8]. The discrepancy between the measured and predicted IE was 0 ± 5.51. The dotted exterior lines are the mean differences ±1.96 standard deviations. It is noted that the central line is located at 0, which indicates that there is no bias in the estimation of the IE, as presented in Fig. 11. In addition, the average IE calculation is in the range of 30–40%, which is in agreement with the literature. Finally, a calculation error of ±5% is found, which we consider low. The RStudio function used for the development of the Bland-Altman plot was "bland.altman.plot."

In the regression plot (Fig. 12), the central red line is the regression line, the black dotted lines close to the regression line correspond to the confidence intervals, and the black dotted exterior lines are the prediction intervals. It is observed that the data are within the prediction interval and only three data are outside it. Moreover, the data are close to the regression line and the confidence interval.

Table 7 presents the 11-variable model that was composed of joint angles of the femur and ankle during the downstroke and knee and ankle angles during the upstroke. Moreover, the joint angular velocities of the femur during the downstroke, of the knee at the bottom dead center, and of the three joints during the upstroke were included. This model considered the mass of the individual and power output. Joint acceleration, pedaling cadences, and KOPS were not considered in this model.

Finally, to evaluate the robustness of the model, a randomized 10-fold cross-validation was performed. Figure 13 presents the results of the randomized tenfold

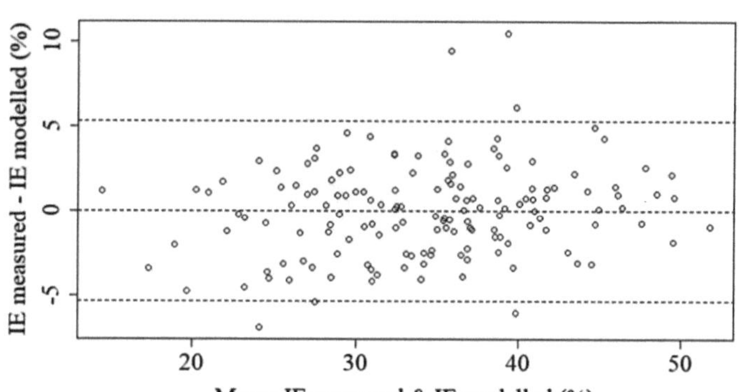

Fig. 11 Bland-Altman plot

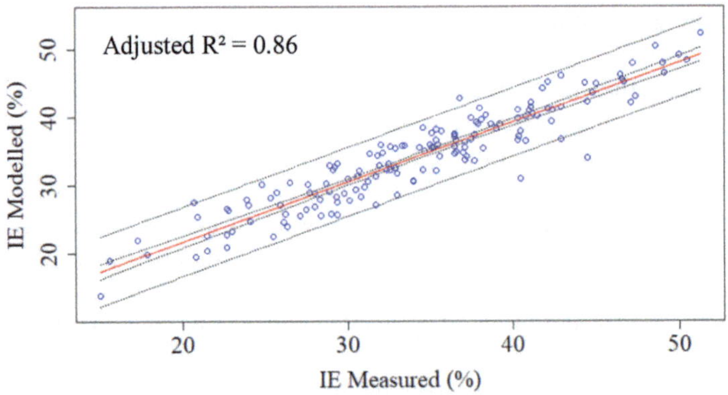

Fig. 12 Regression plot

Table 7 Normalized variables included in the model with the lowest BIC value

Name	Coefficient estimate	Standard error	*p*-value
Intercept	21.66	3.98	2.32e-07
HA45	−23.74	1.99	<2e-16
KA225	8.93	1.504	2.21e-08
AA45	−34.24	2.76	<2e-16
AA270	41.51	2.64	<2e-16
HV90	−10.41	2.82	0.000326
HV270	70.15	7.70	8.42e-16
KV180	−16.31	1.54	<2e-16
KV270	−53.93	6.54	1.18e-13
AV270	19.92	3.03	9.31e-10
POWER	19.09	1.19	<2e-16
MASS	−5.67	1.02	1.45e-07

cross-validation to evaluate the robustness of the model. A routine was developed in RStudio to randomly select the ratio of test data and training data, i.e., x-axis of the graph. Each fold is represented as a fraction, showing in the numerator the training data and in the denominator the test data. Therefore, different test data were randomly chosen from the total dataset, i.e., 10%, 20%, 30%, and 40%. The model was trained with other data, i.e., 90%, 80%, 70%, and 60%, respectively, to predict the IE. Each cross-validation was performed ten times, and the mean ± standard deviation of the RSE and adjusted R^2 was calculated.

In the plots:

X-axis: training data/test data were chosen randomly and performed ten times for each cross-validation.

Y-axis: mean ± standard deviation of the adjusted R^2 and RSE obtained from each condition of the cross-validation.

Dotted lines: results of the model with 100% of the data.

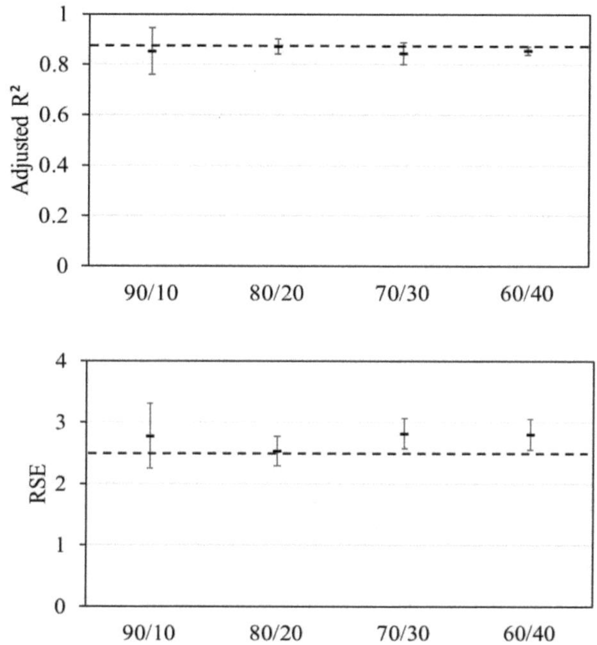

Fig. 13 Randomized 10-fold cross-validation plots

3.1.6 Selected Model

The main objective of this example was to show how to develop a machine learning method for the prediction of the IE. The multiple regression model found is composed of 11 explanatory variables and shows a good correlation, robust agreement, and small error between the measured and modelled IE. With this 11-variable model, it is shown that there is a link between some joint kinematics variables and pedaling kinetics. This variable selection guarantees that the model does not present problems of overfitting because it eliminates the presence of irrelevant variables. In addition, when the tenfold cross validation was done, the 11-variable model maintains stable RSE and R^2 with different ratios of test data and training data, which corroborates the robustness of the model and that the model is generalizable. Furthermore, it is important to note that the multiple linear regression model with the 11 independent variables complies with the assumptions required for this type of model, which makes it reliable for the estimation of the IE. The 11-variable model has an additional advantage to be composed of fewer variables, which implies that its implementation in motion capture systems could be easily achieved with less computational demand.

In the multicollinearity tests, the variables HV270 and KV270 showed an increase in the VIF and a moderate increase in the correlation coefficient, but as justified above, they were included in the model. After the cross-validation procedure, it is observed that the model maintains stable R^2 and RSE, indicating the need to preserve these two variables of the model to guarantee the model generalization. The optimal value of the variables allows estimating the IE with good precision and helps in clearly understanding and visualizing the factors that influence the IE. This indicates that the 11-variable model behaves like a parsimonious model, since when analyzing the models with different variable selection methods, i.e., best subset selection, stepwise forward selection, and stepwise backward selection, it is possible to obtain a good level of precision and a low error with as few variables as possible. It is important to emphasize that the three models with 11 variables were obtained through procedures that differ in the way the subset is found, and therefore, the same independent variables are not obtained. Particularly, the three models share mass and power, which highlights their importance in the IE. Additionally, the possibility of creating a model with ten and fewer variables was also explored, using the best subset selection method. It was found that the minimum number of predictors is ten; the model with ten predictors presents one variable less, i.e., HV90, and a new variable, i.e., KA45, in replacement of KA225. However, the collinearity analysis showed that the variables KA45 and AV270 have a high VIF. It should be noted that this model yields a RSE and R^2 practically equal to the 11-variable model. Thus, this model will not be considered as the selected model.

One of the great advantages of multiple linear regression over other machine learning methods is its interpretability. It is possible to clearly visualize which predictors are directly related to the dependent variable and what their importance in the model is. The multiple linear regression of 11 coefficients, obtained with the best subset selection method, allows us to know exactly which of the 76 independent variables defined at the beginning directly or inversely contribute to the predictor variable [56, 58], which in this case is the IE. The biomechanical factors that influence force production can be clarified by evaluating the regression weights of the coefficients obtained. Figure 14 clarifies which is the most important and the least important predictor that influences the IE.

In this sense the angular velocity of the hip at 270°, coefficient estimate = +70.15, plays the most important role with respect to the other independent variables in the determination of the IE. This is the manifestation of the work done by the hip flexors, i.e., the psoas major, iliacus, and rectus femoris, to quickly raise the leg up to the top dead center [39]. This finding corroborates its impact on the IE during the upstroke phase, which is widely influenced by the hip flexor muscle group. The positive value indicates that the higher the angular velocity, the higher the IE. In addition, subsequently with an inverse relationship, the angular velocity of the knee at 270°, coefficient estimate = −53.93, negatively affects the IE, i.e., a greater knee flexor action decreases the IE, which is also manifested in the low activity that is expected of the knee flexor muscles. Hansen et al. [29] and García-López et al. [23] have also shown the importance of recruiting knee and femur flexor's during the upstroke to optimize the pulling action. The above statement is also related to a study that has shown that the recruitment of knee flexor muscles increases during the upstroke to reduce negative force in elite cyclists compared with non-cyclists [74]. The area around 270° is a critical moment to limit the

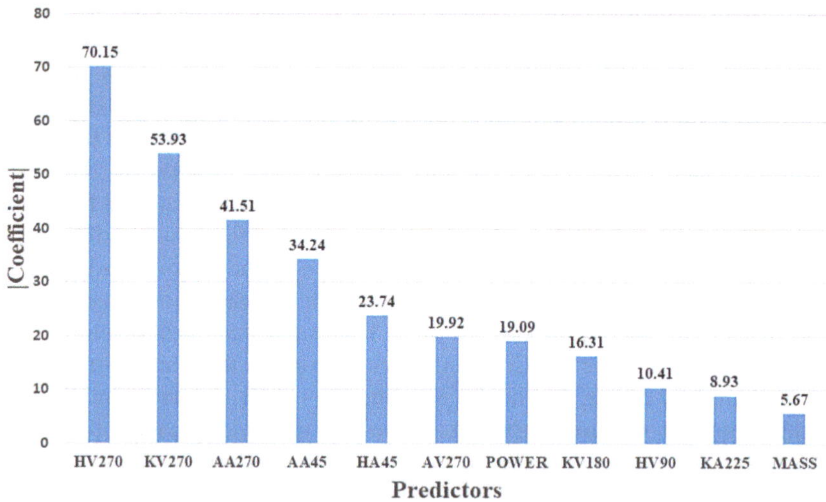

Fig. 14 Regression coefficients and their order of importance

negative force that slows the propulsion [76], which makes this phase of the cycle crucial for optimizing the IE. This is related to the results obtained by García-López et al. [23]. These authors conclude that elite cyclists have a better IE due to a higher pulling force and a greater ankle range of motion. The next most important predictor is ankle angle at 270°, coefficient estimate = +41.51, which indicates that considering the articular limitations of the ankle, i.e., bone blocks, a correct angle influences positively the IE. Our results confirm also the importance of the ankle angle during the upstroke at 270°, where the joint transition from plantar flexion to dorsiflexion plays a role in decreasing the negative torque through the action of the ankle flexor muscles, i.e., the tibialis anterior [51].

Following this point, the angle of the ankle at 45°, coefficient estimate = −34.24, indicates the role of ankle force absorption from the knee and hip extensor muscles, being the most powerful muscles of the lower limb [44, 65]. For this reason, the ankle also plays an important role during the first part of the pedal stroke. The interpretation is that a higher ankle angle at 45° has a negative influence on the IE. Continuing in order of importance of the predictors, the hip angle at 45° also has an inverse relationship with the IE, coefficient estimate = −23.74. This can be interpreted as a correct selection of the saddle height; this angle will be optimal to produce the maximum effective force and, therefore, a greater IE. It should be noted also that this variable is in the first quadrant, i.e., TDC to 90°, where the maximum effective force production is achieved and where muscles such as the gluteus maximus and biceps femoris reach their maximum activation [76]. In addition, it has been shown that during pedaling, the hip is responsible for 50% of the moment of force [50]. Following this predictor is the ankle angular velocity at 270°, coefficient estimate = +19.92, which is linked to the ankle flexor muscles, which effectively control the positioning of the ankle to minimize negative forces, i.e., pedaling pulling phase. Subsequently, the power delivered has a direct relationship with the IE, coefficient estimate = +19.09, which agrees with Zameziati et al. [86] who demonstrated that the power is directly associated with the IE. When the workload increases, especially in the upstroke phase, the IE increases due to the greater recruitment of the knee and hip flexor muscles [6]. The correct saddle selection influences similarly the point where the hip is around the point of maximum extension at 225°, coefficient estimate = +8.93, which will be critical to correctly start the recovery phase. At this point, near the bottom dead center and shortly after, a peak in knee power occurs through the hamstring muscles, which may be associated with the importance of the knee angle and angular velocity at this point in the pedaling cycle [4]. Connecting with the above the knee angular velocity at the bottom dead center, coefficient estimate = −16.31, inversely influences the IE, which directly is a manifestation of the low musculoskeletal activity of the muscles that are involved in the flexion and extension of the knee, since it is a critical point where muscular fibbers are in a maximum stretch. Hip angular velocity at 90°, coefficient estimate = −10.41, has an inverse relationship with the IE and is in the first quadrant, i.e., TDC to 90°, where the maximum effective or tangential force production is achieved. Therefore, the hip moment is particularly high during the first part of the pedal stroke, when the maximum torque and IE are reached [7]. Finally, the individual's mass has an inverse relationship, coefficient estimate = −5.67, which is the coefficient with the lowest weighting with respect to all predictors. It can be associated with the inertial effect of the lower limbs, which represents 40% of the total crank torque [57]. Particularly in the model, no joint angular acceleration appears, and we believe that it may be because these variables affect more the generation of torques in the lower limb. Unfortunately, to the best of our knowledge, few studies relate lower-limb torque with pedaling technique. Surprisingly, the variable KOPS does not appear as a coefficient in the model, which is in controversy with some authors who have found a link between this variable and the IE [46].

In the previous analysis, it is demonstrated that a multiple linear regression is a powerful tool not only for prediction but also for interpretability. Indeed, it allows us to know the most important predictors in the estimation of a variable, which in this case was the IE. In this way, two objectives are achieved: (1) it is possible to determine the IE from the kinematics, without the need to use instrumented pedals, and (2) it clearly supports the hypothesis that IE is directly related to cyclists' lower-limb kinematics.

The results shown are preliminary and promising. The example shown illustrates the methodology for the development of a machine learning method based on multiple linear regression.

4 Conclusions

Multiple linear regression is a machine learning technique that allows full interpretation of any phenomenon, since it allows to visualize the predictors that directly influence the dependent variables.

When a phenomenon to be modelled has a considerable number of predictors, the best subset selection method is a valuable tool of machine learning. This method allows us to select the predictors that are related to the dependent variable.

The example presented in this chapter shows that the IE can be accurately predicted with lower-limb joint kinematics, power output, and individual mass using a multiple linear regression model. This model was composed of 11 explanatory variables, with the lowest BIC value, and showed robustness and generalizability during the randomized tenfold cross-validations. It included angles and angular velocities of the femur, knee, and ankle, demonstrating that both the position and velocity of the joints, particularly at 270° of the pedaling cycle, are important for optimizing the pedaling technique.

This study is highly innovative as it shows that the index of effectiveness can be predicted from variables easily measured by motion capture systems. Therefore, in the near future, this type of model can be integrated into bike fitting systems.

Cross-validation is the important step to determine if the model obtained is generalizable or not and therefore must be implemented in any machine learning methodology.

In the case of biomechanics where multiple kinematic, kinetic, and physical factors are combined, it is important to find a parsimonious model, i.e., as simple as possible, with a smaller number of predictors and a high predictive and explanatory power.

Multiple linear regression is a powerful tool for prediction in sports and for understanding the factors that influence the estimation of kinematic or kinetic variables in any discipline.

Acknowledgments

Acknowledgments The authors thank all the participants who volunteered to participate in the study.

Funding

Acknowledgments This work was supported by the Canada Foundation for Innovation.

Appendix 1

Kinematic equations		
Variables	Equations	Description of the equation terms
Joint angle	$\alpha[n] = \tan^{-1}\frac{Y_a[n]-Y_b[n]}{X_b[n]-X_a[n]}$ (1) $\beta[n] = \alpha[n] + \tan^{-1}\frac{Y_b[n]-Y_c[n]}{X_b[n]-X_c[n]}$ (2) $\theta[n] = \tan^{-1}\frac{Y_b[n]-Y_c[n]}{X_b[n]-X_c[n]} + \tan^{-1}\frac{Y_c[n]-Y_d[n]}{X_d[n]-X_c[n]}$ (3)	n: sample number Marker a: greater trochanter Marker b: femoral lateral epicondyle Marker c: lateral malleolus marker Marker d: foot $Ya[n]$, $Yb[n]$, $Yc[n]$, $Yd[n]$: coordinate Y of marker a, b, c, d $Xa[n]$, $Xb[n]$, $Xc[n]$, $Xd[n]$: coordinated X of marker a, b, c, d $\alpha[n]$, $\beta[n]$, $\theta[n]$: joint angle of femur, knee and ankle $\dot{\alpha}[n]$, $\dot{\beta}[n]$, $\dot{\theta}[n]$: joint angular velocity of femur, knee and ankle
Joint angular velocity	$\dot{\alpha}[n] = \frac{\alpha[n]-\alpha[n-1]}{1/\text{fs}}$ (4) $\dot{\beta}[n] = \frac{\beta[n]-\beta[n-1]}{1/\text{fs}}$ (5) $\dot{\theta}[n] = \frac{\theta[n]-\theta[n-1]}{1/\text{fs}}$ (6)	$\ddot{\alpha}[n]$, $\ddot{\beta}[n]$, $\ddot{\theta}[n]$: joint angular acceleration of femur, knee and ankle fs: sampling frequency
Joint angular acceleration	$\ddot{\alpha}[n] = \frac{\dot{\alpha}[n]-\dot{\alpha}[n-1]}{1/\text{fs}}$ (7) $\ddot{\beta}[n] = \frac{\dot{\beta}[n]-\dot{\beta}[n-1]}{1/\text{fs}}$ (8) $\ddot{\theta}[n] = \frac{\dot{\theta}[n]-\dot{\theta}[n-1]}{1/\text{fs}}$ (9)	

Appendix 2

Kinetic equations		
Total force	$F_{\text{tot}} = \sqrt{\left(F_{\text{T}}^2 + F_{\text{R}}^2 + F_{\text{ML}}^2\right)}$ (10)	F_{tot}: total force (N) F_{T}: tangential force (N) F_{R}: radial force (N) F_{ML}: medio-lateral force (N)
Index of effectiveness	$\text{IE} = \frac{\int_0^{2\pi} F_{\text{T}}(\varphi)d\varphi}{\int_0^{2\pi} F_{\text{tot}}(\varphi)d\varphi} \times 100$ (11)	IE: index of effectiveness (%) φ: crank angle (degree)

Bibliography

1. Aguinis, H., Gottfredson, R. K., & Joo, H. (2013). Best-practice recommendations for defining, identifying, and handling outliers. *Organizational research methods, 16*(2), 270–301.
2. Altelbany, S. (2021). Evaluation of ridge, elastic net and lasso regression methods in precedence of multicollinearity problem: A simulation study. *Journal of Applied Economics and Business Studies, 5*(1), 131–142.
3. Baker, R., Leboeuf, F., Reay, J., & Sangeux, M. (2018). The conventional gait model-success and limitations. In: Springer.
4. Barratt, P. R., Korff, T., Elmer, S. J., & Martin, J. C. (2011). Effect of crank length on joint-specific power during maximal cycling. *Medicine & Science in Sports & Exercise, 43*(9), 1689–1697.
5. Berrar, D. (2019). Cross-Validation. In.
6. Bini, R., Hume, P., Croft, J. L., & Kilding, A. (2013). Pedal force effectiveness in cycling: A review of constraints and training effects.
7. Bini, R. R., & Diefenthaeler, F. (2010). Kinetics and kinematics analysis of incremental cycling to exhaustion. *Sports Biomechanics, 9*(4), 223–235.
8. Bland, J. M., & Altman, D. (1986). Statistical methods for assessing agreement between two methods of clinical measurement. *The lancet, 327*(8476), 307–310.
9. Bogey, R., Perry, J., & Gitter, A. (2005). An EMG-to-force processing approach for determining ankle muscle forces during normal human gait. *IEEE transactions on neural systems and rehabilitation engineering, 13*(3), 302–310.
10. Borg, G. (1982). Psychological bases of physical exertion. *Med Sci Sports Exerc, 14*(5), 377–381.
11. Brooks, G. P., & Ruengvirayudh, P. (2016). Best-subset selection criteria for multiple linear regression. *General Linear Model Journal.*
12. Castle, J. L., Qin, X., & Reed, W. R. (2009). How to pick the best regression equation: A review and comparison of model selection algorithms. *University of Canterbury, Department of Economics and Finance Working Papers in Economics*(09/13).
13. Chowdhury, R. H., Reaz, M. B., Ali, M. A. B. M., Bakar, A. A., Chellappan, K., & Chang, T. G. (2013). Surface electromyography signal processing and classification techniques. *Sensors, 13*(9), 12431–12466.

14. Ciatto, G., Schumacher, M. I., Omicini, A., & Calvaresi, D. (2020). *Agent-based explanations in AI: Towards an abstract framework*. Paper presented at the International workshop on explainable, transparent autonomous agents and multi-agent systems.
15. Cleather, D. J., & Bull, A. M. J. (2010). Influence of inverse dynamics methods on the calculation of inter-segmental moments in vertical jumping and weightlifting. *BioMedical Engineering OnLine, 9*(1), 74. https://doi.org/10.1186/1475-925X-9-74
16. Daoud, J. I. (2017). *Multicollinearity and regression analysis*. Paper presented at the Journal of Physics: Conference Series.
17. Davis III, R. B., Ounpuu, S., Tyburski, D., & Gage, J. R. (1991). A gait analysis data collection and reduction technique. *Human movement science, 10*(5), 575–587.
18. Dorel, S., Couturier, A., Lacour, J.-R., Vandewalle, H., Hautier, C., & Hug, F. (2010). Force-velocity relationship in cycling revisited: benefit of two-dimensional pedal forces analysis. *Med Sci Sports Exerc, 42*(6), 1174–1183.
19. Eckhardt, C. M., Madjarova, S. J., Williams, R. J., Ollivier, M., Karlsson, J., Pareek, A., & Nwachukwu, B. U. (2023). Unsupervised machine learning methods and emerging applications in healthcare. *Knee Surgery, Sports Traumatology, Arthroscopy, 31*(2), 376–381.
20. Faria, E. W., Parker, D. L., & Faria, I. E. (2005). The science of cycling: factors affecting performance—part 2. *Sports medicine, 35*, 313–337.
21. Fonda, B., & Sarabon, N. (2010). Biomechanics of cycling. *Sport Science Review, 19*(1–2), 187.
22. Freijeiro-González, L., Febrero-Bande, M., & González-Manteiga, W. (2022). A critical review of LASSO and its derivatives for variable selection under dependence among covariates. *International Statistical Review, 90*(1), 118–145.
23. García-López, J., Díez-Leal, S., Ogueta-Alday, A., Larrazabal, J., & Rodríguez-Marroyo, J. A. (2016). Differences in pedalling technique between road cyclists of different competitive levels. *Journal of sports sciences, 34*(17), 1619–1626.
24. García García, C., Salmeron Gomez, R., & García Pérez, J. (2022). A review of ridge parameter selection: minimization of the mean squared error vs. mitigation of multicollinearity. *Communications in Statistics-Simulation and Computation*, 1–13.
25. Gatti, A. A., Keir, P. J., Noseworthy, M. D., Beauchamp, M. K., & Maly, M. R. (2021). Hip and ankle kinematics are the most important predictors of knee joint loading during bicycling. *Journal of Science and Medicine in Sport, 24*(1), 98–104.
26. Gatti, A. A., Keir, P. J., Noseworthy, M. D., Beauchamp, M. K., & Maly, M. R. (2022). Equations to prescribe bicycle saddle height based on desired joint kinematics and bicycle geometry. *European journal of sport science, 22*(3), 344–353.
27. Gong, T., Zhao, A., Zhou, J., Chen, Y., Ma, H., Wang, X., & Wang, Y. (2022). A facile inverse dynamics method to study the impacts of fatigue on the lower limb biomechanics. *Biotechnology and Genetic Engineering Reviews*, 1–12.
28. Gupta, S. S., & Yang, H.-M. (1985). Bayes-P_* subset selection procedures for the best population. *Journal of statistical planning and inference, 12*, 213–233.
29. Hansen, E. A., Rønnestad, B. R., Vegge, G., & Raastad, T. (2012). Cyclists' improvement of pedaling efficacy and performance after heavy strength training. *International Journal of Sports Physiology and Performance, 7*(4), 313–321.
30. Hastie, T., Tibshirani, R., & Tibshirani, R. (2020). Best subset, forward stepwise or lasso? Analysis and recommendations based on extensive comparisons.
31. Hirashima, M. (2011). Induced acceleration analysis of three-dimensional multi-joint movements and its application to sports movements. In *Theoretical biomechanics*: IntechOpen.
32. Hoerl, A. E., & Kennard, R. W. (1970). Ridge regression: Biased estimation for nonorthogonal problems. *Technometrics, 12*(1), 55–67.
33. Horvat, T., & Job, J. (2020). The use of machine learning in sport outcome prediction: A review. *Wiley Interdisciplinary Reviews: Data Mining and Knowledge Discovery, 10*(5), e1380.
34. Jiang, X., & Bíró, I. (2023). *Muscle force estimation in running gait analysis after a prolonged running session via OpenSim*. Paper presented at the 2023 4th International Conference on Computer Engineering and Application (ICCEA).

35. Kadaba, M. P., Ramakrishnan, H., & Wootten, M. (1990). Measurement of lower extremity kinematics during level walking. *Journal of orthopaedic research, 8*(3), 383–392.
36. Karagiannopoulos, M., Anyfantis, D., Kotsiantis, S., & Pintelas, P. (2004). Feature selection for regression problems. *Educational Software Development Laboratory, Department of Mathematics, University of Patras, Greece*.
37. Keir, P. J., & Mogk, J. P. (2005). The development and validation of equations to predict grip force in the workplace: contributions of muscle activity and posture. *Ergonomics, 48*(10), 1243–1259.
38. Kidwell, J. S., & Brown, L. H. (1982). Ridge regression as a technique for analyzing models with multicollinearity. *Journal of Marriage and the Family*, 287–299.
39. Lanferdini, F. J., Jacques, T. C., Bini, R. R., & Vaz, M. A. (2014). Cyclists and triathletes do not differ in muscle volume, muscle recruitment or pedal force effectiveness. *Isokinetics and Exercise Science, 22*(4), 279–286.
40. Lebarbier, É., & Mary-Huard, T. (2006). Une introduction au critère BIC: fondements théoriques et interprétation. *Journal de la Société française de statistique, 147*(1), 39–57.
41. Leirdal, S., & Ettema, G. (2011). Pedaling technique and energy cost in cycling. *Medicine and Science in Sports and exercise, 43*(4), 701–705.
42. Lindsey, C., & Sheather, S. (2010). Variable selection in linear regression. *The Stata Journal, 10*(4), 650–669.
43. Martin, J., Brown, N., Anderson, F., & Spirduso, W. (2000). A governing relationship for repetitive muscular contraction. *Journal of Biomechanics, 33*(8), 969–974.
44. Martin, J. C., & Nichols, J. A. (2018). Simulated work loops predict maximal human cycling power. *Journal of Experimental Biology, 221*(13), jeb180109.
45. Menard, M., Domalain, M., Decatoire, A., & Lacouture, P. (2016). Influence of saddle setback on pedalling technique effectiveness in cycling. *Sports Biomechanics, 15*(4), 462–472.
46. Menard, M., Domalain, M., Decatoire, A., & Lacouture, P. (2018). Influence of saddle setback on knee joint forces in cycling. *Sports Biomechanics*.
47. Mesin, L., Merletti, R., & Rainoldi, A. (2009). Surface EMG: the issue of electrode location. *Journal of Electromyography and Kinesiology, 19*(5), 719–726.
48. Mi, J.-X., Li, A.-D., & Zhou, L.-F. (2020). Review study of interpretation methods for future interpretable machine learning. *IEEE Access, 8*, 191969–191985.
49. Miller, T. (2019). Explanation in artificial intelligence: Insights from the social sciences. *Artificial intelligence, 267*, 1–38.
50. Mornieux, G., Guenette, J. A., Sheel, A. W., & Sanderson, D. J. (2007). Influence of cadence, power output and hypoxia on the joint moment distribution during cycling. *European journal of applied physiology, 102*, 11–18.
51. Mornieux, G., Stapelfeldt, B., Gollhofer, A., & Belli, A. (2008). Effects of pedal type and pull-up action during cycling. *International Journal of Sports Medicine*, 817–822.
52. Motio-STT-Systems. (2024). Retrieved from https://motio.stt-systems.com/es/3dma-captura-y-analisis-de-movimiento-optico-3d/
53. Murdoch, W. J., Singh, C., Kumbier, K., Abbasi-Asl, R., & Yu, B. (2019). Interpretable machine learning: definitions, methods, and applications. *arXiv preprint arXiv:1901.04592*.
54. Nakashima, M., Kanie, R., Shimana, T., Matsuda, Y., & Kubo, Y. (2023). Development of a comprehensive method for musculoskeletal simulation in swimming using motion capture data. *Proceedings of the Institution of Mechanical Engineers, Part P: Journal of Sports Engineering and Technology, 237*(2), 85–95.
55. Nasr, A., Bell, S., He, J., Whittaker, R. L., Jiang, N., Dickerson, C. R., & McPhee, J. (2021). MuscleNET: mapping electromyography to kinematic and dynamic biomechanical variables by machine learning. *Journal of Neural Engineering, 18*(4), 0460d0463.
56. Nathans, L. L., Oswald, F. L., & Nimon, K. (2012). Interpreting multiple linear regression: a guidebook of variable importance. *Practical assessment, research & evaluation, 17*(9), n9.
57. Neptune, R., & Herzog, W. (1999). The association between negative muscle work and pedaling rate. *Journal of biomechanics, 32*(10), 1021–1026.

58. Nimon, K. F., & Oswald, F. L. (2013). Understanding the Results of Multiple Linear Regression: Beyond Standardized Regression Coefficients. *Organizational research methods, 16*(4), 650–674. doi:https://doi.org/10.1177/1094428113493929
59. Osborne, J. W., & Waters, E. (2019). Four assumptions of multiple regression that researchers should always test. *Practical assessment, research, and evaluation, 8*(1), 2.
60. Priego Quesada, J. I., Kerr, Z. Y., Bertucci, W. M., & Carpes, F. P. (2018). The categorization of amateur cyclists as research participants: findings from an observational study. *Journal of sports sciences, 36*(17), 2018–2024.
61. Ranstam, J., & Cook, J. (2018). LASSO regression. *Journal of British Surgery, 105*(10), 1348–1348.
62. Ray, S. (2019). *A quick review of machine learning algorithms*. Paper presented at the 2019 International conference on machine learning, big data, cloud and parallel computing (COMITCon).
63. Richter, C., O'Reilly, M., & Delahunt, E. (2021). Machine learning in sports science: challenges and opportunities. *Sports Biomechanics*, 1–7.
64. Roupa, I., da Silva, M. R., Marques, F., Gonçalves, S. B., Flores, P., & da Silva, M. T. (2022). On the modeling of biomechanical systems for human movement analysis: a narrative review. *Archives of Computational Methods in Engineering, 29*(7), 4915–4958.
65. Sachet, I., Brochner Nygaard, N. P., Guilhem, G., Hug, F., & Dorel, S. (2023). Strength capacity of lower-limb muscles in world-class cyclists: new insights into the limits of sprint cycling performance. *Sports Biomechanics, 22*(4), 536–553.
66. Sanderson, D. J., & Black, A. (2003). The effect of prolonged cycling on pedal forces. *Journal of sports sciences, 21*(3), 191–199.
67. Schroeder, M. A., Lander, J., & Levine-Silverman, S. (1990). Diagnosing and dealing with multicollinearity. *Western journal of nursing research, 12*(2), 175–187.
68. Sczepanski, T. L., Gross, M. T., Duncan, P. W., & Chandler, J. M. (1991). Effect of contraction type, angular velocity, and arc of motion on VMO: VL EMG ratio. *Journal of Orthopaedic & Sports Physical Therapy, 14*(6), 256–262.
69. Senthilnathan, S. (2019). Usefulness of correlation analysis. *Available at SSRN 3416918*.
70. Stetter, B. J., Ringhof, S., Krafft, F. C., Sell, S., & Stein, T. (2019). Estimation of knee joint forces in sport movements using wearable sensors and machine learning. *Sensors, 19*(17), 3690.
71. Steyerberg, E. W., van der Ploeg, T., & Van Calster, B. (2014). Risk prediction with machine learning and regression methods. *Biometrical Journal, 56*(4), 601–606.
72. Su, W., Bogdan, M., & Candes, E. (2017). False discoveries occur early on the lasso path. *The Annals of statistics*, 2133–2150.
73. Sudin, S., Aziz, A. N. A., Saad, F. S. A., Khalid, N. S., & Ibrahim, I. I. (2021). *Cycling performance prediction based on cadence analysis by using multiple regression*. Paper presented at the Journal of Physics: Conference Series.
74. Takaishi, T., Yamamoto, T., Ono, T., Ito, T., & Moritani, T. (1998). Neuromuscular, metabolic, and kinetic adaptations for skilled pedaling performance in cyclists. *Medicine and Science in Sports and exercise, 30*, 442–449.
75. Tamura, R., Kobayashi, K., Takano, Y., Miyashiro, R., Nakata, K., & Matsui, T. (2017). Best subset selection for eliminating multicollinearity. *Journal of the Operations Research Society of Japan, 60*(3), 321–336.
76. Turpin, N. A., & Watier, B. (2020). Cycling biomechanics and its relationship to performance. *Applied Sciences, 10*(12), 4112.
77. Uttley, J. (2019). Power analysis, sample size, and assessment of statistical assumptions—Improving the evidential value of lighting research. *Leukos, 15*(2–3), 143–162.
78. Uyanık, G. K., & Güler, N. (2013). A study on multiple linear regression analysis. *Procedia-Social and Behavioral Sciences, 106*, 234–240.
79. Van Eetvelde, H., Mendonça, L. D., Ley, C., Seil, R., & Tischer, T. (2021). Machine learning methods in sport injury prediction and prevention: a systematic review. *Journal of Experimental Orthopaedics, 8*(1), 27.

80. Wang, Y., Luangkesorn, L., & Shuman, L. J. (2011). *Best-subset selection procedure*. Paper presented at the Proceedings of the 2011 Winter Simulation Conference (WSC).
81. Wanich, T., Hodgkins, C., Columbier, J.-A., Muraski, E., & Kennedy, J. G. (2007). Cycling injuries of the lower extremity. *JAAOS-Journal of the American Academy of Orthopaedic Surgeons, 15*(12), 748–756.
82. Winter, D. A. (2009). *Biomechanics and motor control of human movement* [1 ressource en ligne (xiv, 370 pages): illustrations, fichiers HTML et PDF](4th ed.).
83. Wu, G., Siegler, S., Allard, P., Kirtley, C., Leardini, A., Rosenbaum, D., ... Witte, H. (2002). ISB recommendation on definitions of joint coordinate system of various joints for the reporting of human joint motion—part I: ankle, hip, and spine. *Journal of biomechanics, 35*(4), 543–548.
84. Xia, Z., Devaprakash, D., Cornish, B. M., Barrett, R. S., Lloyd, D. G., Hams, A. H., & Pizzolato, C. (2023). Predicting Free Achilles Tendon Strain From Motion Capture Data Using Artificial Intelligence. *IEEE Trans Neural Syst Rehabil Eng, 31*, 3086–3094. doi:https://doi.org/10.1109/tnsre.2023.3296280
85. Xu, L., & Zhang, W.-J. (2001). Comparison of different methods for variable selection. *Analytica Chimica Acta, 446*(1–2), 475–481.
86. Zameziati, K., Mornieux, G., Rouffet, D., & Belli, A. (2006). Relationship between the increase of effectiveness indexes and the increase of muscular efficiency with cycling power. *European journal of applied physiology, 96*, 274–281.
87. Zou, H., & Hastie, T. (2005). Regularization and variable selection via the elastic net. *Journal of the Royal Statistical Society Series B: Statistical Methodology, 67*(2), 301–320.

Machine Learning in Biomechanics: Key Applications and Limitations in Walking, Running and Sports Movements

Carlo Dindorf, Fabian Horst, Djordje Slijepčević, Bernhard Dumphart, Jonas Dully, Matthias Zeppelzauer, Brian Horsak, and Michael Fröhlich

Abstract This chapter provides an overview of recent and promising Machine Learning applications, i.e. *pose estimation*, *feature estimation*, *event detection*, *data exploration and clustering* and *automated classification*, in gait (walking and running) and sports biomechanics. It explores the potential of Machine Learning methods to address challenges in biomechanical workflows; highlights central limitations, i.e. *data and annotation availability* and *explainability*, that need to be addressed; and emphasises the importance of interdisciplinary approaches for fully harnessing the potential of Machine Learning in gait and sports biomechanics.

Keywords Pose estimation · Feature estimation · Event detection · Clustering · Classification · Gait analysis · Sports · Explainability · Kinematics

---

Shared first authorship; authors "C. Dindorf, F. Horst, and D. Slijepčević" contributed equally
Shared senior authorship; authors "M. Zeppelzauer, B. Horsak, and M. Fröhlich" contributed equally

---

C. Dindorf (✉) · J. Dully · M. Fröhlich
Department of Sports Science, University of Kaiserslautern-Landau (RPTU), Kaiserslautern, Germany
e-mail: carlo.dindorf@rptu.de; jonas.dully@rptu.de; michael.froehlich@rptu.de

F. Horst
Institute of Sports Science, Johannes Gutenberg-University Mainz, Mainz, Germany
e-mail: horst@uni-mainz.de

D. Slijepčević · M. Zeppelzauer
Institute of Creative Media Technologies, St. Pölten University of Applied Sciences, Sankt Pölten, Austria
e-mail: djordje.slijepcevic@fhstp.ac.at; matthias.zeppelzauer@fhstp.ac.at

B. Dumphart · B. Horsak
Institute of Health Sciences & Center for Digital Health and Social Innovation, St. Pölten University of Applied Sciences, Saint Pölten, Austria
e-mail: bernhard.dumphart@fhstp.ac.at; brian.horsak@fhstp.ac.at

M. J. Blondin et al. (eds.), *Artificial Intelligence, Optimization, and Data Sciences in Sports*, Springer Optimization and Its Applications 218,
https://doi.org/10.1007/978-3-031-76047-1_4

1 Introduction

Artificial Intelligence (AI), specifically Machine Learning (ML), has attracted increasing interest in many domains, including biomechanics. The biomechanics domain focuses on the quantitative description and analysis of human movements, taking into account aspects such as kinematics, kinetics and electromyography [1]. This comprehensive approach integrates diverse data sources to gain insights into the intricate mechanics of human movement, including gait[1] and sports biomechanics. The motivation for using ML in biomechanics is driven, among other things, by two central challenges in biomechanical workflows:

Laboratory-Bound, Resource and Time-Intensive Data Acquisition and Processing In the biomechanics domain, there is often a need for a precise and comprehensive understanding of human movement, encompassing various body parts and involving a range of biomechanical signals such as joint angles, ground reaction forces and muscular activities. Ensuring high temporal and spatial precision of the data typically requires standardised laboratory-bound set-ups with expensive measurement equipment, e.g. marker-based infrared camera systems and force platforms as well as skilled personnel. As a consequence, acquiring and processing biomechanical data can be resource-intensive and time-consuming. First, ML can be employed to assist domain experts in data acquisition, for example, by estimating biomechanical signals through simple sensor recordings, such as inertial measurement units (IMUs) or video images, even in natural environments, e.g. outside the laboratory. Second, ML can aid in data processing, for example, in data segmentation, as well as identification of measurement errors and outliers. The overarching goal in employing ML is to accelerate biomechanical workflows and make biomechanical data more accessible while simultaneously enhancing data quality.

Complexity of Data Analysis Biomechanical data are extensive and encompass multidimensional and multi-correlated time series signals from various body parts [2]. Biomechanical signals are characterised by a substantial level of both inter-subject [3] and intra-subject [4] variability, alongside the presence of numerous interacting influential factors. For example, when assessing a pathology through ground reaction forces during walking, it is essential to take into account that these forces can be influenced not only by the presence of an underlying pathology, but also by the age, sex, body mass, footwear and walking speed of an individual. Consequently, during data analysis even experienced domain experts face challenges in interpreting the extensive and complex biomechanical data at hand. The integration of ML methods for data exploration, clustering and automated classification seems promising in supporting and enhancing biomechanists' efforts in data analysis and

[1] The term "gait" is utilised when referring to both walking and running movements throughout this chapter.

interpretation. Ultimately, leveraging ML holds the potential to provide data-driven decision support in both research and applied settings.

ML approaches have proven to be effective in handling complex data and automating various aspects of biomechanical data acquisition, processing and analysis. These approaches hold the potential to enhance the efficiency and accuracy of biomechanical workflows and provide biomechanists with valuable tools for gaining deeper insights into human movement. With the integration of ML, the biomechanics domain stands on the brink of substantial progress in advancing our understanding of human movement and its wide-ranging applications.

The present chapter aims to give an overview of promising applications of ML in the biomechanics domain and elaborates central challenges and limitations. In Sect. 2, we provide the reader with background information describing usual biomechanical (Sect. 2.1) and ML (Sect. 2.2) workflows. In Sect. 3, each subsection offers a description and related work of the selected ML applications in biomechanics, while Sect. 4 delves into the overarching central challenges and limitations. Throughout this chapter, we will provide separate descriptions for applications in the biomechanics of walking, running and sports movements. To conclude this chapter (Sect. 5), we consolidate valuable insights from these movements and identify potential synergies for future progress in the domain.

2 Biomechanics and Machine Learning Workflows

2.1 Biomechanics Workflow

Biomechanical testing typically follows a three-phase workflow: data acquisition, data processing and data analysis. In the following, we will outline each of these phases and motivate how ML applications can assist and enhance the capabilities of biomechanists throughout this workflow. Comprehensive background information can be found in biomechanics textbooks [1, 262].

2.1.1 Data Acquisition

Biomechanical data acquisitions involve the use of a variety of measurement devices to measure (i) kinematic, e.g. positions and accelerations of body parts; (ii) kinetic, e.g. ground reaction forces, (iii) electromyographic, e.g. muscle activity; and (iv) anthropometric, e.g. body mass, signals during human movements. In the following, we will outline common biomechanical testing approaches, including laboratory-based, e.g. using marker-based infrared camera systems, force platforms and electromyography, as well as field-based, e.g. using wearable sensors such as IMUs and pressure insoles and portable sensors such as RGB video camera approaches as exemplified in Fig. 1.

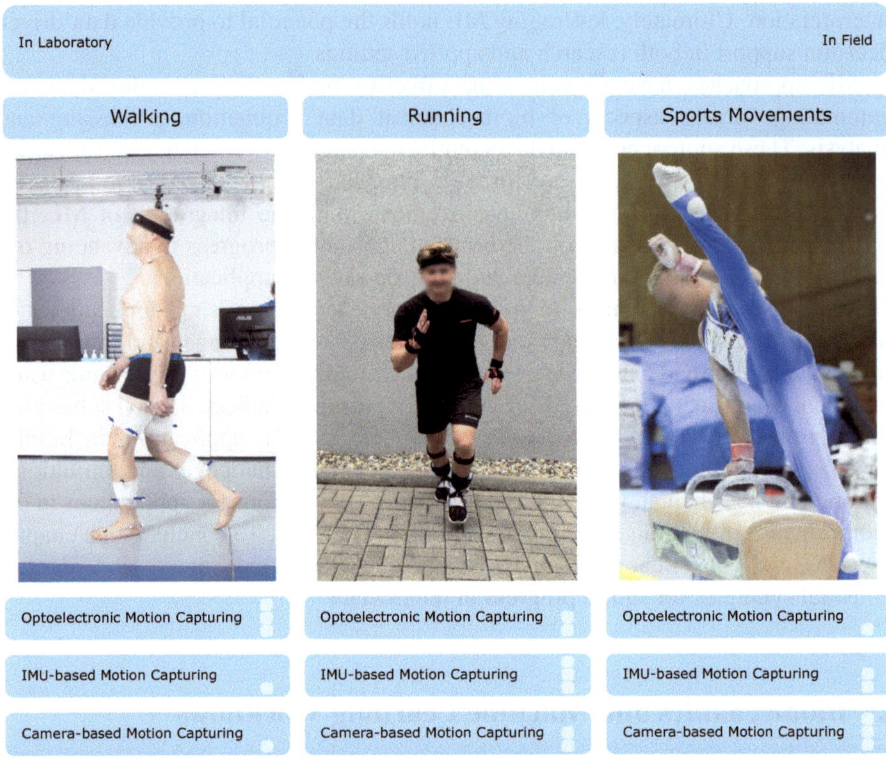

Fig. 1 Overview of three commonly applied biomechanical testing approaches: (i) Biomechanical testing using marker-based infrared camera systems together with force platforms and electromyography in laboratories can be considered the gold standard, where achieving the highest level of precision and accuracy are prioritised over material and personal resources; (ii) biomechanical testing using wearable sensors, e.g. IMUs in natural field-based settings, offering both long-term recordings and real-time feedback; and (iii) biomechanical testing utilising portable sensors, e.g. RGB video cameras is well-suited for conducting assessments in natural field-based settings, where the emphasis is on minimising constraints, even at the expense of spatial and temporal accuracy. The number of squares next to a measurement approaches describes its prevalence for the biomechanical testing of a movement task (walking, running, sports movements). Copyright (left photo): Helene Sorger/St. Pölten University of Applied Sciences

Laboratory-Based Approach In the realm of fundamental and clinical research exploring biomechanical principles in walking, running and sports movements, biomechanists frequently choose to utilise standardised laboratory settings. Biomechanical testing in laboratory settings often involves analysing kinematic signals by tracking markers placed at specific anatomical landmarks on the body with multiple infrared cameras. This approach provides precise 3D position data of the body segments and enables, e.g. the estimation of joint angles using physics-based biomechanical modelling. Force platforms and electromyography are frequently used alongside marker-based approaches. Force platforms provide insight into the magnitude, direction and timing of forces exerted between the body and the platform. These platforms provide high temporal and spatial resolution, but require a proper physical contact with the platform. In certain applications, the electrical activity of muscles during movement is recorded to gain insights into muscle activation patterns using electromyography with electrodes attached to the skin. Comprehensive laboratory-based approaches allow biomechanical modelling that integrates the recorded marker positions, ground reaction forces and muscular activities to estimate biomechanical signals that cannot be directly measured, e.g. joint angles and moments, joint contact forces and muscle forces. The analysis of all these signals enables a comprehensive and accurate understanding of movement mechanics.

The laboratory-based approach provides high accuracy and precision, but also requires high personnel and material resources. Moreover, it comes along with certain experimental limitations. Measurement devices like infrared cameras are often bound to indoor environments without sunlight and cover only a limited spatial area, in which the movement analysis can take place. In addition, the attachment of markers and electrodes to the body may not be feasible for all types of movements and may limit the subjects' range of motion. Thus, measurement methods used in laboratory-based approaches cannot be applied for everyday situations in the field, e.g. standing and walking in public transport, tele-rehabilitation and sports competitions. Nevertheless, the laboratory-based approach can be considered the established standard for biomechanical testing in terms of accuracy, which is used as the best reference when exploring the accuracy of field-based approaches using wearable sensors and video cameras.

Field-Based Approach Field-based methods are used to study biomechanical principles associated with walking, running and sports movements in natural environments. In field settings, biomechanical testing frequently incorporates the use of wearable sensors for prolonged data acquisitions and real-time feedback. Additionally, RGB video cameras are employed, eliminating the need for sensor attachment to the body.

Wearable sensors, e.g. IMUs and pressure insoles, are wireless and lightweight. While these sensors often provide less detailed and more noisy data than laboratory-based approaches, they can be used in various environments with less restricted experimental conditions. IMUs consist of a combination of accelerometers, gyroscopes and sometimes magnetometers. IMUs can be attached to a body segment or equipment, e.g. shoes to measure changes in acceleration and orientation. These data can be used to estimate kinematic signals, e.g. joint angles in real time in non-laboratory settings. This allows, for example, the analysis of running biomechanics in natural environments, e.g. long-distance running on diverse natural terrains such as forest trails. This analysis, however, is limited to kinematic signals like joint angles. Estimating kinetic signals commonly analysed in laboratory-based approaches, such as ground reaction forces and joint moments, presents challenges when relying only on IMU data. ML can offer a promising solution to this challenge, namely, ML-based biomechanical **feature estimation**. For example, an ML method can be used to predict ground reaction forces from available IMU data and therefore significantly enhance the capabilities of IMUs in biomechanical research by providing accurate and reliable estimations of biomechanical signals.

Key Application: Feature Estimation (Sect. 3.2)
Feature estimation employs ML to predict complex biomechanical data, e.g. joint angles and moments from more accessible data sources, including IMUs, pressure insoles and RGB video cameras.

Despite the possibilities that wearable-based approaches offer for assessing movement in the field, several sports-specific movements pose difficulties in terms of attaching hardware to the body. This can include limitations on range of motion, which is especially relevant during competition, and the potential risk of injury to athletes who might fall on the utilised sensors, e.g. high jump.

In situations where sensor and marker attachment to subjects' bodies is not feasible, such as during sports competitions, researchers often resort to acquiring image and video data using cameras, including video or depth cameras, for kinematic analyses in sports practice. In this process, biomechanists manually identify and digitise anatomical landmarks in each frame of a video. However, the accuracy of this manual digitisation depends heavily on factors like camera resolution and lighting conditions. Moreover, this procedure is very time-consuming, so that the analysis of a larger number of recordings of a competition or a training session is only practicable to a limited extent. ML offers a promising solution to address these challenges. Specifically, in the context of human **pose estimation**, ML algorithms can be trained to automatically recognise and track anatomical landmarks in video or image data.

Key Application: Pose Estimation (Sect. 3.1)
Pose estimation is the process of automatically tracking and determining the body's anatomical landmarks, body segments or joint centre locations in video images using ML, enabling the quantification of human movement without marker and sensor attachments to the human body.

Integrating ML into the biomechanical workflow allows for more extensive kinematic analyses, even in scenarios where hardware attachment is not feasible. Thus, the use of ML could significantly reduce the time and effort required to obtain and analyse biomechanical data. This would enable researchers, coaches and clinicians to analyse a more comprehensive set of recordings from natural environments such as sports competitions or training sessions.

2.1.2 Data Processing

Data processing is critical for reducing sensor noise in biomechanical signals and making them comparable across different subjects and recordings. In the following, we briefly mention common processing steps: First, **filtering** is used to smooth out irregularities in the raw signals to enhance the clarity of the movement patterns for subsequent analysis. After filtering, the biomechanical signals undergo an **event detection** to identify and extract movement (sub-)phases in the recordings.

Key Application: Event Detection (Sect. 3.3)
Event detection in time series data is the annotation of certain events that are used to extract useful and vital information or to remove unwanted and unnecessary data for further analysis.

Afterwards, the integrity of the biomechanical signals is checked using **outlier detection** techniques. Identifying and removing outliers or artefacts due to sensor shifts or errors is essential to prevent them from distorting the data analysis. **Normalisation** involves scaling the signals based on specific criteria, often individual characteristics like body size, body mass and speed. This adjustment allows for fair comparisons between different subjects and conditions, as it accounts for variations in size that can influence biomechanical data. Normalisation ensures that data is presented on a consistent scale, making it easier to draw meaningful comparisons and conclusions.

2.1.3 Data Analysis

The analysis of biomechanical data poses challenges due to the extensive volume of data recorded during measurement sessions, which are multidimensional and interrelated [2]. The following two analysis approaches are commonly employed for data analysis in practice:

Qualitative Assessment This approach involves the visual inspection and qualitative assessment of biomechanical signals. Researchers examine the data to identify patterns, anomalies or noteworthy events. While qualitative assessment can provide valuable insights into the data, it often lacks sufficient inter-rater reliability, leading to potential discrepancies in interpretation among different analysts [5, 6].

Extraction of Biomechanical Features Alternatively, biomechanists often choose to extract specific "handcrafted" features from the time series data. For instance, they might extract discrete features such as the maximal and minimal knee angle during walking. Subsequently, inferential statistical analysis is applied to these extracted features. However, the approach of extracting single time-discrete features from several multidimensional time series signals carries certain risks. Reducing complex time series data to a small set of discrete features can potentially lead to information loss, limiting the holistic understanding of biomechanical processes. Additionally, the selection of which features to extract and analyse can introduce subjectivity and bias into the analysis.

To address these challenges and enhance the reliability of biomechanical data analysis, researchers are exploring the integration of ML methods. ML models can be trained using any input data, including handcrafted features, automatically extracted features or the raw time series signals. The latter approach eliminates the need for prior reduction of the available biomechanical time series data. This holistic approach enables the incorporation of comprehensive information, encompassing time series data of whole-body 3D joint angles and ground reaction forces throughout the entire movement cycle.

ML methods, such as Deep Learning models, can automatically identify relevant patterns and features within the multidimensional and interrelated data. This reduces the reliance on manual feature extraction and has the potential to improve the overall accuracy and objectivity of biomechanical analyses. Data exploration and clustering involves the use of unsupervised ML methods to uncover hidden patterns or clusters within the data. This aids in understanding the inherent structure of biomechanical data and can guide subsequent analyses.

Key Application: Data Exploration and Clustering (Sect. 3.4)
Clustering involves grouping instances or individuals with similar biomechanical characteristics using unsupervised ML, thereby revealing underlying structures and subgroups within complex biomechanical data.

In addition to unsupervised ML, supervised ML plays a crucial role in biomechanical data analysis. Supervised ML leverages annotated data to address a wide range of classification and regression tasks in biomechanical settings. Supervised ML can be employed to automatically identify and categorise, e.g. specific pathological conditions.

Key Application: Automated Classification (Sect. 3.5)
Automated classification refers to the process of developing a predictive model that assign input features of data samples to predefined categories or classes using supervised ML.

The use of ML, encompassing data exploration, clustering through unsupervised ML methods and automated classification and regression via supervised ML methods, holds promise for advancing our understanding of human movement.

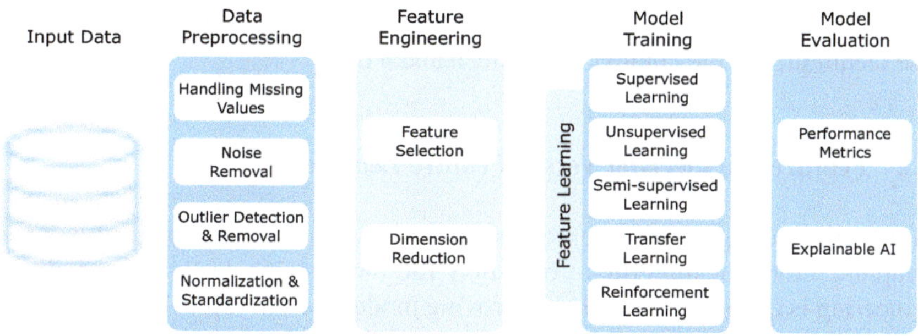

Fig. 2 In the realm of biomechanics, typical ML workflows are characterised by the following phases: data preprocessing, optionally feature engineering, ML model training (which can incorporate feature learning) and ML model evaluation

2.2 *Machine Learning Workflow*

ML approaches have the ability to model relationships between versatile input and output data, enabling various applications throughout the biomechanical workflow. In the following, we will outline the typical ML workflow and emphasise the requirements for applying ML in biomechanics. For more in-depth explanations, refer to relevant sources like [7]. The typical ML workflow can be characterised by the following phases (Fig. 2): data preprocessing, feature engineering, ML model training and ML model evaluation.

2.2.1 Machine Learning-Specific Data Preprocessing

Data preprocessing is an essential step in preparing data for ML models. When dealing with biomechanical data, **data cleaning** is usually omitted, because in the domain of biomechanics, data is already typically filtered, normalised to account for various influencing factors including walking speed and body mass, and outliers are removed, prior to data analysis (Sect. 2.1.2). Hence, these data cleaning steps are only necessary when applying ML to raw sensor signals.

Data normalisation and standardisation become necessary when signals/features in a dataset reside within different value ranges, e.g. different value ranges of joint angles and joint moments. A major motivation for this preprocessing step is to ensure that each signal/feature has an equal opportunity to contribute to the ML model's predictions. Furthermore, in the context of Deep Learning, data normalisation is usually leveraged to foster the convergence of gradient descent during training. Common techniques include min-max normalisation, z-score standardisation and transformations like logarithmic or polynomial transformations. The choice of method can vary based on the respective data and domain. Recent studies showed that these techniques significantly impact model performance [8] and behaviour [9]; however, there are very few studies that systematically compare these techniques in the context of biomechanical data.

2.2.2 Feature Engineering Versus Feature Learning

Feature engineering involves extracting new features from the preprocessed data to capture more meaningful information for the specific task at hand. Feature engineering is crucial not only for improving model accuracy, preventing overfitting and reducing computational demands but also for enhancing interpretability and extracting meaningful information [10]. In conventional methods, feature engineering typically encompasses the crafting of domain-specific features by domain experts. These features are often derived as time-discrete features, such as local and global minima, maxima or other simple descriptive statistics. Moreover, feature selection methods can be employed using ML to automatically select a subset of features from the initially defined feature set. Another commonly applied approach in biomechanics leverages dimension reduction techniques to extract features. Dimension reduction aims to project input data into a lower-dimensional feature space, thereby effectively removing redundant and irrelevant information.

However, it has also been demonstrated that these simple feature engineering approaches can lead to loss of information and performance of the ML models [11]. In Deep Learning, we refer to feature learning, indicating that the process of feature engineering is replaced, and instead, the Deep Learning model takes over the task of learning robust features from the data. As Deep Learning methods can automatically learn task-relevant features from the data during the training process, they hold the potential to outperform the aforementioned traditional approaches [12].

2.2.3 Model Training

Depending on the input data and the objectives of the task, different strategies can be selected to train ML models. Some of the main learning strategies already utilised in biomechanics, or those that may become more significant, encompass:

Supervised Learning An ML algorithm is trained on an annotated, i.e. labelled dataset, where each data sample contains input features and corresponding target labels. The goal is to model relationships between the input features and target labels so that the ML model can make accurate predictions on unseen data for classification (Sect. 3.5) or regression, e.g. feature estimation (Sect. 3.2) tasks.

Unsupervised Learning An ML method utilises non-annotated, i.e. unlabelled data to discover patterns, clusters or relationships within the data based on structural properties of the data. Common applications include clustering similar data points or reducing the dimensionality of available data (Sect. 3.4), e.g. for data exploration.

Semi-Supervised Learning This training approach combines elements of both supervised and unsupervised learning. This approach leverages first a larger amount of unlabelled data to model structural characteristics of the data in an unsupervised way and then utilises a small amount of labelled data to model the task-specific target function.

Transfer Learning This training approach involves pre-training a model on a large dataset and fine-tuning it for a specific task on a smaller dataset. The goal is to leverage patterns learned from one task or domain to improve performance on a related task with limited data.

Reinforcement Learning An agent learns to make a sequence of decisions or actions within an environment through a process of trial and error, aiming to maximise a cumulative reward received as feedback from that environment.

2.2.4 Model Evaluation

Model evaluation is a crucial step in ML, involving the assessment of model performance through performance metrics and enhancing model transparency through explainability methods. These methods enable ML models to be interpretable to users, which is a crucial requirement for many biomechanical applications and regulatory compliance, in addition to their overall performance.

Evaluating model performance primarily involves assessing the generalisation ability of a trained ML model on unseen data. There are several evaluation strategies such as hold-out, k-fold cross-validation and leave-one-out cross-validation. The selection of the approach is commonly influenced by the quantity of available data and the specific domain of application. Regardless of the approach, it is important to maintain the integrity of the test data, e.g. for the task of classifying gait pattern to ensure that each subject's data is exclusively assigned to either the training data or test data. Performance metrics are quantitative measures used to assess how well an ML model performs on a specific task. These metrics vary depending on the type of problem, e.g. classification, regression, clustering and the specific objectives of the task.

Explainable Artificial Intelligence (XAI) involves methods that aim to enhance the transparency and interpretability of ML models. Considering the inherent "black-box" nature of complex ML and Deep Learning models, explainability methods allow users, stakeholders and regulators to gain insight into the model behaviour and decisions. For a comprehensive overview of XAI in biomechanics, refer to Sect. 4.1.

3 Key Applications of Machine Learning in Biomechanics

In the following, we showcase promising ML applications for key tasks in the biomechanical workflow. These applications encompass **pose estimation** (Sect. 3.1), **feature estimation** (Sect. 3.2), **event detection** (Sect. 3.3), **data exploration and clustering** (Sect. 3.4) as well as **automated classification** (Sect. 3.5). Figure 3 provides an overview of the localisation of the key ML applications within the biomechanical workflow, while Fig. 4 illustrates examples of input and output pairs. For each application we present a brief description of the task, followed by prominent research works, and conclude with application-specific limitations and future perspectives. Throughout this chapter, we will provide separate descriptions for applications in biomechanics related to walking, running and sports movements.

3.1 *Pose Estimation*

3.1.1 Problem Description

Over the last three decades, marker-based infrared camera systems have become paramount in accurately quantifying and analysing human movement in sports and

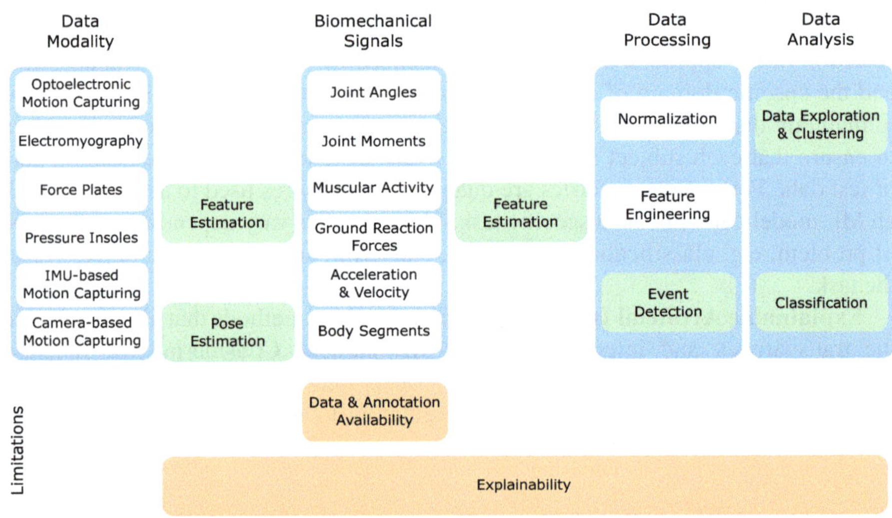

Fig. 3 Overview of the key ML applications (green boxes), i.e. pose estimation, feature estimation, event detection, data exploration and clustering and automated classification, and their integration into the phases of the biomechanical workflow. The key limitations (orange boxes) associated with the use of ML in biomechanics include data and annotation availability and explainability

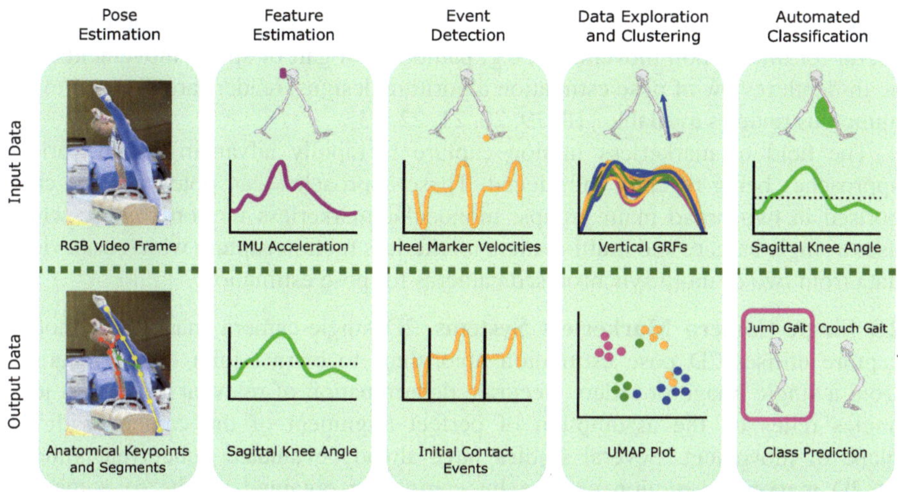

Fig. 4 Examples of input and output pairs for the key ML applications, i.e. pose estimation, feature estimation, event detection, data exploration and clustering as well as automated classification

clinical settings for both research and decision-making purposes. They allow for comprehensive investigation of sports movements, gait and functional activities, providing valuable insights to coaches, clinicians and researchers that significantly advanced our understanding of the biomechanics of human movement. However, despite their value in human movement science and biomechanical research, their accessibility and widespread use are severely limited by high costs, complex hardware, time-consuming laboratory set-ups and the need for highly trained operators. Recent advancements in Deep Learning and computer vision technology offer a potential solution to these problems, i.e. markerless motion capture. This technique allows the two- and three-dimensional tracking of human locomotion from simple videos. Markerless motion capture systems typically utilise Deep Learning models trained on large-scale datasets containing hundred thousands of manually annotated images of individuals engaged in daily activities and sports movements. These so-called pose estimation algorithms can detect anatomical landmarks or joint centre locations automatically in videos that are used to derive the pose of the individual in each video frame or image.

3.1.2 Approaches

Various pose estimation algorithms have been published over the past decade such as OpenPose [13], DeepLabCut [14], DeepPose [15] and DeeperCut [16]. Popular pose estimation algorithms for biomechanical applications especially include OpenPose [17] and DeepLabCut [14]. OpenPose is known for its ability to track multiple individuals in images and its easy handling, while DeepLabCut allows users to retrain pre-trained algorithms for specific tasks, which can be especially useful for uncommon movements, e.g. pathological gait or sports movements. For an in-depth review of pose estimation algorithm designs, readers are directed to the numerous reviews available [18, 19, 20, 21, 22, 23].

The field of markerless motion capture is rapidly advancing, with various approaches being recently introduced. These approaches can potentially be categorised in two broad main groups: monocular markerless motion capture which uses a single camera and multi-camera markerless motion capture which uses video data from two or more synchronised cameras for pose estimation.

2D Single-Camera Markerless Systems 2D single-camera markerless motion capture utilises 2D pose estimation algorithms to extract joint centre locations from a single image or video. Accurate determination of relevant 2D planar joint angles relies on the assumption of perfect alignment of the camera with the plane of movement. Several studies have already evaluated concurrent validity of 2D markerless motion capture by comparing obtained results to a marker-based system [24, 25, 26, 27]. In these studies, activities analysed include walking, underwater running, jumping and cycling. Joint centre locations differed less than 20 mm to the marker-based system and joint angles as well as spatiotemporal parameters were approximately similar. However, depending on the algorithm used, other studies reported greater differences in joint angles and joint centre locations [28, 29]. Most of the studies also reported that in some cases manual corrections were necessary for incorrectly identified joint centre locations due to, e.g. body part occlusion and labelling of the wrong body side. In summary, 2D single-camera methods can provide reasonably accurate spatiotemporal parameters and planar 2D joint angles, but they are limited to the analysis of one body side only and an optimal camera positioning relative to the movement plane is assumed. There are also efforts in using single-camera markerless motion capture to obtain 3D pose estimation. However, errors for these methods are still high, with joint centre differences reported in the literature in the range of 30–60 mm, and, thus, to date do not offer enough accuracy to aid biomechanical applications [30]. For a thorough review of technological aspects for 3D single-camera markerless systems, the reader is referred to recent reviews [19, 31].

3D Multi-Camera Markerless Systems Multi-camera markerless motion capture utilises a minimum of two cameras. The multi-camera approach intends to reduce the problem of partly occluded body parts and typically utilises triangulation methods, such as the direct linear transformation [32], to estimate the 3D pose from the 2D anatomical key features identified in each single video or image. Compared to their single-camera counterparts, multi-camera systems generally seek to offer an alternative method to 3D marker-based state-of-the-art motion capture systems [29]. A few studies have assessed the performance of multi-camera 3D markerless systems utilising, e.g. the OpenPose pose estimation algorithm. These studies, conducted by Nakano et al. [33], Slembrouck et al. [34] and Zago et al. [35], have reported variations in joint centre location ranging from 10 to 50 mm. It was observed that slower movements exhibited more accurate results, with average differences in joint centre location compared to marker-based methods ranging from 10 to 30 mm. Conversely, faster movements such as jumping and throwing yielded differences of 20 to 40 mm [33], which were further exacerbated when captured at lower video frame rates [34, 35]. Recently, Uhlrich et al. [36] have proposed the open-source tool, OpenCap. It combines a multi-camera OpenPose approach with constrained musculoskeletal modelling in a web-based service to quantify human movement dynamics from a minimum of two calibrated smartphone videos. Uhlrich et al. [36] and Horsak et al. [37] reported a difference in joint kinematics between OpenCap and a marker-based motion capturing for various activities including pathological walking patterns of approximately 5–6 degrees on average.

While the aforementioned studies have used open-source algorithms such as OpenPose, in recent years also commercial systems have emerged, such as KinaTrax (KinaTrax, Inc., Boca Raton, FL), SIMI (Simi Reality Motion Systems, Unterschleißheim, Germany), Theia3D (Theia Markerless, Inc., Kingston, Ontario, Canada) and DARI Motion (DARI Motion, Overland Park, Kansas). However, given their very recent availability, there is currently a lack of peer-reviewed research for most of the systems. As of 2024, Theia3D and KinaTrax seem as the most popular system in biomechanical research with at least a few studies evaluating the reliability and concurrent validity for tasks like heel raising, stepping down, walking, hopping, squatting and running-and-cutting manoeuvres [38, 39, 40, 41, 42, 43, 44]. While Kanko et al. [40] and Ripic et al. [44] reported accurate estimations of spatiotemporal parameters for walking using the Theia markerless system, Kanko et al. [38, 40] found differences of 22 to 36 mm for joint centre locations and 3 to 11 degrees for flexion/extension and abduction/adduction in walking kinematics, although rotations about the long axis of a body segment were higher ranging from 7 to 13 degrees. Song et al. [43] investigated eight different functional movement tasks and found their results to be generally in agreement with the results of Kanko et al. [40], but observed increased differences with faster movements, e.g. while running. Wren et al. [41] recently reported errors of approximately 5 to 6 degrees for the lower-limb joint kinematics in patients with a range of diagnose including cerebral palsy and club foot deformities. Further, Kanko et al. [39] also found the reliability of the Theia system being comparable or having even partly slightly superior reliability in walking kinematics compared to previously reported marker-based methods [45, 46]. Lastly, a recent study by Horsak et al. [47] demonstrated that markerless pose estimation pipelines can introduce additionally variability in kinematic data and, therefore, recommended using averaged waveforms rather than single ones to mitigate this problem. Further, they highlighted that caution is advised when using variability-based metrics in gait and human movement analysis based on markerless data as increased inter-trial variability can potentially lead to misleading results.

In summary, multi-camera markerless motion capturing appears to be a highly promising alternative to record 3D human movement without the need for attaching markers. However, while markerless multi-camera systems produce similar kinematic patterns, errors are still above desirable clinical thresholds of two to five degrees in several cases.

Application Examples for Assessing Health and Performance The range of applications for pose estimation is constantly growing. From a human movement science perspective, examples cover the full-range life span, from human development, performance optimisation and injury prevention to motor assessment of persons with motor impairments [48]. For example, Ossmy and Adolph [49] used pose estimation to quantitatively track the development of infant locomotion to monitor body movements. Their results highlighted potential for early detection and intervention for delays in locomotor development. Pose estimation also has been successfully used for the early detection and diagnosis of congenital movement-based disorders such as cerebral palsy. Investigations demonstrate promising results in the early detection of cerebral palsy based on automatic movement assessment from infant video recordings, with performance comparable to standardised cerebral palsy risk measures [50, 51].

Pose estimation also has a wide range of applications that contribute to optimising human health and performance. Examples range from the identification of abnormal gait patterns during walking and running [52, 53, 54], fall detection [55, 56], assessment of movement quality in manual labour work settings [57, 58] and assessment of sports-related injury risk, such as anterior cruciate ligament rupture [59].

While the application of pose estimation in sports seems obvious for performance assessment, injury risk screening and biofeedback training, to date peer-reviewed studies with application examples are still scarce. Recent applications of pose estimation in sports include the estimation of boxing poses [60, 61], the analysis of throwing kinematic during baseball pitching [62], the analysis of ski jumping techniques [63] and the application for various sports-related movements such as squatting, running or jumping [36, 43, 64].

3.1.3 Limitations and Future Perspectives

Markerless motion capture holds exciting potential for advancing real-world biomechanical assessment for both, clinical settings and sports. It has not only the potential to make measurement of human movement substantially more accessible to the broad public, it allows to streamline movement analysis by reducing the time required for data acquisition and processing, without sacrificing reliability [39, 263]. Additionally, markerless techniques offer greater flexibility in analysing the data and can even capture movement data while individuals are wearing clothes [65, 263], an aspect which is of particular interest for various applications in sports.

Accuracy of Kinematic Measurements Research shows that in clinical settings, markerless methods generally yield similar results to marker-based motion capture in terms of spatiotemporal measures, but they still need improvements in accurately determining joint kinematics before they can support decision making.

Noisy Data in Sports Applications While their application in sports seems obvious, the high movement speeds, cluttered background, various lightning conditions and variations in shape due to worn equipment present significant challenges for pose estimation in sports movements.

However, with the currently seen rapid developments in pose estimation algorithms, and the growing amount of annotated data being shared openly, improvements in the accuracy of markerless 3D motion analysis for both, clinical applications and sports, are likely to happen. In some scenarios one could argue that having estimated quantitative data is better than having no data, e.g. in settings where exact kinematic and kinetic measurements are almost impossible to acquire, like in pole vaulting or ski jumping.

3.2 *Feature Estimation*

3.2.1 Problem Description

Advancements in biomechanical modelling have greatly improved our understanding of movement mechanics and dynamics over the past decades. However, achieving comprehensive biomechanical modelling with accurate estimation of joint angles and moments typically requires extensive data acquisitions in standardised laboratory environments, which are both time- and resource-consuming. One goal in the biomechanics domain is the estimation of complex biomechanical signals with minimal experimental constraints and effort, even in natural environments outside of the laboratory. The use of ML approaches for feature estimation of complex biomechanical signals from simplified data acquisition and data processing pipelines has been increasingly explored in recent years.

3.2.2 Approaches

Feature estimation has predominantly been explored in gait biomechanics, examining various possibilities for predicting and utilising biomechanical signals/parameters [7]. Among these, (i) predicting kinetic data, e.g. ground reaction forces, from IMU signals or video images [66, 67, 264]; (ii) predicting kinetic signals, e.g. joint contact forces from ground reaction forces and joint angles [68, 265]; and (iii) forecasting spatiotemporal parameters, e.g. stride length, from IMU or smartwatch signals [69, 70, 71, 72], have garnered significant attention.

Walking Kinetic signals are usually derived from force platform data, which offer high temporal and spatial resolution but require proper physical contact with the platform. However, there are situations where proper contact may not be possible, e.g. when individuals have very small stride lengths and widths or are unable to walk unassisted. In addition, force platforms are often permanently installed in laboratories, which restricts measurements to laboratory environments. The latter limitations have led to efforts to derive kinetic signals from alternative measurement approaches such as IMU or video data. Most studies focused on the estimation of ground reaction forces and joint moments from IMU or video data. These studies reported high correlations to the signals calculated with the use of force platforms and inverse dynamics [73, 74, 75]. Different ML methods have also been investigated and superior performances were reported for Neural Networks and Random Forests [76, 77].

The reported errors in these estimations may be tolerated due to the ability to conduct measurements in natural environments. Moreover, this flexibility can be particularly beneficial in rehabilitation settings, where patients may have pathologies or walking aids that impede clean measurements of ground reaction forces. In such ambulatory tasks, joint moments can be predicted with acceptable error rates using kinematic data collected from IMUs [78]. There are more kinematic variables that can be predicted based on three-dimensional optical measurements, such as centre of gravity [79] and foot clearance, i.e. the distance of the toes to the ground [80]. These studies are mostly conducted with elderly individuals that have a higher risk of falling or tripping because of an altered walking pattern. The recognition of these walking alterations could bring a huge advance to practical walking analysis because at the moment only one step or gait cycle (on the force platform) can be used to calculate the relevant joint moments.

Biomechanical measurements are not only limited to kinetics and kinematics but can also include muscle activity and muscle forces. These forces can also be predicted by kinematic data [81] or used to predict kinetic data such as joint power [82]. Additional optical [83] or sensor-driven measurements [84] can help to improve those predictions. Furthermore, combining ML with pressure insole data offers a promising solution for predicting ground reaction forces [85, 86].

The estimation of spatiotemporal parameters from IMU signals or smart watches has been extensively studied in both healthy and pathological populations, considering various sensor locations [69, 70, 71, 72]. Studies have found that the estimation of temporal walking parameters tends to be more accurate than spatial parameters [72]. With long-term recordings in natural environments and feature estimation techniques, IMUs offer new possibilities for continuously monitoring spatiotemporal walking parameters. These could serve as valuable tool to complement established clinical walking tests or aid in the early detection of diseases. However, it is important to note that further research is required to determine the practical utility of spatiotemporal gait parameters from feature estimation techniques in these applications. This is particularly important given the varying and reduced accuracy compared to the gold standard.

Running Feature estimation has been used in running to estimate kinetic variables that are considered risk factors for running-related injuries. This encompasses investigations that predicted ground reaction force [87, 88], joint force [87] and bone force [89] signals and features from IMU and video data. It is noteworthy that the vertical ground reaction force loading rate could be predicted using only a single IMU sensor [90]. While ongoing advancements hold the potential to improve accuracy and provide valuable tools for injury prevention, it is important to acknowledge that current accuracy levels are often not yet suitable for practical applications.

Sports Movements Feature estimation has been explored in sports movements, such as cutting, side-stepping, squatting or jumping. The predicted features included joint moments and forces [66, 91, 92, 93], ground reaction forces [66], lower-body kinematics [94, 95] and electromyographic variables [96]. Only a few studies have investigated feature estimation techniques in sports-specific movements. For example, arm kinetics have been used to predict fastball velocity [97] and arm kinetics [98] in baseball pitching. In handball, the centre of mass kinematics during a jump shot can be accurately predicted based on force platform signals [99]. Additionally, IMU signals have been employed to predict elbow joint angle during swimming with minor errors compared to marker-based camera systems [100].

3.2.3 Limitations and Future Perspectives

Feature estimation represents a key application of ML approaches in biomechanics, with the aim of enhancing gait and sports biomechanical testing. However, there are challenges and opportunities for further development.

Limited Data Particularly evident in sports movements, feature estimation approaches often encounter limitations due to their dependence on relatively small datasets. The potential of data augmentation techniques to mitigate this issue remains largely unexplored, despite evidence of their effectiveness in enhancing other biomechanical tasks [101]. Additionally, encouraging collaboration among various gait laboratories shows potential for increasing the quantity of available data, which can help improve feature estimation algorithms.

Accuracy in Sports Despite significant advancements, achieving high accuracy in feature estimation, particularly within the context of sports movements, remains a paramount challenge. This challenge can be attributed to two primary factors: (i) Sports movements exhibit a notably elevated level of complexity compared to activities like walking or running. The complexity of movements, varying intensities and dynamic nature of sports activities impose considerable demands on estimation algorithms. Consequently, predicting complex biomechanical parameters, such as muscle forces, becomes inherently challenging. (ii) Another challenge in sports biomechanics is the lack of reference data for training ML models. Unlike the more standardised testing protocols and publicly available datasets in walking and running biomechanics, the diverse and specific nature of sports movements presents obstacles in acquiring data. Consequently, there is typically only limited data accessible for training ML models, particularly for complex parameters like muscle forces.

Challenges in Quantifying Biomechanical Parameters Unlike measurable parameters like joint angles and ground reaction forces, muscle forces pose a unique challenge in biomechanical analysis due to the absence of direct measurement techniques. This inherent complexity complicates the feature estimation process in walking, running and sports biomechanics. Muscle forces are rather approximations than precise measurements, which introduces uncertainties into the training data for feature estimation algorithms. Technological advancements such as instrumented implants offer potential solutions for validating these estimations [102].

Black-Box Models In feature estimation tasks within biomechanical research, there is a prevalent emphasis on predictive performance at the expense of model interpretability. Complex ML models often utilised for feature estimation tasks often lack transparency regarding the input features and their relationships driving these predictions. This does not only hinder biomechanists' ability to extract meaningful insights but also undermines trust in the systems.

These limitations present opportunities for advancement. In particular, the integration of physics-informed ML has shown significant potential in enhancing feature estimations, as exemplified by the work of Taneja et al. [103, 104]. Physics-informed ML entails the incorporation of pre-existing knowledge concerning the fundamental physical principles into the training process of ML models. Through this integration, the model can more effectively capture the underlying dynamics of the system under investigation, thereby yielding more precise and interpretable predictions.

3.3 Event Detection

3.3.1 Problem Description

The goal of event detection in time series data is to annotate certain events to extract useful and vital information or remove unwanted and unnecessary data for further analysis. When analysing movement, recurring cyclical movements like running, swimming, cycling and walking and non-cyclical movements such as football, handball, golf and table tennis can generate a vast amount of time series data. Postprocessing the captured data requires a segmentation of the movement based on certain events. A correct segmentation and detection of events is a crucial task due to the calculation of movement and sports-specific metrics based on the segmentation. With these metrics, a person-specific score or movement pattern can be analysed, which can give information about running performance and injuries [105]; sports-specific game performance of players, e.g. passes, kicks and sprints [106, 107]; and indicating morbidity, mortality or risk of falls in walking [108, 109, 110, 111, 112]. Furthermore, supervised ML approaches can be trained on the segmented data to discover hidden relationships in the underlying multidimensional time series data for further investigations and meaningful insights [106].

3.3.2 Approaches

Both walking and running can be segmented into gait cycles. A single gait cycle can be further divided into a stance and a swing phase. The stance phase starts with the foot's initial contact (IC) on the ground and ends with the same foot leaving the ground, known as the foot off (FO). This FO initiates the swing phase, which ends with the IC of the same foot, simultaneously ending also the gait cycle [113]. This segmentation of the gait cycle allows the calculation and further analysis of spatiotemporal, kinematic and kinetic parameters for each person. The gold standard for detecting IC and FO events in time series data is by defining a certain threshold on the vertical ground reaction force data. However, in gait analysis practice, force platform contacts are often scarce or entirely unavailable due to pathological characteristics or when conducting measurements in the field [114]. For studies focused on running, conventional treadmills are commonly preferred over force-instrumented treadmills due to their cost-effectiveness [105].

In cases where no force platforms are available, a manual identification of gait cycle events is often performed. However, depending on various factors, the identification of gait events can be subjective, tedious and time-consuming work, which is prone to human error [115].

Heuristic-based event detection algorithms have been developed to automate the segmentation process. They are able to automatically detect gait events based on local minima and maxima in kinematic data of markers of the lower extremities. However, heuristic algorithms mostly depend on predefined thresholds, which can vary based on various parameters including walking/running speed, foot strike pattern and marker configuration [114, 105].

Probabilistic-based approaches have been introduced to overcome issues related to heuristic-based algorithms. With recent advancements, ML and Deep Learning methods are able to outperform heuristic algorithms in detecting IC and FO events more robustly and accurately [105, 114, 116, 117]. In the following, different ML-based approaches for walking, running and sports movements data will be outlined.

Walking In walking analysis, an error of below four frames (or 27 ms with 150 Hz capturing frequency) of difference to the ground truth has been used as a common "accuracy threshold" for proposed algorithms [114, 115, 116]. When this threshold is exceeded, data needs to be interpreted with care, as it can potentially result in differences above the minimal detectable change of 5° for certain kinematic parameters [118, 119]. Due to the limiting factors of heuristic approaches, they often tend to exceed this threshold. ML-based event detection algorithms have shown promising results achieving higher performance and more robust event detection compared to heuristic approaches [114, 116, 117].

One of the first ML-based event detection algorithms has been developed by Kidzinski et al. [117]. The authors utilised a stacked Long Short-Term Memory architecture and trained the model on mostly children with neurological disorders. Kidzinski et al. [117] reported a mean accuracy of 10 ms and 13 ms for IC and FO events, respectively, using a capturing frequency of 120 Hz.

An improved event detection algorithm based on a stacked bi-directional Long Short-Term Memory network has been introduced by Lempereur et al. [116]. With a similar patient population ($n = 226$) as Kidzinski et al. [117], Lempereur et al. [116] reported a mean accuracy of 6 ms and 11 ms for IC and FO events, respectively, with a capturing frequency of 120 Hz. These results outperformed the most popular heuristic-based event detection algorithms in the literature.

Another bi-direction Long Short-Term Memory architecture approach was developed by Kim et al. [120] utilising data from children with cerebral palsy ($n = 363$). Additionally, the authors compared results for different marker placements and striking patterns (heel, midfoot and toe). This resulted in a 95% trimmed-average time error of 8 ms and 11 ms for IC and FO, respectively, using a capturing frequency of 150 Hz. Furthermore, their results also showed that the marker set-up can influence the performance up to 10% for different striking patterns and needs to be considered for future approaches.

Filtjens et al. [121] utilised a temporal Convolutional Neural Network to detect events during turning and freezing of walking in patients with Parkinson's disease (data were captured with 100 Hz). The authors reported slightly improved results compared to the Long Short-Term Memory model from Kidzinski et al. [117]. However, as this study utilised a dataset comprising only 15 Parkinson's disease patients, the results need to be interpreted with care. The most recently published automated event detection algorithm by Dumphart et al. [114] trained two separate stacked bi-directional Long Short-Term Memory models for the IC and FO events. The authors utilised the most extensive dataset ($n = 1,211$) containing different pathologies, i.e. malrotation deformities of the lower limbs, club foot, infantile cerebral palsy and healthy participants ($n = 61$). Their approach resulted in slightly improved results compared to the model from Lempereur et al. [116] detecting IC and FO events within 5 ms and 11 ms, respectively, using a capturing frequency of 150 Hz. Furthermore, Dumphart et al. [114] investigated the generalisability across laboratories, suggesting that models trained on data from a different laboratory need to be applied with care due to set-up variations or differences in capturing frequencies.

For field measurements or monitoring daily activities, most of the current literature focuses on the use of single or multiple IMUs placed on the body to detect events. ML-based event detection algorithms have demonstrated superior performance compared to heuristic methods when applied to IMU data. Single-mounted IMU approaches like waist, shank or foot-mounted IMUs have become the preferred option to enhance user compliance, as they require fewer sensors to be attached. Algorithms based on waist-worn [122, 123], shank-worn [124] and foot-worn [125] IMUs show a mean average error of 6 ms to 9 ms for IC and 5 ms to 8 ms for FO detection.

Systems with multiple IMUs are often used to replicate a laboratory analysis when attempting to predict kinematic and kinetic data during walking and other activities [126]. Lin et al. [127] developed a Long Short-Term Memory-based approach to for gait event detection based on five IMUs, which achieved a mean absolute error of 2 ms for IC and 18 ms for FO detection.

Running In running, the error of determining IC and FO events is highly dependent on running speed due to the generally faster movement. A sensitivity study performed by Rivadulla et al. [105] identified changes in the sagittal kinematics of the hip, knee and ankle at IC and FO, categorising them as either acceptable, reasonable or unacceptable. A difference lower than 2° is deemed as acceptable, 2 to 5° as reasonable but requires consideration and above 5° as unacceptable [105]. Heuristic event detection algorithms for running are based on certain thresholds which can be affected by running speed and foot-strike patterns, limiting the accuracy and robustness of such approaches. Osis et al. [128] proposed an ML approach based on Principal Component Analysis to detect IC and FO events during treadmill running using kinematic data acquired from a motion capture system. The results showed that 95% of the IC and FO events were detected within a 20 ms window with a capturing frequency of 200 Hz for both forefoot and heel-strike patterns ($n = 186$). However, 20 ms can lead to differences of up to 7° in the knee flexion angle at relative low running speeds [129], leading to unacceptable errors.

Recently, Rivadulla et al. [105] presented the most accurate and robust event detection approach for running based on a stacked bi-directional Long Short-Term Memory model. The authors utilised five different datasets [130, 131, 132] to overcome issues related to differences in marker configuration, running speed, foot-strike patterns and the limitations arising from relying on data from a single laboratory. These datasets comprised lower limb kinematic data ($n = 80$ runners) that were captured in three different laboratories at different capturing frequencies. This approach outperformed previously published algorithms by detecting IC events with a root-mean-square error of 5 ms and FO events with 6 ms.

IMU-based event detection in running has been researched in and out of the laboratory but mostly with heuristic approaches [133]. Robberechts et al. [134] developed an ML-based event detection algorithm with two accelerometers placed on the tibia. Participants were only rearfoot runners ($n = 93$) and measurements were captured in an indoor running track laboratory (30 m). The algorithm detected IC events with a mean average error of 2.2 ± 3.3 ms and FO events within 4.0 ± 4.52 ms [134]. To address the need for measurements beyond laboratory settings, Donahue et al. [133] developed an IMU-based event detection algorithm, which also estimates kinetic parameters. Data ($n = 16$) was captured using IMUs mounted bilaterally on the dorsal aspect of the foot and on the sacrum, force data from an instrumented insole and GPS data from a sports watch. With a bi-directional Long Short-Term Memory architecture, IC events were detected with a root-mean-square error between 11 and 51 ms and FO events between 20 and 53 ms.

Sports Movements In comparison to cyclic movements like walking and running, sports movements are mostly non-cyclic movements where ground truth events are typically not as distinguishable as in cyclic movements. With current advancements in IMUs, ML and Deep Learning, automated segmentation of sports movements has become efficient and accurate for sports performance analysis and has become a vital tool for athletes and trainers [107, 135, 136]. A recent scoping review [107] analysed 93 studies regarding segmentation of IMU data in sports movements using

ML. Tennis, football and basketball, as well as workout training/exercises, were the most frequently analysed sports movements in the related literature. However, no general conclusion can be formed about weather Deep Learning or traditional ML methods are superior for sports movements segmentation due to a high variety of different tasks analysed.

Pei et al. [137] utilized an IMU embedded in the tennis racket handle to segment tennis strokes into forehand, backhand and serve, achieving an accuracy of 96% for segmentation and 98% for shot detection. In addition to tennis stroke segmentation, Zhao et al. [138] estimated the tennis ball speed for each stroke and Whiteside et al. [139] the hitting load for each stroke for further player performance evaluation. Alobaid et al. [140] segmented different football activities (shooting, passing, heading, running and dribbling) using accelerometer data from a single mobile phone with an accuracy of 90%. Stoeve et al. [141] investigated the shot and pass detection in football with two IMU sensors in each shoe with an F_1 score of 0.93. Acikmese et al. [142] developed a basketball training exercise type segmentation, i.e. forward-backward dribbling, left-right dribbling, regular dribbling, two hands dribbling, shooting and layup, algorithm based on a single wrist-mounted IMU with an accuracy of 99.5%. Jeong et al. [143] proposed an algorithm for exercise segmentation, employing a single wrist-mounted IMU to differentiate between pull-ups, barbell rows, bench presses, dips, squats, deadlifts and military presses, achieving an accuracy of up to 96%.

3.3.3 Limitations and Future Perspectives

Model Generalisability A limitation for ML-based walking and running event detection algorithms is that the proposed algorithms have not been tested on data from different laboratories. Rivadulla et al. [105] used data from different laboratories to train their ML algorithm; however, there was no data present from an unseen laboratory. Dumphart et al. [114] showed the need for caution when using an algorithm trained on data from another laboratory, as it can lead to errors of up to 10 ms due to differences in capturing frequency and recording set-up. These results motivate future research that should focus on developing standardised gait event detection methods that can demonstrate high accuracy and robustness across various laboratories and movement tasks.

Sensitivity for Errors Moreover, conducting sensitivity analyses is crucial to evaluate the degree to which event detection errors impact the computation of kinematic or spatiotemporal parameters. Rivadulla et al. [105] suggested acceptable, reasonable and unacceptable degree thresholds for sagittal knee kinematics at IC and FO that need to be determined for other movements as well.

Transfer to Real-world Settings IMU-based walking and running event detection algorithms are commonly only evaluated in laboratory settings, which does not give a complete picture of how well these methods work in the real world where the movement is performed [144]. Future research should focus on evaluating these approaches also for field measurements beyond the laboratory setting.

Limited Data In sports movements, event detection algorithms are often trained on a relatively low amount of subjects, i.e. mean n = 20 [107, 136]. Collaborations between laboratories could increase the number of subjects as well as further quality testing of the algorithms. Furthermore, data augmentation techniques were rarely used or analysed to mitigate part of this problem. Future research should focus on integrating data augmentation techniques due to the benefits reported in other domains.

Skill Differences The relatively low amount of data could affect the performance of the models in a differently skilled population, e.g. amateur, recreational or professional athletes due to differences in proficiency of movements [107, 136]. This needs to be further investigated and could be an opportunity to explore the use of transfer learning in future work.

3.4 Data Exploration and Clustering

3.4.1 Problem Description

Biomechanical data in both gait and sports biomechanics is characterised by high inter-individual variability, coupled with a frequent lack of data annotations. Hence, employing unsupervised ML methods is crucial for extracting meaningful insights from such data. Cluster analysis, a fundamental component of data exploration, aims to group individuals with similar biomechanical characteristics, thereby revealing the underlying structure of the data. With the inherent complexity of the data, ML-driven approaches can reveal subtle subgroups that may elude human perception, thus uncovering latent patterns. Clustering aids in the discovery of latent groups within non-annotated datasets but also facilitates the process of data annotation and analysis while assisting in the detection of outliers and erroneous data samples.

In biomechanical applications, clustering based on ML techniques is closely associated with the use of visualisations for effective data exploration. This encompasses tasks such as visually identifying clusters within the data [145, 146, 147], e.g. Fig. 5, evaluating class separations in classification tasks [148] and examining different feature extraction methods [148, 149]. Consequently, methods for dimension reduction and visualisation play a pivotal role in presenting high-dimensional biomechanical data in a lower-dimensional space.

3.4.2 Approaches

Walking Less complex methods such as Principal Component Analysis or Multidimensional Scaling have been commonly applied in biomedical data exploration

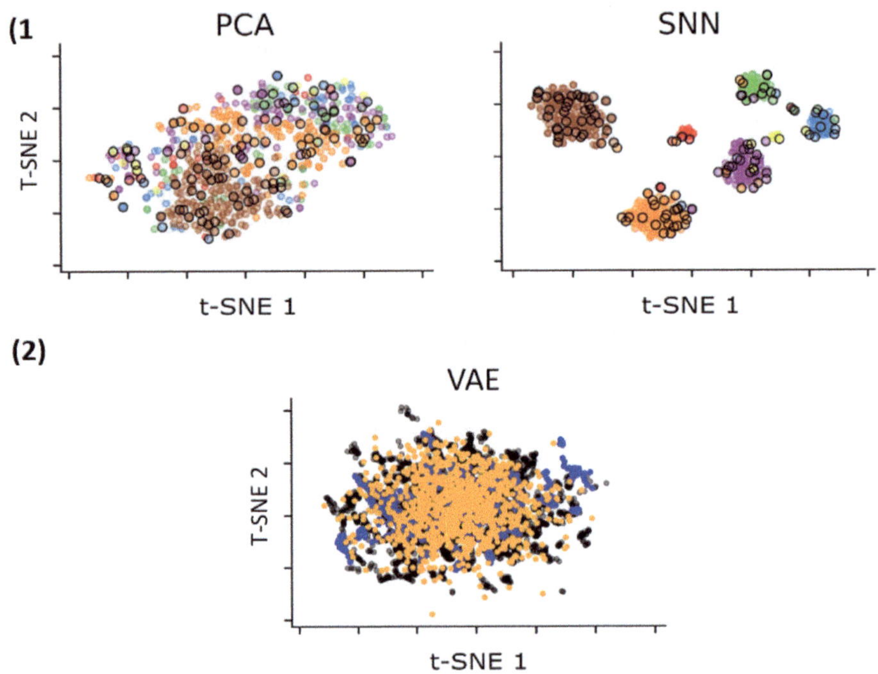

Fig. 5 Overview of different data exploration approaches for biomechanical data. (1) t-Distributed Stochastic Neighbour Embedding visualisations of principal components (left) and embeddings obtained via metric learning using a Siamese Neural Network (right) for various walking patterns associated with cerebral palsy [147]. (2) t-Distributed Stochastic Neighbour Embedding latent space visualisation of a Variational Autoencoder trained on spinal posture data (black circles correspond to training data, blue to testing data and orange to sampled latent vectors from the latent space of the trained Variational Autoencoder for data generation [254])

to uncover latent patterns [150, 151]. Approaches using self-organising maps were also leveraged in the context of walking data and date back to the 1990s [152]. More advanced methods focusing primarily on visualisation as well as learning of latent embeddings were mainly applied to walking data. For visualisation, t-Distributed Stochastic Neighbour Embedding [153, 154, 266] and Uniform Manifold Approximation and Projection [154] were most often used in literature. For learning a latent embedding, which can be used for visualisation purposes, but also for preparatory steps in the context of feature extraction, studies used, e.g. Autoencoder [155] or Siamese Neural Networks [146]. An improvement of separability among subgroups has been observed when t-Distributed Stochastic Neighbour Embedding is applied to learned embeddings, as opposed to its direct application to the raw data [147].

Clustering has been utilised across various applications aimed at enhancing research and practice in walking biomechanics. Most research is conducted on patients with different kinds of pathologies, with orthopaedic problems being the most prevalent. Promising results are observed, particularly when clustering patients with arthritic diseases, for example, distinguishing between symptomatic and asymptomatic patients based on biomechanical parameters [145, 156, 157, 158]. Different clusters based on sex and gait kinematics can be observed in people with total knee arthroplasty and candidates for this kind of surgery [156]. Halim et al. [157] utilised clustering algorithms to identify severity grades in osteoarthritis based on pain scale, kinematic and electromyographic data from patients with osteoarthritis. Leoprace et al. [145] identified four clusters based on kinematic variables for walking data from patients with osteoarthritis. Furthermore, chronic instabilities which are hard to diagnose can be differentiated efficiently with clustering methods based on kinematics [159]. In prosthetic research and practice it is not easy to evaluate how well the prosthesis performs. Clustering has been employed to categorise patients with amputations into homogeneous clusters based on spatiotemporal parameters, revealing distinct patterns within this population [160], which might help to improve the individualisation of prosthetics. Besides orthopaedic diseases, also neurological pathologies, e.g. strokes, show altered kinematics that have been clustered in the literature [161].

Running In running, it is important to detect movements that can lead to injury and optimise the economy of the movement to save energy for longer runs. Hoerzer et al. [162] used self-organising maps to define functional groups with different movement patterns in running. Pardo Albiach et al. [163] evaluated various spatiotemporal parameters and the power output of runners across different running speeds using hierarchical cluster analysis. Further biomechanical variables such as ground reaction forces, bone kinematics and electromyographic data have also been clustered to identify different types of runners [164, 165, 166, 266]. A study by Aljohani et al. [164] made use of various biomechanical variables to train a clustering algorithm that identified four different clusters in runners independent from sex. Senevirathna et al. [165] identified two clusters in runners based on kinetic and spatiotemporal parameters.

Sports Movements Principal Component Analysis and Multidimensional Scaling are also frequently employed in the field of sports biomechanics to uncover hidden patterns in the data [167, 168]. Self-organising maps were also used for kinematic data, e.g. for analysing volleyball spike kinematics [169]. However, in comparison to the analysis of walking and running data, the application of more sophisticated data exploration methods in the field of sports biomechanics is still relatively scarce.

In sports-related movements, clustering is employed to differentiate movement executions from a range of general exercises [170]. Another important application is the detection of pathologies [171] or movements that are linked to injury [154]. Through the identification of particular movement patterns or functional groups within the data, these algorithms have the potential to enhance training programs, particularly for injury prevention and rehabilitation.

Clustering is used in the analysis of sports biomechanics in various kinds of sports. Often it is used to assess the outcome of a special event, e.g. volleyball spike based on upper limb angular velocities [172] or serves in tennis [173]. There are also some applications in individual sports such as the clustering of the canoeing motion from different levels of proficiency based on kinematic data [174] or identifying the phenotype of Olympic wrestlers based on anthropometric data [175]. Giles et al. [176] analysed a large set of spatiotemporal data from tennis matches to identify patterns in the change of players' movement directions.

Clustering has also be used to evaluate coordination patterns in sports movements, e.g. during change of direction using kinematic data [177]. Whiteside et al. [178] used clustering to evaluate the best position for executing a tennis serve with the goal of achieving an ace. Furthermore, clustering has been utilised to analyse stability of hold during shooting, e.g. in biathlon shooting where differences in stability of hold become apparent among athletes with varying levels of performance [179].

3.4.3 Limitations and Future Perspectives

Although cluster analysis hold substantial potentials in gait and sports biomechanics, several limitations and challenges exist.

Noisy Character of Biomechanical Data Biomechanical data is often prone to noise or incomplete data, which can significantly impact the performance and reliability of clustering algorithms.

Complexity of Clustering Time Series Data Biomechanical data is often collected over time, resulting in time series data. Clustering time series data requires specialised techniques that account for temporal dependencies and variations in movement patterns. Choosing appropriate features from the biomechanical data that capture relevant information for clustering is substantial but also challenging. Furthermore, the high dimensionality of the data can lead to computational inefficiency and may cause some clustering algorithms to perform poorly.

Classification of Clusters in Biomechanical Data Cluster analysis can divide the dataset into subsets that share similar structural characteristics. This approach can also be viewed as a preprocessing step for, e.g. subsequent classification tasks. For example, training and evaluating different classifiers on each cluster separately might help to identify the best model for each subgroup of data, leading to more accurate predictions.

Labelling of Biomechanical Data Cluster analysis can aid in semi-supervised learning approaches by assigning cluster annotations to non-annotated (unlabelled) data. This can provide additional training data for the classifier, leading to better performance, especially in scenarios where annotated data is scarce (see also Sect. 4.2). Cluster analyses could therefore also be of great importance in addressing a central challenge of limited data in gait and sports biomechanics research.

3.5 *Automated Classification*

3.5.1 Problem Description

The analysis of biomechanical data poses challenges owing to its multidimensional and interrelated nature [2]. To address these challenges, the application of ML approaches for automated classification has been proposed as a tool to enhance the analysis of biomechanical data. Automated classification refers to the task of predicting the category or class to which a biomechanical data sample belongs. Supervised ML methods, including Support Vector Machines, Random Forests and Convolutional Neural Networks, have been utilised to conduct automated classifications. Supervised ML methods are employed to train classification models based on a collection of samples, known as training data. Each sample within this training dataset comprises specific input features, e.g. 3D ground reaction forces and/or 3D marker trajectories, accompanied by corresponding target labels that denote the respective category or class of the sample, e.g. healthy or pathological gait. ML methods model the training data and learn relationships between the input features and the corresponding target labels. Once ML models are trained, they can be utilised for the automated classification of unseen data samples. By employing these ML models, users can leverage data-driven insights derived from extensive datasets. The literature on the use of supervised ML approaches for the automated classification of biomechanical data is outlined in Sect. 3.5.2.

3.5.2 Approaches

The ability of supervised ML to automatically classify biomechanical data is illustrated by the following examples, separately for the domain of walking, running and sports movements.

Walking In the context of fundamental research on the use of ML for the automated classification in biomechanics, biomechanical data were frequently gathered in laboratory-based settings. In this context, supervised ML methods were employed to test whether biomechanical input features can be classified according to personal or demographic target labels. For instance, supervised ML methods were used to demonstrate that individuals can be accurately identified based on kinetic, e.g. 99.3% in 671 individuals [3]; kinematic, e.g. 100.0% in 57 individuals [180]; and electromyographic, e.g. 98.9% in 79 individuals [181], input features measured during walking. Further studies showed that biomechanical walking patterns allow for identifying individuals across different days [4, 181] and months [182]. The identification of individuals based on their walking patterns can find its applications in personalising therapeutic diagnoses and treatments as well as biometrics and security [183]. Besides personal characteristics, supervised ML methods were used to predict demographic characteristics such age [183, 184] and sex/gender [185, 186] based on biomechanical walking data.

In the realm of clinical gait analysis, typically conducted in laboratory-based settings, studies have primarily focused on distinguishing between pathological and healthy biomechanics for diagnostic purposes [187]. Notable examples include distinguishing patterns in specific conditions, such as stroke [188], Parkinson's disease [189], cerebral palsy [190, 191], multiple sclerosis [190], osteoarthritis [192], anterior cruciate ligament injury [193] and functional walking disorders [11]. The latter studies demonstrate the effective utilisation of ML models for classifying patient groups. These findings suggest that ML models can be developed as decision support tools for diagnostic purposes, aiding clinicians in making informed decisions based on biomechanical data.

Supervised ML methods also demonstrated their potential in predicting fall risk among the elderly [194] and patients with neurological disease [195, 196], contributing to preventive strategies. By analysing long-term data in field-based settings, these models can forecast the likelihood of falls and empower proactive interventions.

Running In laboratory-based settings, supervised ML approaches for automated classification of biomechanical data during running were used for the classification of individual runners [197], age [198, 199], sex/gender [199, 200], running speed [200] and running shoe [200, 201].

A trend that can be observed in recent years is that supervised ML approaches predominantly utilised biomechanical data acquired from wearable IMU sensors in both laboratory- and field-based settings. This trend is linked to the widespread accessibility of such sensors on the consumer market and their extensive utilisation for tracking movements in everyday life and running across performance levels. Supervised ML approaches were able to classify wearable sensor data during running according to different types of outdoor terrains [202], inclinations of surface [203], environmental weather conditions [204], levels of fatigue [205], subjective perceptions of running shoe comfort [206] and runner's performance level [207, 208].

Sports Movements In laboratory-based settings, supervised ML methods were also implemented to accurately identify individuals based on their specific movement patterns in pedalling [181, 209], karate [210] and throwing [211]. Burdack et al. [210] also demonstrated that changes in intra-subject movement patterns were detectable during a karate kicking task, suggesting that adaptations in individual movement patterns occur during the process of fatigue-related accumulation. Novatchkov and Baca [212] explore how supervised ML methods can be used for the automated classification of execution quality in weight training.

In the sports domain, wearable IMU sensors have also been used in field-based settings for activity classification in various sports, e.g. in skateboarding [213], snowboarding [214], beach volleyball [215] and trampoline gymnastics [216].

3.5.3 Limitations and Future Perspectives

Supervised ML approaches have shown promising performance in automated classifications of biomechanical data on various classification tasks, as highlighted in Sect. 3.5.2. Despite promising performances in research, the practical implementation of ML models for automated classification of biomechanical data in clinical settings and athletic routines remains limited. This is related to the fact that many of the ML approaches described in the literature tend to be pilot studies in nature and suffer from the following limitations.

Biased Performance Evaluation Regimes Shortcomings in the evaluation of ML models are commonly observed in the literature. This includes that several studies distributed the available samples randomly across cross-validation folds in k-fold cross-validation configurations, without controlling an equal distribution of classes and confounding factors in each fold and without controlling that the data of one subject is either part of train, validation or test data. Moreover, differences in the number of individuals between categories/classes, i.e. unbalanced datasets, were not taken into account in the performance evaluation of the ML models.

Oversimplified Classification Task Studies have primarily used rather simple, i.e. binary, classification tasks, e.g. healthy control vs. Parkinson's disease. For evaluating the general suitability of ML algorithms for automated classification of biomechanical data, a simplification of the classification task, e.g. to a binary task with two clearly distinct classes is a reasonable research strategy. However, ML models in clinical practice will need to be able to solve advanced tasks, such as early detection, differential stage classification and decision-support for medication adjustment in Parkinson's disease.

Class Dependence During model training, we might have only a limited number of specific classes to work with. However, in real-world applications, the model is likely to encounter additional classes that were not part of its training dataset. Therefore, unsupervised outlier detection methods have recently been proposed which can function as class- or pathology-independent classifiers [217, 218] demonstrating that this approach is preferable to a two- or multi-class classifier in the described scenario [219].

Black-box Models Studies often prioritise predictive performance over interpretability of ML models. While complex, non-linear ML models can accurately classify or predict biomechanical outcomes, the underlying factors and relationships contributing to these predictions may remain opaque. This lack of interpretability hampers the ability of biomechanists to gain meaningful insights and fully understand the mechanisms driving the predictions of the ML models, a fact that may also reduce practitioners' trust in ML models for automated classification of biomechanical data.

The cornerstone for future development of ML models for automated classification is the availability of large-scale, high-quality and multi-modal benchmark dataset from multiple laboratory- and field-based settings for given classification task (see also Sect. 4.2). In addition to ongoing research and development aimed at enhancing the performance, robustness and interpretability of ML models for automatic classification in biomechanics, it is imperative to assess their practical applicability. This entails evaluating whether clinical diagnosis can be enhanced with the assistance of ML models, whether accurate and continuous recording of the ground surface, fatigue and gradient during running using wearable IMU sensors can effectively improve the performance of runners and whether automatic classifications of fall risks can indeed lead to a reduction in the number of falls among elderly individuals or patients with neurological diseases in everyday life.

4 Key Limitations of Machine Learning in Biomechanics

In this section, we will address key limitations of ML in biomechanics that are applicable across the described key applications, including explainability (Sect. 4.1) and data scarcity (Sect. 4.2). Further limitations that pertain primarily to individual applications will be discussed in Sect. 4.3.

4.1 Explainability

4.1.1 Problem Description

In various gait data analysis tasks, researchers frequently employ white-box models like Linear Regression or Decision Trees due to their transparent decision-making processes. In Linear Regression, the relationship between input variables and the output is explicitly defined through coefficients. Similarly, Decision Trees offer a clear decision-making pathway based on a series of if-else conditions, enhancing the interpretability of the model. These white-box models enable researchers to understand the relationships between input features and the predicted outcome effectively. However, as ML approaches in biomechanics evolve, there is a growing trend towards utilising complex, non-linear Deep Learning models such as Convolutional Neural Networks. Despite their promising performance in classifying walking, running and sports movements, these models pose a notable challenge due to their opaque, black-box nature. The lack of transparency presents a challenge in understanding how these models arrive at specific decisions and which underlying patterns and rules they learn from the data.

Explainable Artificial Intelligence (XAI), a field that encompasses different explainability approaches to shed light on the inner workings of complex, non-linear ML models, has gained increasing attention in recent years as an answer to this challenge. In the following, we outline the motivations behind the utilisation of explainability methods in biomechanics. The primary emphasis is on improving the transparency and understanding of the decision-making processes of ML approaches for domain experts, with particular significance in clinical applications. In this context, the critical aspect is to increase trust and acceptance among clinicians and patients into ML-based systems. Explainability methods should, therefore, enable clinical experts to verify decisions and eliminate potential biases by revealing which features they use for predictions. Furthermore, ensuring legal and regulatory compliance is another important aspect. Given the stringent regulations that have been implemented in recent years, especially in relation to the use of automated predictions, particularly within the healthcare sector, e.g. the EU General Data Protection Regulation [220] and the proposed AI Act.[2] Another motivating factor is the desire for personalised medicine. Due to the highly individualised nature of human movement, decision support and treatment planning often need to be customised to meet each patient's specific needs. Explainability methods could offer in-depth insights into the individual movement patterns responsible for particular decisions. The analysis of individual movement patterns is also relevant for running and sports applications. For example, identifying the individual characteristics within biomechanical data concerning athletes and their sports equipment can aid in performance assessment and improvement, the monitoring of training progress and the reduction of injury risks. Finally, explainability methods enable ML engineers to deeper understand the utilised ML methods. This understanding facilitates the evaluation of the effects of data preprocessing procedures and different ML methods. This evaluation considers the classification strategies and utilised input features, extending beyond mere performance assessments. Explainability methods can also help in identifying and mitigating biases, thus optimising ML approaches to align with the specific requirements and tasks within the realm of biomechanics.

Key Limitation: Explainability
Explainability in this context entails providing a human-understandable elucidation of the internal mechanisms and decision-making processes inherent to complex non-linear ML models utilised in biomechanics.

4.1.2 Approaches

According to the taxonomy of Arya et al. [221], three different XAI approaches can be distinguished: data exploration, decision explanation and model explanation.

Data exploration approaches do not explain the ML model, but rather the data that is used to train the model. The goal of these methods is to uncover meaningful data structures and patterns and to help domain experts understand the data distributions within the data. Various methods have been employed for data exploration, e.g. inferential statistical tests, such as Statistical Parametric Mapping [222] for time

[2] https://europa.eu/!PH46HB (accessed on 28.09.2023).

series data [9], visually inspecting data (average waveforms, boxplots and feature space examination) [11, 223] and more advanced data exploration techniques from the field of visual analytics [224].

Decision explanations provide *local* explanations for predictions on individual data samples. These explanations identify which input features contribute to a certain prediction. The majority of these approaches are applied in a post hoc manner to previously trained ML models.

Model explanations explain the *global* operation of a trained model and allow to reveal task- and class-specific prototypes and characteristic patterns. These explanations allow domain experts to identify ambiguous features, detect overlap between classes and assess the correctness of the model.

Walking Decision and model explanations have been explored using propagation- and perturbation-based explainability methods for walking data. Slijepcevic et al. [9] utilised Layer-Wise Relevance Propagation, a propagation-based explainability method, to obtain class and model explanations for the automated classification of functional walking disorders. The authors proposed a model explanation approach based on averaging Layer-Wise Relevance Propagation relevance scores across all samples within a class, alongside an alternative approach relying on SpRAy [225], a method for identifying clusters among explanations. The explainability results highlighted that the investigated ML models (Support Vector Machines, Multilayer Perceptrons and Convolutional Neural Networks) base their predictions on features that align with a statistical and clinical evaluation. In addition, the explainability analysis enabled the identification and comparison of learning strategies employed by the different ML methods. Moreover, the explainability results also confirmed the significance of data normalisation for obtaining unbiased predictions from ML models (addressing bias introduced by different signal amplitudes).

Layer-Wise Relevance Propagation has been further explored to provide insights into sex- and age-dependent walking patterns learned from ML models [226, 227]. Several studies have utilised Layer-Wise Relevance Propagation to identify the features used by ML models for classifying individuals based on their biomechanical walking patterns [3, 180, 209, 228]. The explainability analysis showed that unique person-specific walking characteristics, so-called gait signatures, can be determined very reliably [228].

Rind et al. [229] introduced *gaitXplorer*, a visual analytics approach designed for the classification of walking patterns associated with cerebral palsy. The authors employed the propagation-based method Grad-CAM [230] to provide decision explanations for Convolutional Neural Networks. Slijepcevic et al. [231] proposed model explanations by aggregating individual Grad-CAM explanations (using the median) for classifying walking patterns associated with cerebral palsy. Additionally, the authors compared these explanations with feature importance from tree-based models, i.e. Decision Trees and Random Forests.

Dindorf et al. [232] utilised the perturbation-based explainability method Local Interpretable Model-Agnostic Explanation [233] to explain linear Support Vector Machines trained on kinematic and kinetic waveforms as well as discrete features to differentiate between healthy individuals and patients who underwent total hip arthroplasty. To derive global model explanations, the authors averaged the decision explanations per class. Özates et al. [234] employed Local Interpretable Model-Agnostic Explanation to explore the discrete features that ML models relied on for distinguishing among nine different foot conditions.

Kokkotis et al. [235] leveraged the perturbation-based explainability method Shapley Additive Explanations [236]. The authors applied this method to explain the most influential discrete kinematic and kinetic features for classifying patients with anterior cruciate ligament injuries (both with and without reconstruction surgery) and healthy controls. The authors observed that Shapley Additive Explanations emphasised several discrete features consistent with biomechanical findings in the literature. However, they noted a disparity between the explainability results and the outcomes of traditional statistical analyses.

Furthermore, Dindorf et al. [237] investigated Counterfactual Explanations for the task of identifying the presence of hyperlordosis and hyperkyphosis based on postural data. The authors utilised handcrafted domain-specific features and a Gaussian process classifier.

Running Only a limited number of studies has utilised explainability methods within the domain of running biomechanics. The identified studies all utilised the propagation-based approach Layer-Wise Relevance Propagation. Hoitz et al. utilised Layer-Wise Relevance Propagation to identify running signatures from kinematic and kinetic data from 50 recreational [197] and 20 highly trained runners [238]. Xu et al. [149] utilised Layer-Wise Relevance Propagation to identify differences in running patterns between high- and low-mileage runners based on kinematic and kinetic data. Furthermore, Horst et al. [239] leveraged Layer-Wise Relevance Propagation to identifying running signatures of 30 healthy subjects and evaluate the influence of four footwear conditions on individual running characteristics.

Sports Movements Concerning perturbation-based methods, Lisca et al. [240] employed Shapley Additive Explanations to explain an Extreme Gradient Boost [241] classifier trained for the classification of goalkeeper movements in football based on IMU-derived kinematics of the goalkeeper's hand. Woltmann et al. [242] utilised Shapley Additive Explanations to provide explanations for the prediction of various ML models trained for classifying jumps in trampoline gymnastics. For this purpose, classification relied on time discrete features extracted from the kinetic data of an IMU sensor.

4.1.3 Limitations and Future Perspectives

The utilisation of explainability methods in biomechanics is still relatively in its infancy. Although there is a growing interest in the use of explainability methods in the domain of walking biomechanics, their application in the domain of running, and particularly in the biomechanics of sports movements, remains comparatively unexplored. Initial research, primarily centred on human walking, suggests considerable potential for advancing our understanding of ML in biomechanics and its capacity to enhance ML approaches. For example, explainability methods allow to understand the functioning of ML models and to show which input features were used for their predictions. In addition to performance-based comparisons, explainability methods allow to identify and compare the classification strategies of different ML methods and the effects of different data preprocessing procedures [9, 180]. This can aid in detecting biases in ML models and help ML engineers to mitigate or minimise these biases.

A fundamental challenge in evaluating explainability results is the absence of a ground truth. To address this challenge, Slijepcevic et al. [9] proposed a two-step approach, using Statistical Parametric Mapping and clinical domain experts, for evaluating the explanations. However, there is a need for more extensive examination of different explainability methods to ensure their meaningfulness and validity. At present, different explainability methods have been utilised, e.g. Layer-Wise Relevance Propagation, Grad-CAM, Local Interpretable Model-Agnostic Explanation, Shapley Additive Explanations and Counterfactual Explanations; however, it remains ambiguous which explainability method is most appropriate for certain biomechanical data and tasks. Thus, similar to the lack of a systematic assessment for the elements of the general ML pipeline, there is also a noticeable gap in methodically evaluating various explainability methods for interpreting biomechanical data. A systematic comparison of explainability methods for different biomechanical tasks and data could contribute to making these methods an essential part of ML approaches within the community. These systematic comparisons should also investigate the possibilities of trainable explanation mechanisms incorporated directly into complex Deep Learning models such as the method bounded logit attention [243].

An essential consideration for the usefulness of explainability methods lies in determining the specific form, e.g. interactive counterfactuals in which domain experts require explanations to ensure their practical value. Slijepcevic et al. [9] illustrated that ML models often learn an over-complete set of relevant features, and these features may not always align with all clinically relevant parameters [191]. This discrepancy could potentially diminish experts' trust in the utilised ML methods, even though these methods are effective. Empirical studies assessing the usefulness, trustworthiness and other aspects of explanations in biomechanics are currently lacking and should be addressed in future research.

4.2 Data and Annotation Availability

4.2.1 Problem Description

A prevalent limitation of ML approaches in biomechanics is the reliance on small datasets. ML approaches have been predominantly trained and evaluated on data from a limited amount of individuals, gathered within a single laboratory- or field-based setting. This can be attributed to various factors, including challenges in participant recruitment, the need for specialised expertise and the expensive and complex nature of measurement recording during biomechanical testing. Further limitations emerge due to the inability to record data from entire populations, which can introduce various biases into the datasets. These biases might manifest as imbalanced distributions of sex, age, body mass, body height, speed and performance levels among the categories of a dataset.

The lack of available large-scale benchmark datasets also hinders the comparability and reproducibility of results across studies and limits the widespread adoption of ML-based approaches in biomechanics. The existing literature has a notable lack of comparable classification tasks and datasets. This results in insufficient guidance on appropriate data preprocessing and classification pipelines, including, e.g. feature engineering, scaling and ML methods, suitable for given classification tasks.

A significant limitation for biomechanical data is also the lack of annotations, which are especially necessary in a clinical setting, e.g. pathological gait patterns. Annotations typically require a subjective assessment of the biomechanical data by domain experts. Thus, determining annotations is a labour-intensive and time-consuming task.

Key Limitation: Data and Annotation Availability
Data and annotation availability refers to the limited availability or insufficiency of relevant and comprehensive datasets, task-specific annotations for conducting thorough analyses and research in biomechanics.

4.2.2 Approaches

Walking Public datasets are already accessible for the purpose of data analysis, and a recent systematic review by David et al. [244] presents an overview of existing datasets.

Most of these datasets contain a relatively small number of individuals and walking trials conducted under various settings and using different measurement devices, as documented in previous studies [245, 246, 247, 248, 249, 250]. However, there are a few datasets that offer more potential for ML approaches, such as the GaitRec dataset [251], Gutenberg Gait Database [252] and a dataset published by Derlatka and Parfieniuk [253]. The GaitRec dataset is one of the most comprehensive datasets in the field containing bilateral ground reaction force data of 2,084 patients and 211 healthy controls with a total of 75,732 walking trials. The GaitRec dataset contains clinically relevant annotations determined by a clinical expert. These annotations specify the general anatomical joint level at which an orthopaedic impairment is located, such as at the hip, knee, ankle or calcaneus, and the dataset provides also more detailed annotations about the impairment type. The Gutenberg Gait Database contains also bilateral ground reaction force data of 350 healthy controls with a total of 8,819 walking trials. The dataset by Derlatka and Parfieniuk [253] comprises 13,702 walking trials of bilateral ground reaction force data from 324 healthy controls wearing shoes of various types.

Regarding the standardisation of data preprocessing, the authors of the GaitRec dataset proposed a general preprocessing pipeline for ground reaction force data including thresholding, filtering, outlier detection and an export format. This pipeline was also employed for the Gutenberg Gait Database in order to enable the seamless utilisation of both datasets.

In the ML literature, various strategies have been introduced to address the challenge of dealing with limited data or sparsely labelled data. Few-shot, including variants such as one- and zero-shot, learning approaches have already been explored for the analysis of walking data [147, 254, 255]. These approaches have the advantage of being able to model effectively even limited amounts of data per class. This means that they can extract information from only a few samples per class and create models that are able to detect and classify previously unseen classes. This is especially important in situations where data collection is limited as with biomechanical data, e.g. due to rare pathological walking patterns.

Inter-subject variability is another challenge that can negatively impact model quality. To deal with this challenge, Dindorf et al. [254] were able to achieve the first promising results using Siamese Neural Networks as automated feature extractor for learning features with reduced intra-subject variability between different walking cycles or measurement days.

Running In the context of running, there are also publicly accessible datasets available [244]. However, the number of individuals in these datasets is limited compared to the aforementioned walking datasets. A common characteristic of running datasets is the substantial volume of data gathered for these limited number of individuals, given the recording of numerous running cycles. Running data is frequently a subset of a more extensive dataset that includes both everyday activities, in the context of human activity recognition, and specific sports-related movements. The available running data are commonly recorded in a laboratory-based setting via marker-based infrared camera systems [105, 131, 132, 256] or IMU-based sensors [257, 258] and an instrumented treadmill. The biomechanical assessment also extends beyond the laboratory setting and into the field through the use of IMU-based sensors [257].

Sports Movements Data availability presents an even more pronounced challenge in sports movements compared to walking and running. Collecting comprehensive biomechanical data on sports movements can be challenging and resource-intensive. Moreover, in the realm of high-performance sports, it is common to encounter scenarios in which data collection is restricted to a highly exclusive group of individuals, usually professional athletes. Their data is usually kept confidential to maintain a competitive advantage, among other reasons.

Similar to running data, biomechanical signals extracted from sports movements often form a subset within the broader scope of datasets used in the domain of human activity recognition. These datasets, e.g. HuGaDB dataset [258], MOCAP-ULL Database [256] and SFU Motion Capture Database,[3] cover a wide range of movements, including activities like jumping, parkour rolls over objects, cartwheels, bicycling, martial arts kicking and dance sequences. One of the most extensive datasets that includes sports movements is the MoVi dataset [259] that encompasses records from 90 actors engaging in a total of 21 diverse activities, including both everyday actions and sports-related movements, e.g. crawling, vertical jumping, jumping jacks, kicking, throwing and catching.

4.2.3 Limitations and Future Perspectives

Data availability is a major challenge for ML approaches in biomechanics. To date, datasets have been made publicly available to varying degrees, but there is a clear imbalance between the availability of datasets in the walking, running and sports movement domain. A relatively large number of datasets have been published regarding human walking, while the availability of datasets related to running and sports movements remains comparatively limited. Methods for learning with limited data, e.g. zero-shot learning, have been employed for walking data, but their potential applications in the field of running and sports movements hold promising prospects for the future.

There exists a high demand for high-quality, large-scale benchmark datasets collected in diverse laboratory and field settings. Benchmark datasets would allow for systematic comparisons of various ML methods, as exemplified by recent studies [8, 260]. Since the performance of ML models heavily relies on the quality and representativeness of training data, models trained on specific datasets may not generalise well to new or more diverse datasets, leading to reduced robustness and reliability in real-world applications. Variations in data collection protocols, equipment and participant populations can impact the performance and generalisability of the models. Therefore, benchmark datasets should be controlled for various influencing factors, e.g. age, sex, body height, body mass and speed differences, in order to allow for unbiased training and evaluation of ML models that are able to cope with inter- and intra-subject variability observed in biomechanical data. Large-scale benchmark datasets would allow for a more reliable assessment of the ML model performance on unseen test data, e.g. from another laboratory. With a larger dataset, the utilisation of more complex ML models, such as Deep Learning architectures, holds the potential to yield deeper insights and improve the performance of ML models. Furthermore, large-scale benchmark datasets would allow the application of transfer learning approaches. This is a commonly employed technique, particularly, used in the field of computer vision to achieve high generalisation ability despite limited datasets. Transfer learning involves pre-training ML models on large existing datasets and fine-tuning them

[3] https://mocap.cs.sfu.ca/ (accessed on 28.09.2023).

on a smaller task-relevant dataset. An alternative strategy for addressing the limited size of datasets involves artificially expanding them through the use of Generative Adversarial Networks. These artificial networks can assist in data augmentation by learning the data distribution and enabling the generation of new training data. In the context of spinal posture data, Generative Adversarial Networks have demonstrated potential by enhancing the accuracy of deep Autoencoders through the incorporation of augmented data [101].

Another major limitation relates to the availability of annotations. Accurate and comprehensive annotations of biomechanical data, whether in the context of walking, running or sports movements, is a labour-intensive and subjective process. Annotating data requires human domain experts to identify and label specific events and features in the data. This process can introduce variability. Thus, it is important to create standardised annotation protocols that can be consistently applied to different datasets and research studies. In order to mitigate the requirement for large volumes of annotated data, several promising approaches including unsupervised and self-supervised ML methods could be utilised. These methods allow models to learn a general and transferable representation from large-scale non-annotated data that can be subsequently refined for specific downstream tasks.

The most promising strategy for generating large-scale and comprehensive datasets involves combining existing datasets from various laboratories and studies. Ensuring standardised procedures for data collection, storage and annotation is crucial to enable this integration of datasets acquired from different sources. This standardisation would foster interoperability among the data. A first step in this direction in the context of walking data has been initiated with the GaitRec dataset and the publication of a standardised data preprocessing pipeline [251]. Data sharing initiatives are essential to gather datasets from multiple laboratories and real-world scenarios and thereby promote future research.

4.3 *Further Limitations*

Lack of Reproducibility The current state of practice indicates that the level of reproducibility in the context of ML in biomechanics is restricted. A significant limitation in this regard is the aforementioned challenge regarding the availability of data (Sect. 4.2). Furthermore, ML models are frequently not shared, and the accessibility of code is often lacking. To enhance the reproducibility and transparency of research, the implementation of a badging system for publications in biomechanics similar to the ACM Artifact Review and Badging approach [261] could prove highly beneficial. Such an initiative could play a pivotal role in promoting open science practices and fostering a culture of transparency and collaboration within the biomechanics community, ultimately advancing the field's credibility and impact.

Lack of Comparability Another significant challenge in the domain of ML in biomechanics is the limited comparability among studies. This lack of comparability results from the substantial variation in the evaluation strategies utilised in this community. The assessment of ML model performance in biomechanics can exhibit considerable diversity. This diversity encompasses various aspects, including the selection of evaluation strategies, e.g. hold-out, k-fold cross-validation and leave-one-out cross-validation, the use of different evaluation metrics and the lack of benchmark datasets, which could be used to additionally validate a proposed ML approach. This heterogeneity makes it difficult to draw meaningful comparisons between studies in the field. In response to this challenge, the creation of standardised evaluation protocols and metrics specifically designed for biomechanical applications could be immensely beneficial. To address this challenge, efforts to establish standardised evaluation protocols and metrics tailored to the specific requirements of biomechanical applications, e.g. imbalanced data and limited dataset size, are essential. Such standardisation would not only enhance the credibility of research findings but also facilitate the synthesis of knowledge and the development of more robust and reliable ML solutions.

Neglecting the Biomechanical Context Gait representations learned solely in a data-driven approach have shown certain limitations in accurately capture the gait primitives that underlie fundamental locomotion patterns. Addressing these challenges necessitates a shift towards more comprehensive approaches, using not only data-driven learning but also physics-informed ML approaches.

Physics-informed ML methods integrate domain-specific knowledge, such as biomechanical principles, into the ML model. These methods have the potential to allow a more accurate modelling of the underlying physics of the biomechanical signals. As a result, the ML model could exhibit greater generalisability.

5 Conclusion: The Future of Machine Learning in Biomechanics

The gradual integration of ML into biomechanics marks a significant transformation in the study of human movements. Reflecting on the current research landscape, it is evident that the domains of both gait and sports biomechanics can substantially benefit from the application of ML. While ML has made substantial progress in various key applications, its integration into biomechanics lags behind domains such as healthcare. This disparity is particularly evident in sports biomechanics, where considerably less publications used ML approaches compared to gait biomechanics. Nevertheless, a reciprocal learning dynamic might emerge between gait and sports biomechanics, both in terms of potential insights and methodological advances. However, it should be noted that these two domains are not always easily distinguishable from each other, and there are transitional areas between them.

In certain applications, gait biomechanics is ahead in the development of ML-based approaches, outpacing its implementation in sports biomechanics. Sports biomechanists can obtain guidance on data preprocessing procedures and ML methodologies from gait biomechanics, as a larger number of ML pipelines and systematic comparisons are already available. This includes findings such as the benefits of incorporating entire time series data and dimension reduction techniques such as Principal Component Analysis for input feature generation, as opposed to solely relaying on handcrafted input features. The findings also indicate that ML methods like Support Vector Machines and Random Forests have exhibited comparable, or sometimes even superior, performance to Deep Learning approaches and might be particularly suitable for small-scale datasets in biomechanics. Furthermore, there exists a need to cultivate substantially larger publicly accessible datasets in sports biomechanics compared to already available open source datasets in gait biomechanics.

In comparison, the realm of sports biomechanics has also demonstrated promising advancements in utilising ML, thereby offering potential ML approaches that can be applied to the study of gait biomechanics. For instance, sports-related scenarios frequently involve real-world field conditions, in comparison to the controlled laboratory settings commonly seen in current applications of ML in gait research. Gait biomechanics can get guidance from ML applications in complex sports movements for applications that should work reliably in everyday clinical life.

However, ML in gait and sports biomechanics share similar inherent limitations. Addressing these challenges related to the accessibility of large benchmark datasets or to issues of variability, transparency and trustworthiness will require collaborative efforts. In order to enrich the community in a sustainable and future-oriented way, we need the willingness to share data and work together to advance new developments and achieve greater consensus regarding the application and evaluation of the ML methods. As we navigate this dynamic landscape, interdisciplinary collaboration will serve as the compass guiding us toward unlocking the full potential that ML offers in the multifaceted domains of gait and sports biomechanics.

Funding Information
This work was partly funded by the Research Promotion Agency of Lower Austria (Gesellschaft für Forschungsförderung NÖ) within the Endowed Professorship for Applied Biomechanics and Rehabilitation Research (#SP19-004) and the project DeepForce (#GLF21-1-009), as well as by the Austrian Research Promotion Agency (FFG) within the COIN-program (#898085).

This research was supported by the Central Innovation Programme for Small and Medium-Sized Enterprises (Zentrales Innovationsprogramm Mittelstand, ZIM) of the German Federal Ministry for Economic Affairs and Climate Action, under grant number 16KN113027.

References

1. D. A. Winter, *Biomechanics and Motor Control of Human Movement*. Hoboken, N.J.: John Wiley & Sons, 4th ed. ed., 2009.
2. T. Chau, “A review of analytical techniques for gait data. part 1: Fuzzy, statistical and fractal methods,” *Gait & Posture*, vol. 13, no. 1, pp. 49–66, 2001.
3. F. Horst, D. Slijepcevic, M. Simak, B. Horsak, W. I. Schöllhorn, and M. Zeppelzauer, “Modeling biological individuality using machine learning: A study on human gait,” *Computational and Structural Biotechnology Journal*, vol. 21, pp. 3414–3423, 2023.
4. F. Horst, F. Kramer, B. Schäfer, A. Eekhoff, P. Hegen, B. Nigg, and W. Schöllhorn, “Daily changes of individual gait patterns identified by means of support vector machines,” *Gait & Posture*, vol. 49, pp. 309–314, 2016.
5. M. E. Eastlack, J. Arvidson, L. Snyder-Mackler, J. V. Danoff, and C. L. McGarvey, “Interrater reliability of videotaped observational gait-analysis assessments,” *Physical Therapy*, vol. 71, no. 6, pp. 465–472, 1991.
6. J. J. Brunnekreef, C. J. T. van Uden, S. van Moorsel, and J. G. M. Kooloos, “Reliability of videotaped observational gait analysis in patients with orthopedic impairments,” *BMC Musculoskeletal Disorders*, vol. 6, p. 17, 2005.
7. C. Richter, M. O’Reilly, and E. Delahunt, “Machine learning in sports science: challenges and opportunities,” *Sports Biomechanics*, vol. 23, no. 8, pp. 1–7, 2021.
8. J. Burdack, F. Horst, S. Giesselbach, I. Hassan, S. Daffner, and W. I. Schöllhorn, “Systematic comparison of the influence of different data preprocessing methods on the performance of gait classifications using machine learning,” *Frontiers in Bioengineering and Biotechnology*, vol. 8, p. 260, 2020.
9. D. Slijepcevic, F. Horst, S. Lapuschkin, B. Horsak, A.-M. Raberger, A. Kranzl, W. Samek, C. Breiteneder, W. I. Schöllhorn, and M. Zeppelzauer, “Explaining machine learning models for clinical gait analysis,” *ACM Transactions on Computing for Healthcare (HEALTH)*, vol. 3, no. 2, pp. 1–27, 2021.
10. H. Liu and L. Yu, “Toward integrating feature selection algorithms for classification and clustering,” *IEEE Transactions on Knowledge and Data Engineering*, vol. 17, no. 4, pp. 491–502, 2005.
11. D. Slijepcevic, M. Zeppelzauer, A.-M. Gorgas, C. Schwab, M. Schüller, A. Baca, C. Breiteneder, and B. Horsak, “Automatic classification of functional gait disorders,” *IEEE Journal of Biomedical and Health Informatics*, vol. 22, no. 5, pp. 1653–1661, 2018.
12. C. Dindorf, W. Teufl, B. Taetz, S. Becker, G. Bleser, and M. Fröhlich, “Feature extraction and gait classification in hip replacement patients on the basis of kinematic waveform data,” *Biomedical Human Kinetics*, vol. 13, no. 1, pp. 177–186, 2021.
13. Z. Cao, G. Hidalgo Martinez, T. Simon, S. Wei, and Y. A. Sheikh, “Openpose: Realtime multi-person 2d pose estimation using part affinity fields,” *IEEE Transactions on Pattern Analysis and Machine Intelligence*, vol. 43, pp. 172–186, 2019.
14. A. Mathis, P. Mamidanna, K. M. Cury, T. Abe, V. N. Murthy, M. W. Mathis, and M. Bethge, “DeepLabCut: Markerless pose estimation of user-defined body parts with deep learning,” *Nature Neuroscience*, vol. 21, no. 9, pp. 1281–1289, 2018.
15. A. Toshev and C. Szegedy, “Deeppose: Human pose estimation via deep neural networks,” in *2014 IEEE Conference on Computer Vision and Pattern Recognition*, pp. 1653–1660, 2014.
16. E. Insafutdinov, L. Pishchulin, B. Andres, M. Andriluka, and B. Schiele, “DeeperCut: A Deeper, Stronger, and Faster Multi-person Pose Estimation Model,” in *Computer Vision – ECCV 2016* (B. Leibe, J. Matas, N. Sebe, and M. Welling, eds.), Lecture Notes in Computer Science, (Cham), pp. 34–50, Springer International Publishing, 2016.
17. Z. Cao, T. Simon, S.-E. Wei, and Y. Sheikh, “Realtime multi-person 2d pose estimation using part affinity fields,” in *2017 IEEE Conference on Computer Vision and Pattern Recognition (CVPR)*, pp. 7291–7299, 2017.

18. M. Ben Gamra and M. A. Akhloufi, "A review of deep learning techniques for 2D and 3D human pose estimation," *Image and Vision Computing*, vol. 114, p. 104282, 2021.
19. Y. Chen, Y. Tian, and M. He, "Monocular human pose estimation: A survey of deep learning-based methods," *Computer Vision and Image Understanding*, vol. 192, p. 102897, 2020.
20. L. Song, G. Yu, J. Yuan, and Z. Liu, "Human pose estimation and its application to action recognition: A survey," *Journal of Visual Communication and Image Representation*, vol. 76, p. 103055, 2021.
21. Q. Dang, J. Yin, B. Wang, and W. Zheng, "Deep learning based 2D human pose estimation: A survey," *Tsinghua Science and Technology*, vol. 24, no. 6, pp. 663–676, 2019.
22. S. L. Colyer, M. Evans, D. P. Cosker, and A. I. T. Salo, "A Review of the Evolution of Vision-Based Motion Analysis and the Integration of Advanced Computer Vision Methods Towards Developing a Markerless System," *Sports Medicine-Open*, vol. 4, no. 1, p. 24, 2018.
23. N. J. Cronin, "Using deep neural networks for kinematic analysis: Challenges and opportunities," *Journal of Biomechanics*, vol. 123, p. 110460, 2021.
24. N. J. Cronin, T. Rantalainen, J. P. Ahtiainen, E. Hynynen, and B. Waller, "Markerless 2D kinematic analysis of underwater running: A deep learning approach," *Journal of Biomechanics*, vol. 87, pp. 75–82, 2019.
25. M. Moro, G. Marchesi, F. Odone, and M. Casadio, "Markerless gait analysis in stroke survivors based on computer vision and deep learning: A pilot study," in *Proceedings of the 35th Annual ACM Symposium on Applied Computing*, SAC '20, (New York, NY, USA), pp. 2097–2104, Association for Computing Machinery, 2020.
26. J. F. Drazan, W. T. Phillips, N. Seethapathi, T. J. Hullfish, and J. R. Baxter, "Moving outside the lab: Markerless motion capture accurately quantifies sagittal plane kinematics during the vertical jump," *Journal of Biomechanics*, vol. 125, p. 110547, 2021.
27. G. Serrancolí, P. Bogatikov, J. P. Huix, A. F. Barberà, A. J. S. Egea, J. T. Ribé, S. Kanaan-Izquierdo, and A. Susín, "Marker-Less Monitoring Protocol to Analyze Biomechanical Joint Metrics During Pedaling," *IEEE Access*, vol. 8, pp. 122782–122790, 2020.
28. J. Stenum, C. Rossi, and R. T. Roemmich, "Two-dimensional video-based analysis of human gait using pose estimation," *PLOS Computational Biology*, vol. 17, no. 4, p. e1008935, 2021.
29. L. Wade, L. Needham, P. McGuigan, and J. Bilzon, "Applications and limitations of current markerless motion capture methods for clinical gait biomechanics," *PeerJ*, vol. 10, p. e12995, 2022.
30. N. Sarafianos, B. Boteanu, B. Ionescu, and I. A. Kakadiaris, "3d human pose estimation: A review of the literature and analysis of covariates," *Computer Vision and Image Understanding*, vol. 152, pp. 1–20, 2016.
31. X. Ji, Q. Fang, J. Dong, Q. Shuai, W. Jiang, and X. Zhou, "A survey on monocular 3D human pose estimation," *Virtual Reality & Intelligent Hardware*, vol. 2, no. 6, pp. 471–500, 2020.
32. R. I. Hartley and P. Sturm, "Triangulation," *Computer Vision and Image Understanding*, vol. 68, no. 2, pp. 146–157, 1997.
33. N. Nakano, T. Sakura, K. Ueda, L. Omura, A. Kimura, Y. Iino, S. Fukashiro, and S. Yoshioka, "Evaluation of 3D Markerless Motion Capture Accuracy Using OpenPose With Multiple Video Cameras," *Frontiers in Sports and Active Living*, vol. 2, p. 50, 2020.
34. M. Slembrouck, H. Luong, J. Gerlo, K. Schütte, D. Van Cauwelaert, D. De Clercq, B. Vanwanseele, P. Veelaert, and W. Philips, "Multiview 3D Markerless Human Pose Estimation from OpenPose Skeletons," in *Advanced Concepts for Intelligent Vision Systems* (J. Blanc-Talon, P. Delmas, W. Philips, D. Popescu, and P. Scheunders, eds.), Lecture Notes in Computer Science, (Cham), pp. 166–178, Springer International Publishing, 2020.
35. M. Zago, M. Luzzago, T. Marangoni, M. De Cecco, M. Tarabini, and M. Galli, "3D Tracking of Human Motion Using Visual Skeletonization and Stereoscopic Vision," *Frontiers in Bioengineering and Biotechnology*, vol. 8:181, 2020.
36. S. D. Uhlrich, A. Falisse, Ł. Kidziński, J. Muccini, M. Ko, A. S. Chaudhari, J. L. Hicks, and S. L. Delp, "OpenCap: Human movement dynamics from smartphone videos," *PLOS Computational Biology*, vol. 19, no. 10, p. e1011462, 2023.

37. B. Horsak, A. Eichmann, K. Lauer, K. Prock, P. Krondorfer, T. Siragy, and B. Dumphart, "Concurrent validity of smartphone-based markerless motion capturing to quantify lower-limb joint kinematics in healthy and pathological gait," *Journal of Biomechanics*, vol. 159, p. 111801, 2023.
38. R. M. Kanko, E. K. Laende, G. Strutzenberger, M. Brown, W. S. Selbie, V. DePaul, S. H. Scott, and K. J. Deluzio, "Assessment of spatiotemporal gait parameters using a deep learning algorithm-based markerless motion capture system," *Journal of Biomechanics*, vol. 122, p. 110414, 2021.
39. R. M. Kanko, E. Laende, W. S. Selbie, and K. J. Deluzio, "Inter-session repeatability of markerless motion capture gait kinematics," *Journal of Biomechanics*, vol. 121, p. 110422, 2021.
40. R. M. Kanko, E. K. Laende, E. M. Davis, W. S. Selbie, and K. J. Deluzio, "Concurrent assessment of gait kinematics using marker-based and markerless motion capture," *Journal of Biomechanics*, vol. 127, p. 110665, 2021.
41. T. A. L. Wren, P. Isakov, and S. A. Rethlefsen, "Comparison of Kinematics Between Theia Markerless and Conventional Marker-Based Gait Analysis in Clinical Patients," *Gait & Posture*, vol. 104, pp. 9–14, 2023.
42. N. Ito, H. B. Sigurðsson, K. D. Seymore, E. K. Arhos, T. S. Buchanan, L. Snyder-Mackler, and K. G. Silbernagel, "Markerless motion capture: What clinician-scientists need to know right now," *JSAMS plus*, vol. 1, p. 100001, 2022.
43. K. Song, T. J. Hullfish, R. S. Silva, K. G. Silbernagel, and J. R. Baxter, "Markerless motion capture estimates of lower extremity kinematics and kinetics are comparable to marker-based across 8 movements," *Journal of Biomechanics*, vol. 157, p. 111751, 2023.
44. Z. Ripic, J. F. Signorile, T. M. Best, K. A. Jacobs, M. Nienhuis, C. Whitelaw, C. Moenning, and M. Eltoukhy, "Validity of artificial intelligence-based markerless motion capture system for clinical gait analysis: Spatiotemporal results in healthy adults and adults with Parkinson's disease," *Journal of Biomechanics*, vol. 155, p. 111645, 2023.
45. M. Manca, A. Leardini, S. Cavazza, G. Ferraresi, P. Marchi, E. Zanaga, and M. G. Benedetti, "Repeatability of a new protocol for gait analysis in adult subjects," *Gait & Posture*, vol. 32, no. 2, pp. 282–284, 2010.
46. J. L. McGinley, R. Baker, R. Wolfe, and M. E. Morris, "The reliability of three-dimensional kinematic gait measurements: A systematic review," *Gait & Posture*, vol. 29, no. 3, pp. 360–369, 2009.
47. B. Horsak, K. Prock, P. Krondorfer, T. Siragy, M. Simonlehner, and B. Dumphart, "Inter-trial variability is higher in 3D markerless compared to marker-based motion capture: Implications for data post-processing and analysis," *Journal of Biomechanics*, vol. 166, p. 112049, 2024.
48. J. Stenum, K. M. Cherry-Allen, C. O. Pyles, R. D. Reetzke, M. F. Vignos, and R. T. Roemmich, "Applications of Pose Estimation in Human Health and Performance across the Lifespan," *Sensors*, vol. 21, no. 21, p. 7315, 2021.
49. O. Ossmy and K. E. Adolph, "Real-Time Assembly of Coordination Patterns in Human Infants," *Current Biology: CB*, vol. 30, no. 23, pp. 4553–4562.e4, 2020.
50. L. Adde, J. L. Helbostad, A. R. Jensenius, G. Taraldsen, K. H. Grunewaldt, and R. Støen, "Early prediction of cerebral palsy by computer-based video analysis of general movements: A feasibility study," *Developmental Medicine & Child Neurology*, vol. 52, no. 8, pp. 773–778, 2010.
51. H. Rahmati, O. M. Aamo, Ø. Stavdahl, R. Dragon, and L. Adde, "Video-based early cerebral palsy prediction using motion segmentation," in *2014 36th Annual International Conference of the IEEE Engineering in Medicine and Biology Society*, pp. 3779–3783, 2014.
52. A. A. Chaaraoui, J. R. Padilla-López, and F. Flórez-Revuelta, "Abnormal gait detection with RGB-D devices using joint motion history features," in *2015 11th IEEE International Conference and Workshops on Automatic Face and Gesture Recognition (FG)*, vol. 07, pp. 1–6, 2015.
53. Y. Guo, F. Deligianni, X. Gu, and G.-Z. Yang, "3-D Canonical Pose Estimation and Abnormal Gait Recognition With a Single RGB-D Camera," *IEEE Robotics and Automation Letters*, vol. 4, no. 4, pp. 3617–3624, 2019.

54. J. Kondragunta and G. Hirtz, "Gait Parameter Estimation of Elderly People using 3D Human Pose Estimation in Early Detection of Dementia," in *2020 42nd Annual International Conference of the IEEE Engineering in Medicine & Biology Society (EMBC)*, pp. 5798–5801, 2020.
55. Z.-P. Bian, J. Hou, L.-P. Chau, and N. Magnenat-Thalmann, "Fall Detection Based on Body Part Tracking Using a Depth Camera," *IEEE Journal of Biomedical and Health Informatics*, vol. 19, no. 2, pp. 430–439, 2015.
56. Y. Chen, R. Du, K. Luo, and Y. Xiao, "Fall detection system based on real-time pose estimation and SVM," in *2021 IEEE 2nd International Conference on Big Data, Artificial Intelligence and Internet of Things Engineering (ICBAIE)*, pp. 990–993, 2021.
57. S. Han, M. Achar, S. Lee, and F. Peña-Mora, "Empirical assessment of a RGB-D sensor on motion capture and action recognition for construction worker monitoring," *Visualization in Engineering*, vol. 1, no. 1, p. 6, 2013.
58. S. Han and S. Lee, "A vision-based motion capture and recognition framework for behavior-based safety management," *Automation in Construction*, vol. 35, pp. 131–141, 2013.
59. N. Blanchard, K. Skinner, A. Kemp, W. Scheirer, and P. Flynn, ""keep me in, coach!": A computer vision perspective on assessing acl injury risk in female athletes," in *2019 IEEE Winter Conference on Applications of Computer Vision (WACV)*, pp. 1366–1374, 2019.
60. J. Lin, X. Xie, W. Wu, S. Xu, C. Liu, T. Hudoyberdi, and X. Chen, "Model transfer from 2D to 3D study for boxing pose estimation," *Frontiers in Neurorobotics*, vol. 17, p. 1148545, 2023.
61. B. K. Lahkar, A. Muller, R. Dumas, L. Reveret, and T. Robert, "Accuracy of a markerless motion capture system in estimating upper extremity kinematics during boxing," *Frontiers in Sports and Active Living*, vol. 4, p. e939980, 2022.
62. G. S. Fleisig, J. S. Slowik, D. Wassom, Y. Yanagita, J. Bishop, and A. Diffendaffer, "Comparison of marker-less and marker-based motion capture for baseball pitching kinematics," *Sports Biomechanics*, vol. 19, no. 0, pp. 1–10, 2022.
63. W. Bao, T. Niu, N. Wang, and X. Yang, "Pose estimation and motion analysis of ski jumpers based on ECA-HRNet," *Scientific Reports*, vol. 13, no. 1, p. 6132, 2023.
64. D. Pagnon, M. Domalain, and L. Reveret, "Pose2Sim: An End-to-End Workflow for 3D Markerless Sports Kinematics-Part 2: Accuracy," *Sensors*, vol. 22, no. 7, p. 2712, 2022.
65. V. T. Keller, J. B. Outerleys, R. M. Kanko, E. K. Laende, and K. J. Deluzio, "Clothing condition does not affect meaningful clinical interpretation in markerless motion capture," *Journal of Biomechanics*, vol. 141, p. 111182, 2022.
66. W. R. Johnson, J. Alderson, D. Lloyd, and A. Mian, "Predicting athlete ground reaction forces and moments from spatio-temporal driven cnn models," *IEEE Transactions on Bio-Medical Engineering*, vol. 66, no. 3, pp. 689–694, 2019.
67. M. S. B. Hossain, Z. Guo, and H. Choi, "Estimation of lower extremity joint moments and 3d ground reaction forces using imu sensors in multiple walking conditions: A deep learning approach," *IEEE Journal of Biomedical and Health Informatics*, vol. 27, no. 6, 2023.
68. C. M. Saliba, A. L. Clouthier, S. C. E. Brandon, M. J. Rainbow, and K. J. Deluzio, "Prediction of knee joint contact forces from external measures using principal component prediction and reconstruction," *Journal of Applied Biomechanics*, vol. 34, no. 5, pp. 419–423, 2018.
69. C. A. Bailey, A. Mir-Orefice, T. K. Uchida, J. Nantel, and R. B. Graham, "Smartwatch-based prediction of single-stride and stride-to-stride gait outcomes using regression-based machine learning," *Annals of Biomedical Engineering*, vol. 51, no. 11, pp. 2504–2517, 2023.
70. J. R. Verbiest, B. Bonnechère, W. Saeys, P. van de Walle, S. Truijen, and P. Meyns, "Gait stride length estimation using embedded machine learning," *Sensors*, vol. 23, no. 16, 2023.
71. S. Vandermeeren, H. Bruneel, and H. Steendam, "Feature selection for machine learning based step length estimation algorithms," *Sensors*, vol. 20, no. 3, 2020.
72. M. Sharifi Renani, C. A. Myers, R. Zandie, M. H. Mahoor, B. S. Davidson, and C. W. Clary, "Deep learning in gait parameter prediction for oa and tka patients wearing imu sensors," *Sensors*, vol. 20, no. 19, p. 5553, 2020.

73. M. A. Boswell, S. D. Uhlrich, Ł. Kidziński, K. Thomas, J. A. Kolesar, G. E. Gold, G. S. Beaupre, and S. L. Delp, "A neural network to predict the knee adduction moment in patients with osteoarthritis using anatomical landmarks obtainable from 2d video analysis," *Osteoarthritis and Cartilage*, vol. 29, no. 3, pp. 346–356, 2021.
74. A. J. Aljaaf, A. J. Hussain, P. Fergus, A. Przybyla, and G. J. Barton, "Evaluation of machine learning methods to predict knee loading from the movement of body segments," in *2016 International Joint Conference on Neural Networks (IJCNN)*, IEEE, pp. 5168–5173, 2016.
75. M. Mundt, W. Thomsen, T. Witter, A. Koeppe, S. David, F. Bamer, W. Potthast, and B. Markert, "Prediction of lower limb joint angles and moments during gait using artificial neural networks," *Medical & Biological Engineering & Computing*, vol. 58, no. 1, pp. 211–225, 2020.
76. S. M. Moghadam, T. Yeung, and J. Choisne, "A comparison of machine learning models' accuracy in predicting lower-limb joints' kinematics, kinetics, and muscle forces from wearable sensors," *Scientific Reports*, vol. 13, no. 1, p. 5046, 2023.
77. G. Giarmatzis, E. I. Zacharaki, and K. Moustakas, "Real-time prediction of joint forces by motion capture and machine learning," *Sensors*, vol. 20, no. 23, p. 6933, 2020.
78. J. Camargo, D. Molinaro, and A. Young, "Predicting biological joint moment during multiple ambulation tasks," *Journal of Biomechanics*, vol. 134, p. 111020, 2022.
79. J. Moon, D. Lee, H. Jung, A. Choi, and J. H. Mun, "Machine learning strategies for low-cost insole-based prediction of center of gravity during gait in healthy males," *Sensors*, vol. 22, no. 9, 2022.
80. D. T. H. Lai, S. B. Taylor, and R. K. Begg, "Prediction of foot clearance parameters as a precursor to forecasting the risk of tripping and falling," *Human Movement Science*, vol. 31, no. 2, pp. 271–283, 2012.
81. T. T. Dao, "From deep learning to transfer learning for the prediction of skeletal muscle forces," *Medical & Biological Engineering & Computing*, vol. 57, no. 5, pp. 1049–1058, 2019.
82. O. Heeb, A. Barua, C. Menon, and X. Jiang, "Building effective machine learning models for ankle joint power estimation during walking using fmg sensors," *Frontiers in Neurorobotics*, vol. 16, p. 836779, 2022.
83. L. Zhang, X. Zhu, E. M. Gutierrez-Farewik, and R. Wang, "Ankle joint torque prediction using an nms solver informed-ann model and transfer learning," *IEEE Journal of Biomedical and Health Informatics*, vol. 26, no. 12, pp. 5895–5906, 2022.
84. N. Sun, M. Cao, Y. Chen, Y. Chen, J. Wang, Q. Wang, X. Chen, and T. Liu, "Continuous estimation of human knee joint angles by fusing kinematic and myoelectric signals," *IEEE Transactions on Neural Systems and Rehabilitation Engineering : A Publication of the IEEE Engineering in Medicine and Biology Society*, vol. 30, pp. 2446–2455, 2022.
85. V. A. de Jesus Oliveira, D. Slijepčević, B. Dumphart, S. Ferstl, J. Reis, A.-M. Raberger, M. Heller, B. Horsak, and M. Iber, "Auditory feedback in tele-rehabilitation based on automated gait classification," *Personal and Ubiquitous Computing*, vol. 27, no. 5, pp. 1–14, 2023.
86. M. Iber, B. Dumphart, V.-A. de Jesus Oliveira, S. Ferstl, J. M. Reis, D. Slijepčević, M. Heller, A.-M. Raberger, and B. Horsak, "Mind the steps: Towards auditory feedback in tele-rehabilitation based on automated gait classification," in *Proceedings of the 16th International Audio Mostly Conference*, pp. 139–146, 2021.
87. W. R. Johnson, A. Mian, C. J. Donnelly, D. Lloyd, and J. Alderson, "Predicting athlete ground reaction forces and moments from motion capture," *Medical & Biological Engineering & Computing*, vol. 56, no. 10, pp. 1781–1792, 2018.
88. R. K. Begg, "Neural network-based prediction of missing key features in vertical grf-time recordings," *Journal of Medical Engineering & Technology*, vol. 30, no. 5, pp. 315–322, 2006.
89. E. S. Matijevich, L. R. Scott, P. Volgyesi, K. H. Derry, and K. E. Zelik, "Combining wearable sensor signals, machine learning and biomechanics to estimate tibial bone force and damage during running," *Human Movement Science*, vol. 74, p. 102690, 2020.

90. R. Derie, P. Robberechts, P. van den Berghe, J. Gerlo, D. de Clercq, V. Segers, and J. Davis, "Tibial acceleration-based prediction of maximal vertical loading rate during overground running: A machine learning approach," *Frontiers in Bioengineering and Biotechnology*, vol. 8, p. 33, 2020.
91. B. J. Stetter, S. Ringhof, F. C. Krafft, S. Sell, and T. Stein, "Estimation of knee joint forces in sport movements using wearable sensors and machine learning," *Sensors*, vol. 19, no. 17, p. 3690, 2019.
92. C. R. Chaaban, N. T. Berry, C. Armitano-Lago, A. W. Kiefer, M. J. Mazzoleni, and D. A. Padua, "Combining inertial sensors and machine learning to predict vgrf and knee biomechanics during a double limb jump landing task," *Sensors*, vol. 21, no. 13, p. 4383, 2021.
93. M. Zago, C. Sforza, C. Dolci, M. Tarabini, and M. Galli, "Use of machine learning and wearable sensors to predict energetics and kinematics of cutting maneuvers," *Sensors*, vol. 19, no. 14, 2019.
94. X. Chen, K. Zhang, H. Liu, Y. Leng, and C. Fu, "A probability distribution model-based approach for foot placement prediction in the early swing phase with a wearable imu sensor," *IEEE Transactions on Neural Systems and Rehabilitation Engineering : A Publication of the IEEE Engineering in Medicine and Biology Society*, vol. 29, pp. 2595–2604, 2021.
95. Z. Choffin, N. Jeong, M. Callihan, E. Sazonov, and S. Jeong, "Lower body joint angle prediction using machine learning and applied biomechanical inverse dynamics," *Sensors*, vol. 23, no. 1, p. 228, 2022.
96. A. Moniri, D. Terracina, J. Rodriguez-Manzano, P. H. Strutton, and P. Georgiou, "Real-time forecasting of semg features for trunk muscle fatigue using machine learning," *IEEE Transactions on Bio-medical Engineering*, vol. 68, no. 2, pp. 718–727, 2021.
97. K. F. Nicholson, G. S. Collins, B. R. Waterman, and G. S. Bullock, "Machine learning and statistical prediction of fastball velocity with biomechanical predictors," *Journal of Biomechanics*, vol. 134, p. 110999, 2022.
98. K. F. Nicholson, G. S. Collins, B. R. Waterman, and G. S. Bullock, "Machine learning and statistical prediction of pitching arm kinetics," *The American Journal of Sports Medicine*, vol. 50, no. 1, pp. 238–247, 2022.
99. A.-R. I. Akl and A. A. Hassan, "An artificial neural network approach for predicting kinematics in handball throws," *American Journal of Sports Science*, vol. 5, no. 5, pp. 35–39, 2017.
100. A. Macaro, M. J. Connick, E. Beckman, and S. M. Tweedy, "Using machine learning techniques and wearable inertial measurement units to predict front crawl elbow joint angle: a pilot study," *36th Conference of the International Society of Biomechanics in Sports*, vol. 36, no. 1, p. 366, 2018.
101. C. Dindorf, J. Dully, J. Konradi, C. Wolf, S. Becker, S. Simon, J. Huthwelker, F. Werthmann, J. Kniepert, P. Drees, U. Betz, and M. Fröhlich, "Enhancing biomechanical machine learning with limited data: generating realistic synthetic posture data using generative artificial intelligence," *Frontiers in Bioengineering and Biotechnology*, vol. 12, p. 1350135, 2024.
102. A. Zargham, M. Afschrift, J. de Schutter, I. Jonkers, and F. de Groote, "Inverse dynamic estimates of muscle recruitment and joint contact forces are more realistic when minimizing muscle activity rather than metabolic energy or contact forces," *Gait & Posture*, vol. 74, pp. 223–230, 2019.
103. K. Taneja, X. He, Q. He, X. Zhao, Y.-A. Lin, K. J. Loh, and J.-S. Chen, "A feature-encoded physics-informed parameter identification neural network for musculoskeletal systems," *Journal of Biomechanical Engineering*, vol. 144, no. 12, 2022.
104. J. Zhang, Y. Zhao, F. Shone, Z. Li, A. F. Frangi, S. Q. Xie, and Z.-Q. Zhang, "Physics-informed deep learning for musculoskeletal modeling: Predicting muscle forces and joint kinematics from surface emg," *IEEE Transactions on Neural Systems and Rehabilitation Engineering*, vol. 31, pp. 484–493, 2022.
105. A. Rivadulla, X. Chen, G. Weir, D. Cazzola, G. Trewartha, J. Hamill, and E. Preatoni, "Development and validation of FootNet; a new kinematic algorithm to improve foot-strike and toe-off detection in treadmill running," *PLOS ONE*, vol. 16, p. e0248608, 2021.

106. R. Komitova, D. Raabe, R. Rein, and D. Memmert, "Time Series Data Mining for Sport Data: A Review," *International Journal of Computer Science in Sport*, vol. 21, pp. 17–31, 2022.
107. C. Kranzinger, S. Bernhart, W. Kremser, V. Venek, H. Rieser, S. Mayr, and S. Kranzinger, "Classification of Human Motion Data Based on Inertial Measurement Units in Sports: A Scoping Review," *Applied Sciences*, vol. 13, p. 8684, 2023.
108. Y. Zhang, B. V. Taylor, S. Simpson, L. Blizzard, J. A. Campbell, A. J. Palmer, and I. van der Mei, "Feelings of depression, pain and walking difficulties have the largest impact on the quality of life of people with multiple sclerosis, irrespective of clinical phenotype," *Multiple Sclerosis Journal*, vol. 27, no. 8, pp. 1262–1275, 2020.
109. A. V. Patel, J. S. Hildebrand, C. R. Leach, P. T. Campbell, C. Doyle, K. Shuval, Y. Wang, and S. M. Gapstur, "Walking in Relation to Mortality in a Large Prospective Cohort of Older U.S. Adults," *American Journal of Preventive Medicine*, vol. 54, pp. 10–19, 2018.
110. M. A. Brodie, S. R. Lord, M. J. Coppens, J. Annegarn, and K. Delbaere, "Eight-Week Remote Monitoring Using a Freely Worn Device Reveals Unstable Gait Patterns in Older Fallers," *IEEE Transactions on Biomedical Engineering*, vol. 62, pp. 2588–2594, 2015.
111. J. Verghese, R. Holtzer, R. B. Lipton, and C. Wang, "Mobility stress test approach to predicting frailty, disability, and mortality in high-functioning older adults," *Journal of the American Geriatrics Society*, vol. 60, pp. 1901–1905, 2012.
112. J. M, S. C, M. A, T. M, S.-T. A, and S. D, "Daily walking intensity as a predictor of quality of life in patients with chronic obstructive pulmonary disease.," *Medicine and Science in Sports and Exercise*, vol. 44, pp. 1212–1218, 2012.
113. R. Baker, *Measuring Walking: A Handbook of Clinical Gait Analysis*. London: Mac Keith Press, 2013.
114. B. Dumphart, D. Slijepcevic, M. Zeppelzauer, A. Kranzl, F. Unglaube, A. Baca, and B. Horsak, "Robust deep learning-based gait event detection across various pathologies," *PLOS ONE*, vol. 18, p. e0288555, 2023.
115. R. M. S. Visscher, S. Sansgiri, M. Freslier, J. Harlaar, R. Brunner, W. R. Taylor, and N. B. Singh, "Towards validation and standardization of automatic gait event identification algorithms for use in paediatric pathological populations," *Gait & Posture*, vol. 86, pp. 64–69, 2021.
116. M. Lempereur, F. Rousseau, O. Rémy-Néris, C. Pons, L. Houx, G. Quellec, and S. Brochard, "A new deep learning-based method for the detection of gait events in children with gait disorders: Proof-of-concept and concurrent validity," *Journal of Biomechanics*, vol. 98, p. 109490, 2020.
117. Ł. Kidziński, S. Delp, and M. Schwartz, "Automatic real-time gait event detection in children using deep neural networks," *PLOS ONE*, vol. 14, p. e0211466, 2019.
118. J. M. Wilken, K. M. Rodriguez, M. Brawner, and B. J. Darter, "Reliability and minimal detectible change values for gait kinematics and kinetics in healthy adults," *Gait & Posture*, vol. 35, pp. 301–307, 2012.
119. M. Geiger, A. Supiot, D. Pradon, M. C. Do, R. Zory, and N. Roche, "Minimal detectable change of kinematic and spatiotemporal parameters in patients with chronic stroke across three sessions of gait analysis," *Human Movement Science*, vol. 64, pp. 101–107, 2019.
120. Y. K. Kim, R. M. S. Visscher, E. Viehweger, N. B. Singh, W. R. Taylor, and F. Vogl, "A deep-learning approach for automatically detecting gait-events based on foot-marker kinematics in children with cerebral palsy-Which markers work best for which gait patterns?," *PLOS ONE*, vol. 17, p. e0275878, 2022.
121. B. Filtjens, A. Nieuwboer, N. D'cruz, J. Spildooren, P. Slaets, and B. Vanrumste, "A data-driven approach for detecting gait events during turning in people with Parkinson's disease and freezing of gait," *Gait & Posture*, vol. 80, pp. 130–136, 2020.
122. M. Z. Arshad, D. Lee, D. Jung, A. Jamsrandorj, J. Kim, and K.-R. Mun, "Deep Learning-Based Gait Event Prediction through a Single Waist-worn Wearable Sensor," in *2023 IEEE International Conference on Consumer Electronics (ICCE)*, pp. 1–6, 2023.
123. M. Z. Arshad, A. Jamsrandorj, J. Kim, and K.-R. Mun, "Gait Events Prediction Using Hybrid CNN-RNN-Based Deep Learning Models through a Single Waist-Worn Wearable Sensor," *Sensors*, vol. 22, p. 8226, 2022.

124. R. Romijnders, E. Warmerdam, C. Hansen, G. Schmidt, and W. Maetzler, "A Deep Learning Approach for Gait Event Detection from a Single Shank-Worn IMU: Validation in Healthy and Neurological Cohorts," *Sensors*, vol. 22, p. 3859, 2022.
125. M. Zago, M. Tarabini, M. Delfino Spiga, C. Ferrario, F. Bertozzi, C. Sforza, and M. Galli, "Machine-Learning Based Determination of Gait Events from Foot-Mounted Inertial Units," *Sensors*, vol. 21, p. 839, 2021.
126. E. Dorschky, M. Nitschke, A.-K. Seifer, A. J. van den Bogert, and B. M. Eskofier, "Estimation of gait kinematics and kinetics from inertial sensor data using optimal control of musculoskeletal models," *Journal of Biomechanics*, vol. 95, p. 109278, 2019.
127. P.-H. Lin, C.-L. Shih, D. P. Y. Wong, and P. H. Chou, "Gait Parameters Analysis Based on Leg-and-shoe-mounted IMU and Deep Learning," in *2021 International Symposium on VLSI Design, Automation and Test (VLSI-DAT)*, pp. 1–4, 2021.
128. S. T. Osis, B. A. Hettinga, and R. Ferber, "Predicting ground contact events for a continuum of gait types: An application of targeted machine learning using principal component analysis," *Gait & Posture*, vol. 46, pp. 86–90, 2016.
129. S. T. Osis, B. A. Hettinga, J. Leitch, and R. Ferber, "Predicting timing of foot strike during running, independent of striking technique, using principal component analysis of joint angles," *Journal of Biomechanics*, vol. 47, pp. 2786–2789, 2014.
130. G. Weir, C. Jewell, H. Wyatt, M. B. Trudeau, E. Rohr, G.-P. Brüggemann, and J. Hamill, "The influence of prolonged running and footwear on lower extremity biomechanics," *Footwear Science*, vol. 11, pp. 1–11, 2019.
131. E. S. Matijevich, L. M. Branscombe, L. R. Scott, and K. E. Zelik, "Ground reaction force metrics are not strongly correlated with tibial bone load when running across speeds and slopes: Implications for science, sport and wearable tech," *PLOS ONE*, vol. 14. no. 1, p. e0210000, 2019.
132. R. K. Fukuchi, C. A. Fukuchi, and M. Duarte, "A public dataset of running biomechanics and the effects of running speed on lower extremity kinematics and kinetics," *PeerJ*, vol. 5, p. e3298, 2017.
133. S. R. Donahue and M. E. Hahn, "Estimation of gait events and kinetic waveforms with wearable sensors and machine learning when running in an unconstrained environment," *Scientific Reports*, vol. 13, p. 2339, 2023.
134. P. Robberechts, R. Derie, P. Van den Berghe, J. Gerlo, D. De Clercq, V. Segers, and J. Davis, "Predicting gait events from tibial acceleration in rearfoot running: A structured machine learning approach," *Gait & Posture*, vol. 84, pp. 87–92, 2021.
135. J. McGrath, J. Neville, T. Stewart, and J. Cronin, "Upper body activity classification using an inertial measurement unit in court and field-based sports: A systematic review," *Proceedings of the Institution of Mechanical Engineers, Part P: Journal of Sports Engineering and Technology*, vol. 235, pp. 83–95, 2021.
136. R. Hart, H. Smith, and Y. Zhang, "Systematic review of automatic assessment systems for resistance-training movement performance: A data science perspective," *Computers in Biology and Medicine*, vol. 137, p. 104779, 2021.
137. W. Pei, J. Wang, X. Xu, Z. Wu, and X. Du, "An embedded 6-axis sensor based recognition for tennis stroke," in *2017 IEEE International Conference on Consumer Electronics (ICCE)*, pp. 55–58, 2017.
138. H. Zhao, S. Wang, G. Zhou, and W. Jung, "TennisEye: Tennis ball speed estimation using a racket-mounted motion sensor," in *Proceedings of the 18th International Conference on Information Processing in Sensor Networks*, IPSN '19, (New York, NY, USA), pp. 241–252, 2019.
139. D. Whiteside, O. Cant, M. Connolly, and M. Reid, "Monitoring Hitting Load in Tennis Using Inertial Sensors and Machine Learning," *International Journal of Sports Physiology and Performance*, vol. 12, pp. 1212–1217, 2017.
140. O. Alobaid, L. Ramaswamy, and K. Rasheed, "A Machine Learning Approach for Identifying Soccer Moves Using an Accelerometer Sensor," in *2018 International Conference on Computational Science and Computational Intelligence (CSCI)*, pp. 1168–1174, 2018.

141. M. Stoeve, D. Schuldhaus, A. Gamp, C. Zwick, and B. M. Eskofier, "From the Laboratory to the Field: IMU-Based Shot and Pass Detection in Football Training and Game Scenarios Using Deep Learning," *Sensors*, vol. 21, p. 3071, 2021.
142. Y. Acikmese, B. C. Ustundag, and E. Golubovic, "Towards an artificial training expert system for basketball," in *2017 10th International Conference on Electrical and Electronics Engineering (ELECO)*, pp. 1300–1304, 2017.
143. P. Jeong, M. Choe, N. Kim, J. Park, and J. Chung, "Physical Workout Classification Using Wrist Accelerometer Data by Deep Convolutional Neural Networks," *IEEE Access*, vol. 7, pp. 182406–182414, 2019.
144. L. C. Benson, A. M. Räisänen, C. A. Clermont, and R. Ferber, "Is This the Real Life, or Is This Just Laboratory? A Scoping Review of IMU-Based Running Gait Analysis," *Sensors*, vol. 22, p. 1722, 2022.
145. G. Leporace, F. Gonzalez, L. Metsavaht, M. Motta, F. P. Carpes, J. Chahla, and M. Luzo, "Are there different gait profiles in patients with advanced knee osteoarthritis? a machine learning approach," *Clinical Biomechanics*, vol. 88, p. 105447, 2021.
146. C. Dindorf, J. Konradi, C. Wolf, B. Taetz, G. Bleser, J. Huthwelker, F. Werthmann, E. Bartaguiz, P. Drees, U. Betz, and M. Fröhlich, "Visualization of interindividual differences in spinal dynamics in the presence of intraindividual variabilities," in *LEVIA'22 : Leipzig Symposium on Visualization in Applications 2022* (C. Gillmann, J. Schmidt, S. Jänicke, and D. Wiegreffe, eds.), (Leipzig), Leipzig University, 2022.
147. P. Krondorfer, D. Slijepčević, F. Unglaube, A. Kranzl, C. Breiteneder, M. Zeppelzauer, and B. Horsak, "Deep learning-based similarity retrieval in clinical 3d gait analysis," *Gait & Posture*, vol. 90, pp. 127–128, 2021.
148. X. Li, J. Liu, Y. Huang, D. Wang, and Y. Miao, "Human motion pattern recognition and feature extraction: An approach using multi-information fusion," *Micromachines*, vol. 13, no. 8, 2022.
149. D. Xu, W. Quan, H. Zhou, D. Sun, J. S. Baker, and Y. Gu, "Explaining the differences of gait patterns between high and low-mileage runners with machine learning," *Scientific Reports*, vol. 12, no. 1, p. 2981, 2022.
150. A. Hartebrodt and R. Röttger, "Federated horizontally partitioned principal component analysis for biomedical applications," *Bioinformatics Advances*, vol. 2, no. 1, pp. 1–10, 2022.
151. H. Uğuz, "A biomedical system based on artificial neural network and principal component analysis for diagnosis of the heart valve diseases," *Journal of Medical Systems*, vol. 36, no. 1, pp. 61–72, 2012.
152. M. Köhle and D. Merkl, "Identification of gait patterns with self-organizing maps based on ground reaction force," *Proc. of the European Symposium on Artificial Neural Networks (ESANN'96)*, pp. 73–78, 1996.
153. G. Hajian, A. Etemad, and E. Morin, "An investigation of dimensionality reduction techniques for emg-based force estimation," in *2019 41st Annual International Conference of the IEEE Engineering in Medicine and Biology Society (EMBC)*, IEEE, pp. 698–701, 2019.
154. C. de La Fuente, R. Silvestre, R. Yañez, M. Roby, M. Soldán, W. Ferrada, and F. P. Carpes, "Preseason multiple biomechanics testing and dimension reduction for injury risk surveillance in elite female soccer athletes: short-communication," *Science & Medicine in Football*, vol. 7, no. 2, pp. 183–188, 2023.
155. V. Hernandez, D. Kulić, and G. Venture, "Adversarial autoencoder for visualization and classification of human activity: Application to a low-cost commercial force plate," *Journal of Biomechanics*, vol. 103, p. 109684, 2020.
156. K. L. Young-Shand, P. C. Roy, M. J. Dunbar, S. S. R. Abidi, and J. L. Astephen Wilson, "Gait biomechanics phenotypes among total knee arthroplasty candidates by machine learning cluster analysis," *Journal of Orthopaedic Research : Official Publication of the Orthopaedic Research Society*, vol. 41, no. 2, pp. 335–344, 2023.
157. H. N. A. Halim and A. Azaman, "Clustering-based support vector machine (svm) for symptomatic knee osteoarthritis severity classification," in *Proceedings of the 2022 9th International Conference on Biomedical and Bioinformatics Engineering*, pp. 140–146, 2022.

158. F. Su, W. Wu, Y. Cheng, and Y. Chou, “Fuzzy clustering of gait patterns of patients after ankle arthrodesis based on kinematic parameters,” *Medical Engineering & Physics*, vol. 23, no. 2, pp. 83–90, 2001.
159. S. Qian, S.-C. Yen, E. Folmar, and C.-A. Chou, “Self-expressive subspace clustering to recognize motion dynamics for chronic ankle instability,” *IISE Transactions on Healthcare Systems Engineering*, vol. 10, no. 1, pp. 60–73, 2020.
160. D. Ichimura, R. Amma, G. Hisano, H. Murata, and H. Hobara, “Spatiotemporal gait patterns in individuals with unilateral transfemoral amputation: A hierarchical cluster analysis,” *PloS one*, vol. 17, no. 12, p. e0279593, 2022.
161. K. Kaczmarczyk, A. Wit, M. Krawczyk, and J. Zaborski, “Gait classification in post-stroke patients using artificial neural networks,” *Gait & Posture*, vol. 30, no. 2, pp. 207–210, 2009.
162. S. Hoerzer, V. von Tscharner, C. Jacob, and B. M. Nigg, “Defining functional groups based on running kinematics using self-organizing maps and support vector machines,” *Journal of Biomechanics*, vol. 48, no. 10, pp. 2072–2079, 2015.
163. J. Pardo Albiach, M. Mir-Jimenez, V. Hueso Moreno, I. Nácher Moltó, and J. Martínez-Gramage, “The relationship between vo2max, power management, and increased running speed: Towards gait pattern recognition through clustering analysis,” *Sensors*, vol. 21, no. 7, 2021.
164. M. Aljohani and K. Kipp, “Use of self-organizing maps to study sex- and speed-dependent changes in running biomechanics,” *Human Movement Science*, vol. 72, p. 102649, 2020.
165. A. M. Senevirathna, A. J. Pohl, M. J. Jordan, W. B. Edwards, and R. Ferber, “Differences in kinetic variables between injured and uninjured rearfoot runners: A hierarchical cluster analysis,” *Scandinavian Journal of Medicine & Science in Sports*, vol. 33, no. 2, pp. 160–168, 2023.
166. S. Jauhiainen, A. J. Pohl, S. Äyrämö, J.-P. Kauppi, and R. Ferber, “A hierarchical cluster analysis to determine whether injured runners exhibit similar kinematic gait patterns,” *Scandinavian Journal of Medicine & Science in Sports*, vol. 30, no. 4, pp. 732–740, 2020.
167. K. J. Deluzio and J. L. Astephen, “Biomechanical features of gait waveform data associated with knee osteoarthritis: an application of principal component analysis,” *Gait & Posture*, vol. 25, no. 1, pp. 86–93, 2007.
168. N. Welch, C. Richter, A. Franklyn-Miller, and K. Moran, “Principal component analysis of the biomechanical factors associated with performance during cutting,” *Journal of Strength and Conditioning Research*, vol. 35, no. 6, pp. 1715–1723, 2021.
169. B. Serrien, M. Goossens, and J.-P. Baeyens, “Issues in using self-organizing maps in human movement and sport science,” *International Journal of Computer Science in Sport*, vol. 16, no. 1, pp. 1–17, 2017.
170. S. M. Remedios, D. P. Armstrong, R. B. Graham, and S. L. Fischer, “Exploring the application of pattern recognition and machine learning for identifying movement phenotypes during deep squat and hurdle step movements,” *Frontiers in Bioengineering and Biotechnology*, vol. 8, p. 364, 2020.
171. T. C. Phan, A. Pranata, J. Farragher, A. Bryant, H. T. Nguyen, and R. Chai, “Machine learning derived lifting techniques and pain self-efficacy in people with chronic low back pain,” *Sensors*, vol. 22, no. 17, 2022.
172. J. Sarvestan, Z. Svoboda, F. Alaei, and F. Mulloy, “Analysis of whole-body coordination patterning in successful and faulty spikes using self-organising map-based cluster analysis: A secondary analysis,” *Sensors*, vol. 21, no. 4, 2021.
173. S. Kovalchik and M. Reid, “A shot taxonomy in the era of tracking data in professional tennis,” *Journal of Sports Sciences*, vol. 36, no. 18, pp. 2096–2104, 2018.
174. L. Liu, S. Qiu, Z. Wang, J. Li, and J. Wang, “Canoeing motion tracking and analysis via multi-sensors fusion,” *Sensors*, vol. 20, no. 7, p. 2110, 2020.
175. D. A. Bonilla, J. O. Peralta-Alzate, J. A. Bonilla-Henao, W. Urrutia-Mosquera, R. Cannataro, J. Kočí, and J. L. Petro, “Unsupervised machine learning analysis of the anthropometric characteristics and maturity status of young colombian athletes,” *Journal of Physical Education and Sport*, vol. 22, no. 1, pp. 256–265, 2022.

176. B. Giles, P. Peeling, S. Kovalchik, and M. Reid, "Differentiating movement styles in professional tennis: A machine learning and hierarchical clustering approach," *European Journal of Sport Science*, vol. 23, no. 1, pp. 44–53, 2023.
177. J. Sarvestan, E. Shirzad, A. R. Arshi, and F. Mousavian, "Analysis of change of direction performance with dominant and non-dominant leg using linear and nonlinear approaches," *Sports Biomechanics*, pp. 1–17, 2022.
178. D. Whiteside and M. Reid, "Spatial characteristics of professional tennis serves with implications for serving aces: A machine learning approach," *Journal of Sports Sciences*, vol. 35, no. 7, pp. 648–654, 2017.
179. A. Baca and P. Kornfeind, "Stability analysis of motion patterns in biathlon shooting," *Human Movement Science*, vol. 31, no. 2, pp. 295–302, 2012.
180. F. Horst, S. Lapuschkin, W. Samek, K.-R. Müller, and W. I. Schöllhorn, "Explaining the unique nature of individual gait patterns with deep learning," *Scientific Reports*, vol. 9, no. 1, p. 2391, 2019.
181. F. Hug, C. Vogel, K. Tucker, S. Dorel, T. Deschamps, E. Le Carpentier, and L. Lacourpaille, "Individuals have unique muscle activation signatures as revealed during gait and pedaling," *Journal of Applied Physiology*, vol. 127, no. 4, pp. 1165–1174, 2019.
182. F. Horst, M. Mildner, and W. Schöllhorn, "One-year persistence of individual gait patterns identified in a follow-up study—a call for individualised diagnose and therapy," *Gait & Posture*, vol. 58, pp. 476–480, 2017.
183. R. Begg and J. Kamruzzaman, "A machine learning approach for automated recognition of movement patterns using basic, kinetic and kinematic gait data," *Journal of Biomechanics*, vol. 38, no. 3, pp. 401–408, 2005.
184. Y. Zhou, R. Romijnders, C. Hansen, J. van Campen, W. Maetzler, T. Hortobágyi, and C. J. C. Lamoth, "The detection of age groups by dynamic gait outcomes using machine learning approaches," *Scientific Reports*, vol. 10, no. 1, p. 4426, 2020.
185. J.-H. Yoo, D. Hwang, and M. S. Nixon, "Gender classification in human gait using support vector machine," in *Advanced Concepts for Intelligent Vision Systems* (J. Blanc-Talon, W. Philips, D. Popescu, and P. Scheunders, eds.), pp. 138–145, Springer, Berlin, Heidelberg, 2005.
186. R. K. Pathan, M. A. Uddin, N. Nahar, F. Ara, M. S. Hossain, and K. Andersson, "Gender classification from inertial sensor-based gait dataset," in *Intelligent Computing and Optimization* (P. Vasant, I. Zelinka, and G.-W. Weber, eds.), (Cham), pp. 583–596, Springer International Publishing, 2021.
187. J. Figueiredo, C. P. Santos, and J. C. Moreno, "Automatic recognition of gait patterns in human motor disorders using machine learning: A review," *Medical Engineering & Physics*, vol. 53, pp. 1–12, 2018.
188. H. Y. Lau, K. Y. Tong, and H. Zhu, "Support vector machine for classification of walking conditions of persons after stroke with dropped foot," *Human Movement Science*, vol. 28, no. 4, pp. 504–514, 2009.
189. F. Wahid, R. K. Begg, C. J. Hass, S. Halgamuge, and D. C. Ackland, "Classification of parkinson's disease gait using spatial-temporal gait features," *IEEE Journal of Biomedical and Health Informatics*, vol. 19, no. 6, pp. 1794–1802, 2015.
190. M. Alaqtash, T. Sarkodie-Gyan, H. Yu, O. Fuentes, R. Brower, and A. Abdelgawad, "Automatic classification of pathological gait patterns using ground reaction forces and machine learning algorithms," in *2011 Annual International Conference of the IEEE Engineering in Medicine and Biology Society*, pp. 453–457, 2011.
191. D. Slijepcevic, M. Zeppelzauer, F. Unglaube, A. Kranzl, C. Breiteneder, and B. Horsak, "Explainable machine learning in human gait analysis: A study on children with cerebral palsy," *IEEE Access*, vol. 11, pp. 65906–65923, 2023.
192. C. Nüesch, V. Valderrabano, C. Huber, V. von Tscharner, and G. Pagenstert, "Gait patterns of asymmetric ankle osteoarthritis patients," *Clinical Biomechanics*, vol. 27, no. 6, pp. 613–618, 2012.

193. J. Christian, J. Kröll, G. Strutzenberger, N. Alexander, M. Ofner, and H. Schwameder, "Computer aided analysis of gait patterns in patients with acute anterior cruciate ligament injury," *Clinical Biomechanics*, vol. 33, pp. 55–60, 2016.
194. T. E. Lockhart, R. Soangra, H. Yoon, T. Wu, C. W. Frames, R. Weaver, and K. A. Roberto, "Prediction of fall risk among community-dwelling older adults using a wearable system," *Scientific Reports*, vol. 11, no. 1, p. 20976, 2021.
195. B. Filtjens, P. Ginis, A. Nieuwboer, M. R. Afzal, J. Spildooren, B. Vanrumste, and P. Slaets, "Modelling and identification of characteristic kinematic features preceding freezing of gait with convolutional neural networks and layer-wise relevance propagation," *BMC Medical Informatics and Decision Making*, vol. 21, no. 1, p. 341, 2021.
196. G. Shalin, S. Pardoel, E. D. Lemaire, J. Nantel, and J. Kofman, "Prediction and detection of freezing of gait in parkinson's disease from plantar pressure data using long short-term memory neural-networks," *Journal of NeuroEngineering and Rehabilitation*, vol. 18, p. 167, 2021.
197. F. Hoitz, V. von Tscharner, J. Baltich, and B. M. Nigg, "Individuality decoded by running patterns: Movement characteristics that determine the uniqueness of human running," *PloS one*, vol. 16, no. 4, p. e0249657, 2021.
198. R. K. Fukuchi, B. M. Eskofier, M. Duarte, and R. Ferber, "Support vector machines for detecting age-related changes in running kinematics," *Journal of Biomechanics*, vol. 44, no. 3, pp. 540–542, 2011.
199. A. Phinyomark, B. A. Hettinga, S. T. Osis, and R. Ferber, "Gender and age-related differences in bilateral lower extremity mechanics during treadmill running," *PLOS ONE*, vol. 9, pp. 1–10, 08 2014.
200. C. Maurer, P. Federolf, V. von Tscharner, L. Stirling, and B. M. Nigg, "Discrimination of gender-, speed-, and shoe-dependent movement patterns in runners using full-body kinematics," *Gait & Posture*, vol. 36, no. 1, pp. 40–45, 2012.
201. M. B. Trudeau, V. Von Tscharner, J. Vienneau, S. Hoerzer, and B. M. Nigg, "Assessing footwear effects from principal features of plantar loading during running," *Medicine & Science in Sports & Exercise*, vol. 47, no. 9, 2015.
202. P. Dixon, K. Schütte, B. Vanwanseele, J. Jacobs, J. Dennerlein, J. Schiffman, P.-A. Fournier, and B. Hu, "Machine learning algorithms can classify outdoor terrain types during running using accelerometry data," *Gait & Posture*, vol. 74, pp. 176–181, 2019.
203. N. U. Ahamed, D. Kobsar, L. C. Benson, C. A. Clermont, S. T. Osis, and R. Ferber, "Subject-specific and group-based running pattern classification using a single wearable sensor," *Journal of Biomechanics*, vol. 84, pp. 227–233, 2019.
204. N. U. Ahamed, D. Kobsar, L. Benson, C. Clermont, R. Kohrs, S. T. Osis, and R. Ferber, "Using wearable sensors to classify subject-specific running biomechanical gait patterns based on changes in environmental weather conditions," *PLOS ONE*, vol. 13, pp. 1–15, 2018.
205. L. Marotta, J. H. Buurke, B.-J. F. van Beijnum, and J. Reenalda, "Towards machine learning-based detection of running-induced fatigue in real-world scenarios: Evaluation of imu sensor configurations to reduce intrusiveness," *Sensors*, vol. 21, no. 10, 2021.
206. D. Koska and C. Maiwald, "A time-series based framework for exploring the unknown effects of shoe comfort-induced biomechanical adaptations," *Footwear Science*, vol. 12, no. 2, pp. 113–122, 2020.
207. C. A. Clermont, L. C. Benson, S. T. Osis, D. Kobsar, and R. Ferber, "Running patterns for male and female competitive and recreational runners based on accelerometer data," *Journal of Sports Sciences*, vol. 37, no. 2, pp. 204–211, 2019.
208. Q. Liu, S. Mo, V. C. K. Cheung, B. M. F. Cheung, S. Wang, P. P. K. Chan, A. Malhotra, R. T. H. Cheung and R. H. M. Chan, "Classification of runners' performance levels with concurrent prediction of biomechanical parameters using data from inertial measurement units," *Journal of Biomechanics*, vol. 112, p. 110072, 2020.
209. J. Aeles, F. Horst, S. Lapuschkin, L. Lacourpaille, and F. Hug, "Revealing the unique features of each individual's muscle activation signatures," *Journal of The Royal Society Interface*, vol. 18, no. 174, p. 20200770, 2021.

210. J. Burdack, F. Horst, D. Aragonés, A. Eekhoff, and W. I. Schöllhorn, "Fatigue-related and timescale-dependent changes in individual movement patterns identified using support vector machine," *Frontiers in Psychology*, vol. 11, p. 551548, 2020.
211. F. Horst, D. Janssen, H. Beckmann, and W. I. Schöllhorn, "Can individual movement characteristics across different throwing disciplines be identified in high-performance decathletes?," *Frontiers in Psychology*, vol. 11, p. 2262, 2020.
212. H. Novatchkov and A. Baca, "Artificial intelligence in sports on the example of weight training," *Journal of Sports Science & Medicine*, vol. 12, no. 1, p. 27, 2013.
213. B. H. Groh, T. Kautz, D. Schuldhaus, and B. M. Eskofier, "Imu-based trick classification in skateboarding," in *KDD Workshop on Large-scale Sports Analytics*, vol. 17, 2015.
214. B. H. Groh, M. Fleckenstein, and B. M. Eskofier, "Wearable trick classification in freestyle snowboarding," in *2016 IEEE 13th International Conference on Wearable and Implantable Body Sensor Networks (BSN)*, pp. 89–93, 2016.
215. T. Kautz, B. H. Groh, J. Hannink, U. Jensen, H. Strubberg, and B. M. Eskofier, "Activity recognition in beach volleyball using a deep convolutional neural network," *Data Mining and Knowledge Discovery*, vol. 31, pp. 1678–1705, 2017.
216. L. Woltmann, C. Hartmann, W. Lehner, P. Rausch, and K. Ferger, "Sensor-based jump detection and classification with machine learning in trampoline gymnastics," *German Journal of Exercise and Sport Research*, vol. 53, pp. 187–195, 2023.
217. C. Dindorf, J. Konradi, C. Wolf, B. Taetz, G. Bleser, J. Huthwelker, F. Werthmann, E. Bartaguiz, J. Kniepert, P. Drees, U. Betz, and M. Fröhlich, "Classification and automated interpretation of spinal posture data using a pathology-independent classifier and explainable artificial intelligence (xai)," *Sensors*, vol. 21, no. 18, p. 6323, 2021.
218. W. Teufl, B. Taetz, M. Miezal, C. Dindorf, M. Fröhlich, U. Trinler, A. Hogan, and G. Bleser, "Automated detection and explainability of pathological gait patterns using a one-class support vector machine trained on inertial measurement unit based gait data," *Clinical Biomechanics (Bristol, Avon)*, vol. 89, p. 105452, 2021.
219. K. Hempstalk and E. Frank, "Discriminating against new classes: One-class versus multi-class classification"," in *AI 2008: Advances in Artificial Intelligence* (W. Wobcke and M. Zhang, eds.), pp. 325–336, Springer, Berling, Heidelberg, 2008.
220. European Union, "Regulation (EU) 2016/679 of the european parliament and of the council of 27 april 2016 on the protection of natural persons with regard to the processing of personal data and on the free movement of such data, and repealing directive 95/46/ec (General Data Protection Regulation)," *Official Journal of the European Union*, vol. L 119, pp. 1–88, 2016. Available at: https://eur-lex.europa.eu/eli/reg/2016/679/oj.
221. V. Arya, R. K. Bellamy, P.-Y. Chen, A. Dhurandhar, M. Hind, S. C. Hoffman, S. Houde, Q. V. Liao, R. Luss, A. Mojsilović, *et al.*, "One explanation does not fit all: A toolkit and taxonomy of ai explainability techniques," arXiv, e1909.03012, 2019.
222. T. C. Pataky, "Generalized n-dimensional biomechanical field analysis using statistical parametric mapping," *Journal of Biomechanics*, vol. 43, no. 10, pp. 1976–1982, 2010.
223. S. B. Hausmann, A. M. Vargas, A. Mathis, and M. W. Mathis, "Measuring and modeling the motor system with machine learning," *Current Opinion in Neurobiology*, vol. 70, pp. 11–23, 2021.
224. M. Wagner, D. Slijepcevic, B. Horsak, A. Rind, M. Zeppelzauer, and W. Aigner, "Kavagait: Knowledge-assisted visual analytics for clinical gait analysis," *IEEE Transactions on Visualization and Computer Graphics*, vol. 25, no. 3, pp. 1528–1542, 2018.
225. S. Lapuschkin, S. Wäldchen, A. Binder, G. Montavon, W. Samek, and K.-R. Müller, "Unmasking clever hans predictors and assessing what machines really learn," *Nature Communications*, vol. 10, no. 1, p. 1096, 2019.
226. F. Horst, D. Slijepcevic, M. Zeppelzauer, A.-M. Raberger, S. Lapuschkin, W. Samek, W. I. Schöllhorn, C. Breiteneder, and B. Horsak, "Explaining automated gender classification of human gait," *Gait & Posture*, vol. 81, pp. 159–160, 2020.

227. D. Slijepcevic, F. Horst, M. Simak, S. Lapuschkin, A.-M. Raberger, W. Samek, C. Breiteneder, W. I. Schöllhorn, M. Zeppelzauer, and B. Horsak, "Explaining machine learning models for age classification in human gait analysis," *Gait & Posture*, vol. 97, pp. S252–S253, 2022.
228. D. Slijepcevic, F. Horst, M. Simak, W. I. Schöllhorn, M. Zeppelzauer, and B. Horsak, "Towards personalized gait rehabilitation: How robustly can we identify personal gait signatures with machine learning?," *Gait & Posture*, vol. 106, pp. S192–S193, 2023.
229. A. Rind, D. Slijepčević, M. Zeppelzauer, F. Unglaube, A. Kranzl, and B. Horsak, "Trustworthy visual analytics in clinical gait analysis: A case study for patients with cerebral palsy," in *2022 IEEE Workshop on TRust and EXpertise in Visual Analytics (TREX)*, pp. 8–15, IEEE, 2022.
230. R. R. Selvaraju, M. Cogswell, A. Das, R. Vedantam, D. Parikh, and D. Batra, "Grad-cam: Visual explanations from deep networks via gradient-based localization," in *Proceedings of the IEEE International Conference on Computer Vision*, pp. 618–626, 2017.
231. D. Slijepcevic, M. Zeppelzauer, F. Unglaube, A. Kranzl, C. Breiteneder, and B. Horsak, "Explainable machine learning in human gait analysis: A study on children with cerebral palsy," *IEEE Access*, vol. 10, pp. 1–10, 2023.
232. C. Dindorf, W. Teufl, B. Taetz, G. Bleser, and M. Fröhlich, "Interpretability of input representations for gait classification in patients after total hip arthroplasty," *Sensors*, vol. 20, no. 16, p. 4385, 2020.
233. M. T. Ribeiro, S. Singh, and C. Guestrin, ""why should i trust you?" explaining the predictions of any classifier," in *Proceedings of the 22nd ACM SIGKDD international conference on knowledge discovery and data mining*, pp. 1135–1144, 2016.
234. M. Özates, F. Salami, S. Wolf, S. Campos, S. Hagmann, U. Schneider, and A. Yaman, "Application of an explainable ai method in classification of functional foot conditions," *Gait & Posture*, vol. 90, pp. 172–173, 2021.
235. C. Kokkotis, S. Moustakidis, T. Tsatalas, C. Ntakolia, G. Chalatsis, S. Konstadakos, M. E. Hantes, G. Giakas, and D. Tsaopoulos, "Leveraging explainable machine learning to identify gait biomechanical parameters associated with anterior cruciate ligament injury," *Scientific Reports*, vol. 12, no. 1, p. 6647, 2022.
236. S. M. Lundberg and S.-I. Lee, "A unified approach to interpreting model predictions," *Advances in Neural Information Processing Systems*, vol. 30, 2017.
237. C. Dindorf, O. Ludwig, S. Simon, S. Becker, and M. Fröhlich, "Machine learning and explainable artificial intelligence using counterfactual explanations for evaluating posture parameters," *Bioengineering*, vol. 10, no. 5, p. 511, 2023.
238. F. Hoitz, L. Fraeulin, V. von Tscharner, D. Ohlendorf, B. M. Nigg, and C. Maurer-Grubinger, "Isolating the unique and generic movement characteristics of highly trained runners," *Sensors*, vol. 21, no. 21, p. 7145, 2021.
239. F. Horst, F. Hoitz, D. Slijepcevic, N. Schons, H. Beckmann, B. M. Nigg, and W. I. Schöllhorn, "Identification of subject-specific responses to footwear during running," *Scientific Reports*, vol. 13, no. 1, p. 11284, 2023.
240. G. Lisca, C. Prodaniuc, T. Grauschopf, and C. Axenie, "Less is more: learning insights from a single motion sensor for accurate and explainable soccer goalkeeper kinematics," *IEEE Sensors Journal*, vol. 21, no. 18, pp. 20375–20387, 2021.
241. T. Chen and C. Guestrin, "Xgboost: A scalable tree boosting system," in *Proceedings of the 22nd Acm Sigkdd International Conference on Knowledge Discovery and Data Mining*, pp. 785–794, 2016.
242. L. Woltmann, C. Hartmann, W. Lehner, P. Rausch, and K. Ferger, "Sensor-based jump detection and classification with machine learning in trampoline gymnastics," *German Journal of Exercise and Sport Research*, vol. 53, no. 2, pp. 187–195, 2023.
243. T. Baumhauer, D. Slijepcevic, and M. Zeppelzauer, "Bounded logit attention: Learning to explain image classifiers," in *NeurIPS'22 Workshop on All Things Attention: Bridging Different Perspectives on Attention*, 2022.

244. P.-F. David, R.-C. David, M. C. Juan, and T. Diego, “Human locomotion databases. a systematic review.,” *IEEE Journal of Biomedical and Health Informatics*, vol. 28. no. 3, pp. 1716–1729, 2023.
245. J. K. Moore, S. K. Hnat, and A. J. van den Bogert, “An elaborate data set on human gait and the effect of mechanical perturbations,” *PeerJ*, vol. 3, p. e918, 2015.
246. Luo Yue, Sarah Coppola, Philippe Dixon, Song Li, Jack Dennerlein, and Boyi Hu, “A database of human gait performance on irregular and uneven surfaces collected by wearable sensors.” *Scientific Data*, vol. 7, no. 1, p. 219, 2020.
247. C. A. Fukuchi, R. K. Fukuchi, and M. Duarte, “A public dataset of overground and treadmill walking kinematics and kinetics in healthy individuals,” *PeerJ*, vol. 6, p. e4640, 2018.
248. T. Lencioni, I. Carpinella, M. Rabuffetti, A. Marzegan, and M. Ferrarin, “Human kinematic, kinetic and emg data during different walking and stair ascending and descending tasks,” *Scientific Data*, vol. 6, no. 1, p. 309, 2019.
249. C. Schreiber and F. Moissenet, “A multimodal dataset of human gait at different walking speeds established on injury-free adult participants,” *Scientific Data*, vol. 6, no. 1, p. 111, 2019.
250. D. Camara Miraldo, R. Naville Watanabe, and M. Duarte, “An open data set of inertial, magnetic, foot-ground contact, and electromyographic signals from wearable sensors during walking,” *Motor Control*, vol. 24, no. 4, pp. 558–570, 2020.
251. B. Horsak, D. Slijepcevic, A.-M. Raberger, C. Schwab, M. Worisch, and M. Zeppelzauer, “GaitRec, a large-scale ground reaction force dataset of healthy and impaired gait,” *Scientific Data*, vol. 7, p. 143, 2020.
252. F. Horst, D. Slijepcevic, M. Simak, and W. I. Schöllhorn, “Gutenberg gait database, a ground reaction force database of level overground walking in healthy individuals,” *Scientific Data*, vol. 8, p. 232, 2021.
253. M. Derlatka and M. Parfieniuk, “Real-world measurements of ground reaction forces of normal gait of young adults wearing various footwear,” *Scientific Data*, vol. 10, no. 1, p. 60, 2023.
254. C. Dindorf, J. Konradi, C. Wolf, B. Taetz, G. Bleser, J. Huthwelker, F. Werthmann, P. Drees, M. Fröhlich, and U. Betz, “Machine learning techniques demonstrating individual movement patterns of the vertebral column: the fingerprint of spinal motion,” *Computer Methods in Biomechanics and Biomedical Engineering*, vol. 25, no. 7, pp. 821–831, 2022.
255. K. A. Duncanson, S. Thwaites, D. Booth, G. Hanly, W. S. Robertson, E. Abbasnejad, and D. Thewlis, “Deep metric learning for scalable gait-based person re-identification using force platform data,” *Sensors*, vol. 23, no. 7, p. 3392, 2023.
256. R. Estévez-García, J. Martín-Gutiérrez, S. M. Mendoza, J. R. Marante, P. Chinea-Martín, O. Soto-Martín, and M. Lodeiro-Santiago, “Open data motion capture: Mocap-ull database,” *Procedia Computer Science*, vol. 75, pp. 316–326, 2015.
257. S. Khandelwal and N. Wickström, “Evaluation of the performance of accelerometer-based gait event detection algorithms in different real-world scenarios using the marea gait database,” *Gait & Posture*, vol. 51, pp. 84–90, 2017.
258. R. Chereshnev and A. Kertész-Farkas, “Hugadb: Human gait database for activity recognition from wearable inertial sensor networks,” in *Analysis of Images, Social Networks and Texts: 6th International Conference, AIST 2017, Moscow, Russia, July 27–29, 2017, Revised Selected Papers 6*, pp. 131–141, Springer, 2018.
259. S. Ghorbani, K. Mahdaviani, A. Thaler, K. Kording, D. J. Cook, G. Blohm, and N. F. Troje, “Movi: A large multi-purpose human motion and video dataset,” *PLOS ONE*, vol. 16, no. 6, p. e0253157, 2021.
260. D. Slijepcevic, M. Zeppelzauer, C. Schwab, A.-M. Raberger, C. Breiteneder, and B. Horsak, “Input representations and classification strategies for automated human gait analysis,” *Gait & Posture*, vol. 76, pp. 198–203, 2020.
261. Association for Computing Machinery, “Artifact review and badging—current,” 2020.
262. Uchida, Thomas K and Delp, Scott L, “Biomechanics of movement: the science of sports, robotics, and rehabilitation,” Mit Press, 2021.

263. Horsak, Brian and Kainz, Hans and Dumphart, Bernhard, “Repeatability and minimal detectable change including clothing effects for smartphone-based 3D markerless motion capture,” *Journal of Biomechanics*, vol. 175, p. 112281, Elsevier, 2024.
264. Slijepcevic, Djordje and Krondorfer, Philipp and Unglaube, Fabian and Kranzl, Andreas and Zeppelzauer, Matthias and Horsak, Brian, “Predicting ground reaction forces in overground walking from gait kinematics using machine learning,” *Gait & Posture*, vol. 113, pp. 214–215, Elsevier, 2024.
265. Krondorfer, Philipp and Slijepčević, Djordje and Unglaube, Fabian and Kranzl, Andreas and Zeppelzauer, Matthias and Kainz, Hans and Horsak, Brian, “Predicting knee contact forces in walking: A comparative study of machine learning models including a physics-informed approach,” *Gait & Posture*, vol. 113, pp. 125–126, Elsevier, 2024.
266. Slijepcevic, Djordje and Horst, Fabian and Zeppelzauer, Matthias and Schöllhorn, Wolfgang Immanuel, “Individual Responses to Running Shoes: An Investigation Using Unsupervised Machine Learning,” *ISBS Proceedings Archive*, vol. 42, no. 1, p. 858, 2024.

Artificial Intelligence and Machine Learning-Based Data Analytics for Sports: General Overview and NBA Case Study

Akemi Gálvez, Vei S. Chan, Sara Pérez-Carabaza, and Andrés Iglesias

Abstract Artificial intelligence and machine learning are two of the most groundbreaking and disruptive computing technologies in today's digital world. Recent developments in these fields, such as deep learning, reinforcement learning, and others, are having a profound impact in many domains, including sports. On the other hand, data analytics represents a new paradigm for processing and analyzing the massive amount of data on sports that can be obtained today through cameras, sensors, GPS, wearables, and other devices. Combining these three technologies is revolutionizing the way we approach sports, from analyzing player performance to developing new training methods, preventing injuries, optimizing game strategies, engaging spectators, and many others. The objective of this chapter is twofold: on the one hand, it provides a general overview about the historical evolution, uses, tools, and applications of artificial intelligence and machine learning-based data analytics to the world of sports in a comprehensive way; on the other hand, it presents a more detailed analysis of a case study—the world of NBA professional basketball. Several illustrative examples about the use of these technologies in sports are also presented. As these technologies continue to evolve, we can expect to see even more innovative applications of these fields in sports for the years to come.

Keywords Artificial intelligence · Machine learning · Data analytics · Sports analytics · Game strategy Optimization · NBA gameplay optimization

A. Gálvez · S. Pérez-Carabaza · A. Iglesias (✉)
Department of Applied Mathematics and Computational Sciences, University of Cantabria, Santander, Spain
e-mail: galveza@unican.es; sara.perezcarabaza@unican.es; iglesias@unican.es

V. S. Chan
Faculty of Computing, University of Technology Malaysia, Johor Bahru, Malaysia
e-mail: vschan2@live.utm.my

M. J. Blondin et al. (eds.), *Artificial Intelligence, Optimization, and Data Sciences in Sports*, Springer Optimization and Its Applications 218,
https://doi.org/10.1007/978-3-031-76047-1_5

1 Introduction

Artificial intelligence (AI) and machine learning (ML) are two of the most groundbreaking and disruptive computing technologies in today's digital world. They are paving the way for some of the most impressive technological advances arising during the last two decades. Recent developments in these fields, such as deep learning, reinforcement learning, and others, are having a profound impact in many domains, including sports. On the other hand, data analytics (DA) represents a new paradigm for processing and analyzing the massive amount of data that can be obtained from a myriad of different data-acquiring devices, including sensors, cameras, GPS (Global Positioning System), wearables, and many others. The combination of these three powerful areas, artificial intelligence, machine learning, and data analytics, is revolutionizing our approach to sports in many different ways, opening the door to breakthrough developments, advances, and applications that were unimaginable just a few short years ago. From analyzing player's performance to developing new training methods, preventing injuries, optimizing game strategies, engaging spectators, and many others, the world of sports is changing drastically as these new digital tools are increasingly being introduced into sports practice at both amateur and professional level, as well as in the media and sports broadcasting.

This chapter discusses how artificial intelligence, machine learning, and data analytics are currently being applied in the field of sports and how these changes are modifying the current landscape of sports. The objective of this chapter is twofold: on the one hand, it provides a general overview about the historical evolution, uses, tools, and applications of artificial intelligence and machine learning-based data analytics to the world of sports in a comprehensive way; on the other hand, it presents a more detailed analysis of a case study—the world of NBA (National Basketball Association) professional basketball. Several illustrative examples about the use of these technologies in the NBA and in other team sports are also presented.

The structure of this chapter is as follows: in Sect. 2 the computational tools pertaining to sports analytics (artificial intelligence, machine learning, and data analytics) are briefly discussed. To enrich the discussion, our description includes more than 100 bibliographic entries reporting relevant real-world applications of AI, ML, and DA in several fields other than sports. Then, their applications to the world of sports, commonly referred to as sports analytics, are presented in Sect. 3. Some challenges and limitations of sports analytics are also discussed in that section. Section 4 presents an in-depth analysis of a case study: the world of sports analytics in the American professional basketball league NBA. The chapter closes with the main conclusions in Sect. 5.

2 The Computational Tools

The computational tools considered in this chapter are artificial intelligence, machine learning, and data analytics. They are briefly described in this section. In addition, some important applications of these techniques in several fields are also outlined. This description is in no way intended to be exhaustive. For further details, the reader is kindly referred to the bibliographic entries included in the section.

Artificial intelligence (AI) consists of the development of algorithms, methods, and computer systems that can perform tasks that typically require human intelligence, such as learning, reasoning, problem-solving, perception, language understanding, and decision-making [97, 147, 167]. Currently, AI is widely used for many advanced applications, playing a significant and evolving role in our current digital society across various sectors and industries. At its turn, *machine learning* (ML), while generally considered to be a subset of artificial intelligence, is specifically focused on the development of algorithms and statistical models that enable computers to improve their performance on a specific task through learning from data, without being explicitly programmed [3, 43, 66]. In other words, instead of being programmed with explicit instructions, a machine learning system learns from examples and experiences to make predictions or decisions [125, 126, 161]. Nowadays, machine learning is increasingly used owing to its ability to handle complex, data-driven challenges, automate decision-making, and provide knowledge or insights that were previously impossible or very difficult to achieve [88, 189].

Within the domain of ML, there are primarily two paradigms: supervised and unsupervised learning. Supervised learning operates with labeled data, where both the inputs and the desired outputs are known. This allows the model to learn to predict the outcome based on the given input. For instance, it can predict the success of a basketball shot based on data from previous shots. On the other hand, unsupervised learning works with unlabeled data, seeking to uncover hidden structures or patterns. In a sports context, it could be used to identify clusters of players with similar playing styles without having predefined categories, or clusters of fans to enhance their experience and engagement. Both paradigms offer invaluable tools to gain insights and enhance decision-making in sports.

Data analytics (DA) is an important and increasingly popular field of research that consists of examining and interpreting large sets of data to uncover meaningful insights, patterns, and trends. This field, widely used in science, social science, and business areas, typically involves many different techniques, tools, and methodologies to transform raw data into actionable information that can be used for enhanced decision-making, problem-solving, and system optimization. Data analytics encompasses a wide range of tasks and activities, such as data collection, data cleaning and preparation, exploratory data analysis, data modeling and visualization, data interpretation, predictive analysis, and decision-making, to mention just a few. Through these tasks, data analytics has emerged as an integral tool for companies and organizations to extract valuable insights from data, enabling them to make informed decisions, solve complex problems, increase the productivity and performance, and drive innovation.

In our current digital era AI and ML have become pervasive in many scientific, technological, and industrial domains, and new, exciting applications are arising in almost all aspects of our daily life [163, 171]. Similarly, DA is used across several industries and interest areas. The most popular ones include finance, business, marketing, healthcare, logistics and manufacturing, and social sciences. Just for illustration, we include here some interesting applications of artificial intelligence, machine learning, and data analytics:

- *Automation, quality control, logistics, and manufacturing*: AI-powered systems automate repetitive and labor-intensive tasks, leading to increased efficiency and productivity in areas such as human resources, automation, and logistics [2, 29, 34, 162, 194]. Industries such as manufacturing, logistics, and customer service have benefited from AI-driven automation [36, 37, 104]. Also, ML is increasingly used in manufacturing processes to monitor equipment health, predict maintenance and logistic needs, and perform quality assessment to ensure that manufactured goods meet the expected standards for quality and safety [5, 159]. Currently, DA is applied to supply chain optimization in order to improve inventory management, demand forecasting, and logistics for efficient supply chain operations [32, 180].
- *Healthcare*: AI and ML are revolutionizing healthcare with applications in many different medical areas [116]. Powerful AI and ML algorithms have been developed for medical diagnosis of several diseases, very often relying on medical imaging, e.g., MRI (magnetic resonance imaging), CT (computer tomography), or X-rays. An illustrative example is given by the recent COVID-19 pandemic, for which advanced AI and ML-driven methods have been obtained for reliable detection of the disease from chest CT scans. Also, new AI-based optimization methods have been developed for designing social distancing enhanced physical spaces [173, 174, 175]. AI algorithms can assist doctors in making more accurate diagnoses and treatment decisions and as a support tool for decision-making. AI has also been successfully applied to accelerate drug discovery for COVID-19 and other diseases, while ML is applied to screen thousands of potential drug candidates for rapid selection of the most promising drugs against viruses and other microorganisms. AI-driven applications have been developed for personalized medical treatment plans and patient monitoring as well as for contact tracing and tracking of the COVID-19 pandemic [78]. ML methods have been developed for health management-related decision-making plans, for instance, for prediction of those patients most likely to require hospitalization or intensive care, allowing for more efficient allocation of human and material resources. Furthermore, AI-powered robots are currently in use in hospitals for tasks such as patient monitoring and transfer, cleaning and disinfection of rooms and units affected by infectious diseases, or admission procedures. Several DA techniques are currently being applied in the field of healthcare analytics to

analyze patient data and medical records to improve patient outcomes, optimize resource allocation, and enhance clinical processes [8, 45].

- *Autonomous vehicles*: AI and ML technologies are crucial for the development of self-driving cars and other autonomous vehicles [12, 136, 192]. These systems use sensors and AI algorithms to navigate and make real-time decisions on the road [52, 129]. They also make use of the large number of data generated by self-driving cars, drones, and other autonomous systems to learn from real-world data in order to enhance the operational decision-making in dynamic environments [67, 99, 156].
- *Natural language processing (NLP)*: NLP is a subfield of AI focused on enabling computers to understand, interpret, and generate human language [35, 82, 90, 191]. Virtual assistants like Siri, Alexa, and Google Assistant use AI-based NLP techniques to process and respond to user queries and communicate with humans in a rather natural way. In addition, ML powers language understanding and generation in applications like chatbots, virtual assistants, language translation, sentiment analysis, and text summarization [87, 95, 134, 181].
- *Finance and business intelligence*: AI is used in financial trading to develop and optimize many financial decision-making processes, such as investment strategies, credit scoring, and financial risk assessment [74, 109]. The huge amount of financial data available paired with the impressive actual computing power enables the application of sophisticated ML algorithms to identify patterns and trends in market data for better investment decisions [42]. ML is also applied to detect fraudulent activities in financial transactions, credit card usage, and online transactions by identifying unusual patterns or behaviors. Data analytics is used for risk assessment of market trades, to design and optimize investment strategies and as a supporting tool for the financial decision-making, a field commonly known as financial analytics [25, 122]. Data analytics is also a primary tool for analyzing sales data, determining main customer's behaviors, and predicting market trends in order to design better pricing strategies, improve customer experience, and identify growth opportunities in the market [79, 140].
- *Marketing, E-commerce, and recommendation systems*: AI-driven recommendation engines analyze user's behavior to suggest products, services, or content that users are likely to be interested in [13, 77, 154]. Companies like Amazon and Netflix use AI for personalized recommendations. E-Commerce, content streaming platforms, and social media are good examples of the ability of ML to provide customers and other end users with individualized suggestions based on their preferences and observed behaviors [39, 81]. Data analytics tools are also applied to identify behavioral patterns of customers and market segmentation to boost campaign performance, enhance marketing efforts, and target specific audiences and market niches [30, 151].
- *Computer vision and pattern recognition*: AI and ML are used in image and video analysis, enabling computers to recognize objects, people, scenes, emotions, and others in images and videos [22, 71, 108, 112, 165], even falling objects or people [183, 188]. This technology is used in facial recognition, object

detection, environment exploration, autonomous vehicles, and manufacturing [1, 89, 115, 182, 199]. Optical recognition systems are widely used for text scanning and interpretation and language translation [14, 197].

- *Entertainment and content creation*: AI is used in creating digital content, such as generating art, music, and even writing articles or stories [7, 24, 31, 86, 105]. AI can simulate human creativity to produce new and unique outputs [40, 62, 133, 155]. Also, ML is widely used to create realistic simulations, generate plausible behaviors for digital characters [76], and develop new interactive experiences in virtual/augmented reality environments and video games [83], image and video [19, 27, 33, 113], and other fields.
- *Environmental monitoring and energy management*: AI is used in monitoring and managing environmental conditions, such as analyzing satellite data to track deforestation, predicting natural disasters, managing climate change-related challenges, and others [4, 18, 50]. Also, ML models are used to optimize energy consumption by analyzing usage patterns and making real-time adjustments to reduce waste and costs [48, 91, 144]. In addition, DA is widely used in interpreting air pollution and how gas emissions from factories affect the environment, tracking the environmental impact, using the monitoring results to comply with environmental regulations, and optimizing the management of resources such as energy and emissions controls.
- *Cybersecurity and risk assessment*: AI helps in detecting and responding to cybersecurity threats by analyzing network traffic, identifying unusual patterns, and predicting other potential digital threats [69, 148, 166, 176, 185]. ML is also used to learn behavioral patterns for intrusion detection [44, 68] and risk prediction not only in digital networks [38, 64, 152, 195] and Internet of Things [6] but also in real-world physical networks, such as traffic highways, airplane routes, pedestrian crossings, bicycle paths, water networks, oil pipelines, and so on.
- *Agriculture*: AI and ML help optimize crop yields by analyzing factors like weather, soil conditions, and pest data to make informed decisions about planting, irrigation, and harvesting [85]. They are also widely used to monitor crops [138] and automate various tasks (e.g., precision irrigation, plant nutrition), leading to considerable savings of key resources (water, energy, fertilizers, human labor workforce).
- *Education*: AI is being integrated into educational systems to provide personalized learning experiences for students [84]. AI-powered tools can adapt to individual learning styles and pace. ML models can identify patterns of poor or irregular performance and modify the learning pathway and approach through data-based informed decision-making to adapt to different learners' profile and abilities. Applying data analytics to social data is becoming increasingly popular as an effective social network supporting tool to analyze social human behavior, interpret and predict opinion trends, and understand complex societal patterns.
- *Research and innovation*: AI accelerates scientific research by analyzing complex data, simulating experiments, and discovering new patterns or insights that may not be apparent through traditional methods [120]. The ability of ML models

to uncover hidden patterns in data may lead to out-of-the-box thinking, thus helping the researchers to modify their research plans on the fly yielding to unexpected discoveries that would not be possible otherwise.

3 Applications of Artificial Intelligence, Machine Learning, and Data Analytics to Sports

Artificial intelligence, machine learning, and data analytics have revolutionized many industries, with sports being a very remarkable example. With the increasing amount of data generated by sports teams and individuals, AI tools are being used to gain insights and make data-driven decisions. AI and ML are being used in a variety of sports applications, such as athlete performance analysis, injury prevention, talent identification, and fan engagement. This section discusses the historical evolution of the field, the various tools and applications of AI and ML in sports, and the benefits and challenges associated with its implementation.

3.1 The Moneyball Effect in Sports

Sports analytics, often referred to as Moneyball, involves making sport decisions based on data. The term *Moneyball* was coined after the famous Michael Lewis's book *"Moneyball: The Art of Winning an Unfair Game"*, published in 2003 [103]. The book came after the writer knew about the Oakland Athletics general manager Billy Beane, who consistently built successful baseball teams year after year, breaking away from the long-standing tradition of relying on the instinctual judgments of seasoned scouts. Lewis secured exclusive access to Billy Beane, the mastermind behind the Athletics' distinctive approach to valuing players, which involved reevaluating offensive efficiency, prioritizing metrics like on-base percentage and power while downplaying the importance of stolen bases and focusing on putting the ball in play. This revolutionary strategy shifted the paradigm of talent acquisition, emphasizing statistical achievements as much as, if not more than, traditional scouting methods. That strategy paid off. Despite being a low-budget MLB (Major League Baseball) franchise, the Oakland Athletics sought to maintain a high level of success through an analytics-driven strategy, even in the face of financial constraints. This approach resonated with some franchises, changing the way many major league front offices did business. They started to hire full-time sport analysts and, later, to create sports analytics departments in the franchises, which are now pervasive in all major American professional team sport organizations. The Moneyball effect also influenced other professional sports teams, including European club association football. Reportedly, famous Premier League club managers such as Arsene Wenger (Arsenal), Alex Ferguson

(Manchester United), or Sam Allardyce (Bolton) and even club owners such as John W. Henry (Liverpool F.C.) were ardent supporters of the use of data analytics for sport decision-making.

While the traditional approach of talent evaluation through scouts using the "eye test" still exists, through sports analytics it got an unexpected ally. Undoubtedly, as "big data" continues to shape decisions both on and off the court, the entire team operations departments of professional team sports, including the players, will need to embrace and adapt to the influence of analytics.

3.2 *Sports Analysis*

One of the main areas where AI, ML, and DA are being intensively used in sports is in data analysis [20, 141]. With the vast amounts of data that are usually generated in sports today, it is becoming increasingly difficult for coaches and analysts to sift through it all collected data and extract meaningful insights. Powerful computational intelligence techniques can help to automate this process, allowing coaches and analysts to focus on interpreting the data rather than collecting and organizing it [17, 21, 54, 59].

One example of how AI and ML are being used in sports analysis is in the field of player tracking. For decades, professional sports relied on basic statistics and the conventional wisdom of the main actors of the game (i.e., players, coaches, managers, scouts, advisors, front office members) to drive the team plans, shape a new roster, and develop team play strategies and, in general, for all major decision-making involving the team. Illustrative examples from team sports abound. In American football, simple statistics such as completed yards, touchdowns, and interceptions were typically considered the state of the art when it comes to statistical analysis of the game. Similarly, box-score statistics for professional basketball players were for many years merely a compilation of simple data such as scored points, rebounds and assists, and, more recently, other data such as steals, turnovers, and blocks. In professional baseball, statistics relied on stolen bases, runs batted in, and batting average as the ground truth for player's assessment, as it used to happen in baseball dating back from the nineteenth century. Another illustrative example appears in football (called soccer in the USA and other countries), where for years the basic statistical sheets contained little more than the goals, penalties, faults, yellow and red cards, substitutions, and injuries. This is in sharp contrast with the current data-driven approach of modern football, where thousands of data are routinely collected from every single match or player and made available to everyone either for free or for a very affordable fee.

The use of data in sports can mean a very important boost in the performance of teams in a competition—provided they are used appropriately—but also to poorer performance otherwise. A remarkable example of this latter case is the pioneering use of statistics in English football carried out by Charles Reep, a dedicated supporter of Herbert Chapman's Arsenal, a team that revolutionized football with its very successful and highly influential W-M formation in the 1930s. In the 1950s, disappointed that Chapman's football system had not been fully embraced, he began recording actions in professional matches, leading to his pioneering role as a performance analyst for some clubs in professional English football.

Reep's early findings highlighted that teams often scored from moves with three or fewer received passes. He advocated for a direct style of play with long balls forward. His partnership with Wolverhampton Wanderers' manager Stan Cullis resulted in three First Division titles. However, Reep's simplistic approach did more harm than good for English football, as it was based on erroneous conclusions. After analyzing hundreds of games between 1953 and 1967, he found that only 5% of all moves consist of four or more received passes and only 1% of six or more. He also introduced the position of maximum opportunity (POMO) zones, particular areas of the game field from where goals were most often scored. According to Reep, too many passes were not only ineffective but also harmful to the teams making them, because many goals are conceded from ball lost situations derived from excessive passing. As a result, Wolves players were asked to avoid lateral passes and instead move the ball to the POMO zone quickly while encouraging the wingers to eagerly run to such zones to maximize the chances of getting the ball and shoot. These concepts became foundational in English football for decades.

Jonathan Wilson's analysis, nicely explained in his book *"Inverting the Pyramid"* [186], debunked Reep's ideas, revealing that moves with more passes were often more effective. Reep's calculations indicate that 91.5% of moves in the games consisted of three received passes or fewer, but goals from these moves accounted for less than 80% of goals, suggesting that the long-ball play tactic is not as effective as claimed. Also, teams that make a lot of passes tend to exhaust the opposing team by forcing them to chase their players to regain possession of the ball, something they can capitalize on to their advantage in the final stages of matches. This misconception of the game became the keystone of English football, affecting the way football was played in England dramatically. It is significant that the brief stint of Reep with Sheffield Wednesday faced challenges due to players resisting his primitive and direct style.

But still, Reep's methods continued to influence managers across England. In the 1980s, Charles Hughes become the director of coaching for the Football Association and embraced Reep's ideas, vehemently advocating for long-ball play. In his book *"The Winning Formula"*, Hughes supported the long-ball play that characterized English football for decades, under the premise that the players would score if the ball was played into the POMO zone as soon and often as possible and that long-ball play is the optimal way to achieve this goal. The English national team head coach Graham Taylor and others adopted these principles, but their teams failed to succeed against more advanced game tactics, and by the 1990s, Reep's methods were considered outdated, marking the end of long-ball play era in English football.

With the founding of the Premier League in 1992 and the influence of different football cultures, performance analysis moved towards a more reliable direction based on enhanced data analytics. In 1996 a new British sports analytics company called Opta Sports was founded to analyze Premier League matches, signing a contract with television broadcaster Sky Sports for the 1996–1997 season. The following year, Opta turned into the official statistics provider for the Premier League and became sponsored by Carling, which paid for the Opta Index to be freely offered to the clubs and media. Back then, statistics were comparatively less complex than they are in the present day, yet they remained groundbreaking by the prevailing standards of that era. In just a short span of time, the sport transitioned from having limited data to being inundated with an excess of information. In no time, both Opta and its competitors were disseminating thousands of new data items every week. Clubs began uncovering insights they had never even considered previously, such as the number of kilometers each player covered during a match and their statistics on tackles and passes made.

A fresh wave of club managers enthusiastically adopted and utilized this available technology. Popular club managers such as Manchester United's Alex Ferguson, Arsenal's Arsene Wenger, and many others started using match data to develop new game plans; redefine their team rosters and scouting policies; change the eating, resting, and training habits of their players; and apply advanced statistics to improve the team performance during matches.

Surprisingly, the process of collecting these data in football is mostly manual even today. Two individuals analyze every match live, with each responsible for recording a team's data, which is subsequently reviewed by two separate analysts the day after, thus repeating the whole process to see the game again with different eyes. These analysts employ keyboards to document what they refer to as "events," categorizing them based on four variables: the type of action (e.g., pass, goal, assist, dribble), the field location, the moment in the game, and the involved player. Approximately 1,800 events are documented per game. Before starting match analysis, each analyst undergoes an intensive three-month training period. Initially challenging, this process becomes almost automatic with practice, similar to playing a video game console like PlayStation. About 2,000 live game events were initially collected in 1996 for any given game, each meticulously labelled with its time code and pitch coordinate. This task holds significant allure for young enthusiasts, excited by the opportunity to earn money while watching football matches. However, staff changes notably high, as many individuals experience burnout within just a couple of years due to its repetitive nature.

The quantity of collected data points experienced a substantial surge when larger teams of analysts were brought into the equation. In recent years, each member of the team has been tasked with monitoring, tracking, and coding the individual performance of a specific player on the field. This comprehensive assessment encompasses recording a player's movements both with and without the ball, tallying how often they penetrate defensive lines, and measuring the pressure they exert on opposing ball carriers. This thorough analysis resulted in a remarkable increase, with over 15,000 data points gathered from each game, offering team analysts, coaches, players, media, and fans fresh insights to enhance their comprehension and enjoyment of the game.

During the 2000s and 2010s, numerical data became an integral component of performance analysis and opposition assessment within football clubs, resulting in tangible advantages on the field. A notable example lies in the management of data pertaining to penalty shootouts. Understanding the shooting tendencies of both field players and goalkeepers facilitated the development of winning strategies that proved pivotal in numerous critical playoffs and finals, both at the national and international levels. To harness these advantages to the fullest, dedicated data analytics departments and roles were established, previously unheard in the sport. Teams could gain a significant edge over their competitors, particularly by employing artificial intelligence techniques to analyze vast data volumes, a practice now commonplace in sports clubs and associations worldwide. Thanks to technological advancement, there now exist millions of data points, necessitating club discernment in selecting those that confer genuine benefits, whether on the football field or in player transfer negotiations.

This example in English football can be readily extended to other leagues and sports. In many sports leagues, player-tracking data are available, where the locations of all players and the ball or puck are recorded several times per second throughout the game using camera-based or chip-based technologies. Using computer vision and machine learning algorithms, it is now possible to track the movements of players on the field or court and apply data analytics on the collected data to analyze and improve players performance in real time. This data can be used to identify patterns in player's behavior, such as running speed, acceleration, and direction changes, which can be used to optimize training and performance. Other data, such as heart rate, speed, distance covered, and time spent on the pitch, are analyzed to identify what makes a player perform well and to design optimal plans to enhance players' physical conditioning.

3.3 Applications of Sports Analytics

As indicated above, AI, ML, and DA have found a number of remarkable applications in sports. In this section, we briefly outline some of the most relevant applications of these technologies for team sports.

3.3.1 Sports Training

An important area where AI and sports analytics are being used is in training [106, 196]. AI methods can be used to develop customized training programs for individual athletes based on their performance data [51, 145]. For example, if an athlete is struggling with a particular skill, AI can analyze their performance data and develop a training program that is tailored to their specific needs [150].

AI and ML can also be used to simulate game scenarios and provide feedback to athletes in real time. For example, a basketball player could practice shooting free throws in a virtual environment, with AI providing feedback on their technique and form. This can help athletes to improve their skills more quickly and effectively than traditional training methods. Besides, player tracking technology, such as GPS and RFID (radio-frequency identification) devices, provides data on player movements, speed, distance covered, and acceleration [142]. Coaches can analyze this data to monitor players' physical condition and ensure they are meeting training goals [53, 56, 63]. This information helps in designing tailored training regimens to improve players' fitness and endurance. Powerful computational methods based on metaheuristics and swarm intelligence have been developed for planning training sessions [55, 58, 98]. As a further step in the process, AI-powered artificial sport trainers have also been developed [57, 60].

Also, biomechanics data is used to analyze an athlete's movement patterns and techniques [26, 193]. This can help in refining techniques to optimize performance and reduce the risk of injury [96]. For example, in sports like golf, tennis, or baseball, computational intelligence-based automated assessment and biomechanical analysis can help improve swing mechanics [11, 119], while in basketball, it can help to optimize the motion routines for shooting, dribbling, jumping, or dunking [149, 198]. During training sessions, wearable technology and sensors can provide real-time feedback to athletes. For example, a football player might wear a sensor in their shoe to measure the power and accuracy of their shots. Immediate feedback helps athletes make adjustments and improve faster. Sports analytics can also help in managing the workload of athletes. By tracking body movements and training loads and monitoring player fatigue levels, coaches can adjust training intensity and rest periods to prevent overtraining and reduce the risk of injuries or poor performance due to fatigue [70].

3.3.2 Player Health and Injury Prevention

A primary application of sports analytics is the player health and injury prevention, which poses a significant challenge for sports teams due to the multifactorial and somewhat unpredictable nature of injuries. Nevertheless, cutting-edge technologies like biometric data analysis, in-depth biomechanical assessments, and video data analysis have the potential to reduce injury risk and expedite recovery by implementing appropriate treatment protocols. Several AI tools are also being used to predict and prevent injuries by deriving good practices for player eating, resting, and training routines [131]. ML algorithms can be used to identify patterns of previous injuries and how to mitigate them in the future. Wearable devices like smartwatches and fitness trackers are used to track the players' movement and collect data. These data are then analyzed to identify injury risks and to prevent them. Systematic video analysis studies have been performed to identify and prevent injuries in professional sports [41, 92, 158]

In addition to the player health and injury prevention considerations, sports analytics in this field is aimed at mitigating the substantial costs associated with player injuries. These costs encompass not only player salaries during injury but also the decline in team performance during that period. Foundational analytics in this domain involve using descriptive analytics to assess player activity, with the objective of evaluating physical condition and the impact of different training and game sessions [132]. By analyzing factors like heart rate and blood pressure, conclusions about athletes' physical well-being and stress levels can be drawn [127, 178].

3.3.3 Player and Game Performance

Sports analytics for player and game performance has two primary areas of focus: player scouting and lineup/team strategy optimization. Traditionally, player valuation relied heavily on human-based scouting, which was costly, limited in scalability, and prone to subjective biases. However, technology and data-driven processes have become more prevalent in supporting scouting efforts and player selection and for player performance analysis [110, 111, 177]. Performance data, including advanced statistics, are used to calculate key performance indicators (which, of course, vary depending on the particular team sport) and determine a player's market value [80, 137]. Data-driven market values tend to be more accurate and cost-effective than human-based assessments (provided that reliable performance indicators are actually used), enabling better budget management for the front office and team managers of the clubs, franchises, and sport organizations. Advanced analytical methods in this area encompass video analysis, fan-generated open data analysis, and the use of proprietary data for competitive advantages. Individual player performance analytics also aid in salary optimization for sports teams by quantifying each player's importance to the team's success [179].

The second area pertains to selecting matchday lineups and strategies. This involves simulating games or analyzing different tactical approaches. Deciding on team formations during matches is crucial, and analytical techniques, including Decision Trees (DTs), are employed to assess various formation options. Other techniques such as data clustering are used in sports like football to perform data analysis on different parts of the game field and to arrange players in the most suitable team formation for the match. Frontier analytics in this field utilize data from sources like video footage, locational data, and biometric data. For instance, player tracking data is used to dynamically measure, classify, and study team formations during matches. Locational and biometric devices, such as GPS devices, RFID, and accelerometers, are gaining prominence and can enhance player performance analysis. Wearable technologies like the adidas MiCoach system provide real-time performance data to coaches, facilitating team improvement. Other applications of these technologies may include assessing player interactions. The use of big data analytics is expected to play a significant role in the future of tactical analysis in many team sports.

3.3.4 Game Strategy Optimization

Sports analytics plays a crucial role in optimizing game strategies for sports teams. Sports analytics helps teams thoroughly scout their opponents [16, 170]. Coaches and analysts review past game footage, player statistics, and team tendencies to identify patterns and weaknesses [47, 157]. This valuable input informs game plans and strategy adjustments to exploit opponent vulnerabilities.

Advanced AI and ML methods can also help identify favorable player matchups and for tactical adjustments [160]. Coaches can determine which players have an advantage over their counterparts on the opposing team and devise strategies to exploit these matchups [101]. Advanced statistical and AI models and computational algorithms, such as Bayesian networks and Markov chains, can be used to analyze player and team performance. These models provide insights into player's contributions, team dynamics, and optimal strategies [73]. Coaches can also receive data on their team's performance and the opponent's performance and make tactical adjustments accordingly. For example, if a certain play or formation is consistently failing, coaches can quickly switch to a different strategy. Sports analytics also provides data on in-game situations, such as when to go for it on fourth down in American football or when to make substitutions or switch defenses in basketball. Coaches use these data to make informed decisions that maximize their team's chances of success.

In team sports like football, basketball, handball, rugby, American football, or others, set pieces and special teams plays are crucial [94, 187]. Sports analytics helps teams optimize these plays by analyzing data on positioning, timing, and execution. In baseball, pitchers and batters use analytics to study the tendencies of their opponents. Pitchers may adjust their pitch selection based on the batter's weaknesses, and batters may anticipate the types of pitches they are likely to face. In basketball, analytics is used to determine the most efficient shot selection. Teams analyze shot charts to identify high-percentage scoring areas and encourage players to take shots from those locations.

Analytics can help manage player fatigue during games. Teams use data on player workload and performance to make substitutions and ensure that players are performing at their best throughout the game. Teams track various performance metrics during games, such as player positioning, ball possession, and passing accuracy. These data help coaches assess their team's overall performance and make strategic adjustments as needed.

Teams can use analytics to simulate various game scenarios and strategies and for predicting outcomes. By running simulations, coaches can determine the most effective strategies for different situations, such as late-game scenarios or penalty situations. Machine learning algorithms can be trained on large amounts of historical data to identify patterns and trends that can be used to predict the outcome of a game or match. This can be used by coaches and analysts to develop strategies for upcoming games, or by fans to place bets on the outcome of a game.

3.3.5 Business Analytics for Sports

In the professional sport industry, sports analytics finds significant application in business-oriented operations [10, 72]. These operations encompass areas like ticket pricing, fan loyalty, promotions, social media, and fan engagement, resembling typical business analytics but often less sophisticated due to the primary focus on on-field performance in sports organizations [128].

One prevalent application involves strategically adjusting ticket prices to maximize revenue. Variable ticket pricing, where prices remain static but vary for specific games against popular opponents, is widespread [200]. A more advanced approach, dynamic pricing, adjusts ticket prices throughout the season based on factors like team performance, league standings, and even weather conditions [172]. AI-driven projects, like Vitesse Arnhem's pilot in the Dutch football league Eredivisie, in which ticket prices are based on several data and the willingness of the club fans to pay and where the seat allocation in the stadium was unknown for fans at the ticket purchase time, have shown very revealing results: almost 52% of ticket buyers did not have plans to attend the match but changed their plans attracted by personal offers, with 86% declaring to be highly satisfied with their ticket price. Also, although 66% preferred a different seat location prior to the game, 78% were satisfied with their assigned seat after the match. As a result of this dynamic pricing approach, even with a decrease of 7% of accepted tickets, the total revenue increased about 48%. These surprising outcomes have demonstrated the value of dynamic pricing, leading to the increased adoption of similar policies by other clubs worldwide.

Sports analytics also plays a role in enhancing various interactions between sports consumers and their teams. While basic efforts involve displaying statistics and content on websites, more advanced applications personalize online content using fan data from social media and membership records [143]. This enables clubs to develop highly targeted marketing approaches, tailoring promotions and activities before and after games based on a fan's history and interactions. For instance, FC Bayern Munich employs digitally generated fan data to enhance the offline fan shop experience and boost revenue through personalized offerings, facilitated by software solutions from software company SAP, an excellent example of the possibilities that sports analytics brings to motivate and captivate fans on a global scale [153].

3.3.6 Fan Engagement

The advent of sports analytics data has brought about a revolutionary change for both sports teams and enthusiasts, reshaping the way they connect with their beloved sports in many different ways. AI and ML are being used to enhance the fan experience in sports. AI can be used to provide real-time analysis and insights to fans watching a game, helping to increase engagement and provide a more immersive experience. For example, it can be used to provide real-time statistics and analysis of a game, or to generate personalized highlight reels based on a fan's interests and preferences [65, 146]. Owing to its ability to extract meaningful patterns from data, ML can be used to enhance fans' experience by identifying hidden patterns in the game play, provide the fans with challenging quizzes about play predictions, or offer similar engaging content [102, 114].

AI is being used to increase fan engagement by digital interaction. Nowadays, chatbots and virtual assistants are used to interact with fans, answer their queries, and provide them with information about the team and the players. AI tools are also used to personalize the fan experience, recommend products, provide personalized content, and make reliable predictions about sport game results [100, 121, 123].

Other important ways to boost fans engagement and enhance their overall sports experience are through fantasy sports, betting, and social media engagement. Fantasy sports have experienced an explosive surge in popularity, and data analytics is at the heart of this phenomenon. Enthusiasts who partake in fantasy leagues rely on sports analytics data to craft their dream teams [135, 190]. They meticulously scrutinize player statistics, past performances, and upcoming matchups to strategically select players for their fantasy roster. By harnessing data-driven insights, fans endeavor to outwit their competitors and compete for virtual championships within their fantasy leagues. This data-based approach infuses an extra layer of thrill and competition into the fan experience.

Data analysis through AI and ML has also left a significant mark on sports betting practices. Sports enthusiasts with a penchant for betting utilize sports analytics data to make well-informed choices when laying their wagers. They can assess the likelihood of various outcomes by examining player and team performances, injury reports, weather conditions, and historical data. Sports analytics provides a scientific and data-driven foundation for betting, enhancing the odds of successful bets and the potential for winning [28].

Finally, sports analytics data find widespread dissemination on social media platforms, establishing an interactive arena where fans can engage with insights and dialogues about their cherished sports. Sports organizations, analysts, and fans alike share statistical tidbits, player comparisons, and data-driven narratives, sparking spirited debates and discussions among fans. Social media has evolved into a central hub for fan interaction, with data-driven insights stoking enthusiasm and anticipation for forthcoming games and events. Enthusiasts have the opportunity to join conversations, contribute their analyses, and further enrich the sports community on social media.

3.3.7 Game Analysis

Sports analytics plays a pivotal role in game analysis by providing data-driven insights into player and team performance, tactics, and strategies [9, 139]. Different techniques have been devised for game analysis from video sequences in football and other sports using AI and ML-based methods such as deep learning, support vector machines, and semantic analysis [46, 107, 184].

Another illustrative example is given by the analysis of passing in football. As expected, passing is the most frequent event in football. A proficient passing game is crucial for success in football because it empowers a team to control the game by maintaining possession, conserving energy, and consequently enhancing their scoring opportunities. Hence, it is important to analyze the player performance in this particular game feature. The conventional metric is the number of completed passes. However, categorizing passes solely as completed or not overlooks the vast array of pass styles employed by different players and fails to consider the overall flow of a team. So, new AI-based models have been devised for measuring the pass quality on a continuous spectrum, assessing the level of risk associated with each pass based on historical pass data [23]. An AI algorithm was trained using data from 358,753 passes observed across 380 games involving 20 participating teams. Numerous factors were considered during this training process. First and foremost, it took into account whether the players were positioned in their own field or within the opponent's territory. Additionally, minute-by-minute game results were analyzed throughout the entire match duration. Moreover, the passes were graphically plotted on the field, with the ends of the field serving as the X and Y axes. Lastly, the algorithm considered the type of pass, categorizing them as normal, header, cross, corner, shot, goal kick, or free kick.

With all these input data, AI and ML methods (specifically, K-means clustering and XGBoost) were applied to discern patterns that correlated specific types of passes by players with the most favorable outcomes. Notably, the analysis uncovered insights such as that a significant proportion of passes are failed when the distance was very short, less than 5 meters. This is probably accounting for interceptions at an early stage of the pass rather than due to unforced errors. Furthermore, it revealed that passes in the range of 15 to 30 meters were more likely to be completed than failed, whereas beyond 30 meters, the completion rate sharply declined while failed passes stabilized. Another crucial factor that emerged was the location on the field where passes were executed. For instance, as players approached the opponent's goal, the frequency of failed passes increased.

The algorithm also accounts for the distinct roles of individual footballers. If the model predicts a pass with a probability of 0.8 and it is successfully completed, 0.2 points are credited to the footballer's passing score. Conversely, when the pass is not completed, 0.8 points are deducted from the player's passing score. These scores are subsequently averaged over the player's total number of passes, resulting in an average risk-over score. This score enables player comparisons based on the risks they assume and their ability to manage them and succeed in their passing game. Considering that both risk management and the volume of passes are critical metrics for evaluating player performance, multiplying these two factors together yields a comprehensive passing score, which in turn facilitates more in-depth analysis. This offers coaches the chance to assemble teams and reserves composed of players who, statistically, are likely to outperform their market valuation and identify underperforming players.

Some other examples of sports analytics for game analysis for the particular case of the professional basketball in the NBA league will be discussed in the second part of this chapter. Many other examples from other sports (not discussed here owing to limitations of space) can also be found in the available literature in the field.

3.4 Challenges and Limitations of Sports Analytics

As any other technology at our disposal, sports analytics comes not only with advantages but also with challenges and limitations. A major limitation of sports analytics is that it often overlooks the human element of the game. Factors such as player motivation, injuries, interpersonal relationships, team chemistry, and others can have a significant impact on the outcome of a game, but they are challenging to quantify and predict. Consequently, they are not usually considered in current sport metrics, even though they can have actually a strong impact on the player and team performance if not considered and properly addressed. How to embed this information into the current sports analytics is still an open problem in the field.

Another limitation of sports analytics is that coaches and players may not always trust the data, leading to a reluctance to change strategies or make decisions based on analytics. Currently, there are many fierce advocates of the anti-analytics movement. As an illustrative example, popular former NBA All-Star players such as Charles Barkley, Shaquille O'Neal, or Isiah Thomas, or even current NBA stars such as Kevin Durant, Zach Levine, Bradley Beal, or Paul George, dislike analytics and disdain its use to a significant extent. Other similar claims against sports analytics have emerged for other team sports as well [169].

Another challenge of particular concern is the issue of biometrics and data privacy along with data ownership and usage. Like anyone else, players do not like the notion of their biometric data being disclosed publicly to the media. They also fear that their health and performance data could be used by the teams in contract negotiations where previous health issues, bad training or eating habits, or other data might be used against them. Nowadays, privacy, ownership, and usage of data generated from these new systems and technologies is a matter of high concern for players and their respective associations.

4 Case Study: National Basketball Association (NBA)

Any NBA fan or follower is well aware of the significant changes taking place over the years within the league. NBA has been the most innovative American professional sports league (for instance, in 2013 it became the first American sports league to use player tracking in every game) and is always evolving as teams explore innovative strategies to rise above mediocrity and compete to get access to the playoffs and ultimately, for championships. This transformative journey towards championship contention is driven by (among other factors) sports analytics, which has proven highly effective in transforming once-average teams into serious contenders.

Nowadays, nearly all NBA franchises employ data analysts. The use of sports analytics is becoming a pervasive trend across all aspects of the sport, from scouting rookies and identifying undervalued players to optimizing shot selection and the growing prominence of the three-point shot. But this journey took a long time to become a reality.

4.1 The NBA Journey to Sports Analytics

4.1.1 The Old Era

In the early twentieth century, forward-thinking individuals began exploring systematic data analysis in basketball. Howard Hobson, who led Oregon to the first NCAA (National Collegiate Athletic Association) championship in 1939, emerged as one of the pioneering coaches to statistically analyze the game. He authored *"Scientific Basketball"* in 1949, focusing on shooting percentages and other aspects of the sport [75]. While his innovative concepts initially met resistance, Hobson's research ultimately contributed to significant rule changes like widening the free-throw lane, introducing the shot clock (adopted by the NBA in 1954), and implementing the 3-point line (firstly introduced in the short-lived American Basketball League-ABL in 1961 and, then, reintroduced by the now-defunct American Basketball Association-ABA in 1967, the first year of that league).

In the mid-1950s, Air Force head coach Bob Spear and his assistant, Dean Smith, who later achieved Hall of Fame status during his remarkable 36-year tenure at North Carolina, developed a possession evaluation system. This system assessed teams' performance both offensively and defensively, recognizing that traditional box score metrics did not capture the full essence of a game, particularly its pace and number of possessions. Another basketball statistics' pioneer, Paul Keller, who believed that *"points in a game actually mean nothing, but points per possession mean everything,"* devised in 1960 his offensive efficiency rating (OER) system [93] and applied it to track Ohio State's success during its national title season in 1960 and runner-up finishes in 1961 and 1962.

However, despite efforts by Smith and other forward-thinkers to delve deeper into statistical analysis during the following decades, conventional box score statistics, including points, rebounds, and assists, remained prevalent in basketball analysis throughout the 1990s. At the dawn of the NBA, the teams and media used a very basic statistical sheets containing the team lineups, the score for quarters and final score, and some basic stuff about the players: points, rebounds, and assists. Also, the field goal percentage (FG%) and the free-throw percentage (FT%) are given by $FG\% = FG/FGA$ and $FT\% = FT/FTA$, respectively, where FG and FT indicate the field goals and free throws converted respectively, and the final A in the denominator stands for attempts.

These statistics provided a basic overview of a player's raw production, but they did not accurately reflect their true impact on the team or allow for meaningful comparisons with their teammates. As a result, additional metrics like WOWY (with or without you) and WOWYR (WOWY regressed) were developed years later to assess a player's contribution by examining how the team's performance differed when the player was on the court compared to when they were on the bench (see [168] for details).

4.1.2 The Box Score Era

In the 1970s, new statistics were added to the mixture, starting with rebound percentages in 1971 and then defensive and offensive rebounds, steals, blocks, and team turnovers first for the season 1973–1974 and individual turnovers since the 1977–1978 season, subsequent to the NBA-ABA merger. In the 1980s, more complete game logs were collected, and for the 2000s we also had the plus-minus values, which keep track of the net changes in the score when a given player is either on the court or on the bench and were published in official score boxes in NBA since the 2007–2008 season. However, this single metrics can be strongly misleading, so more sophisticated plus-minus indices (net, adjusted, regularized-adjusted, augmented-box, and others) were subsequently developed.

About the mid-1990s a new set of valuable data, called Play-by-Play (PBP) data, were provided. The rationale behind is that conventional box score statistics, while providing per-game summaries for both individual players and teams, offer only a partial glimpse of the intricate dynamics within a game. Moreover, these statistics can frequently lead to misconceptions, particularly when assessing defensive performance. In contrast, PBP data emerged as the primary data source for numerous advanced statistics, including metrics like adjusted plus-minus. In short, a PBP record presents a chronological account of a game through discrete events. A standard PBP dataset typically includes the following details:

- Timestamp indicating when the possession occurred
- The player responsible for commencing the possession (e.g., for steals or defensive rebounds)
- The opposing player responsible for initiating the possession (e.g., for missed shots or turnovers), along with the court location from which the shot was taken, and various unique identifiers used to categorize the nature of the possession

Following the widespread success of Moneyball, Dean Oliver embarked on a similar mission in 2004, pioneering *"Basketball in Paper"* during an era when advanced metrics were not yet commonplace in NBA front offices and among fans. The book focuses on the role that statistical data play in understanding the game by employing statistical analysis to assess players and teams, gaining insights into their performance variations, influences, and the underlying reasons behind them. With that goal in mind, the book introduced possession-based statistics and the so-called four factors, which encompass a set of metrics (effective field goal percentage, turnover percentage, offensive rebound percentage, and free-throw rate) derived from box score statistics. These metrics demonstrate a robust correlation with winning basketball games and assist in unveiling a team's strategic strengths and weaknesses. As a consequence, the book quickly became an essential resource for basketball analysis. This led to a surge in the creation of analytical blogs and websites highly popular today in the NBA fan community such as *Basketball Reference* [15], *FiveThirtyEight* [61], and *NBAstuffer* [130].

The book also proposed other formulas such as WinShares as a mean to quantify players' contribution to their team wins. This index was inspired by the popular Win shares index proposed by Bill James for baseball as a comprehensive metric designed to facilitate the comparison of players across various positions and from different historical periods and was later adapted to the professional basketball and eventually widely accepted in the NBA world. Years later, *Basketball Reference* developed its unique iteration of win shares. The emergence of advanced box score metrics such as effective field-goal percentage (EFG); usage rate; true shooting percentage; adjusted plus/minus; value over replacement player (VORP), which is basketball's counterpart to baseball's wins above replacement (WAR); and John Hollinger's player efficiency rating (PER) also played a prominent role in this statistical revolution.

4.1.3 The Play Tracking Era

In the first decade of the twenty-first century, a new technological innovation appeared: Synergy play tracking. The practice of analyzing NBA game footage had remained constant since the advent of these videos. Coaches and technical staff from teams invested extensive hours in this endeavor, enabling them to pinpoint areas for improvement and craft more effective team strategies and tactics. Additionally, it served as a valuable tool for showcasing specific video clips to players, aiding in refining their movements, learning new offensive techniques, and studying the offensive tendencies of opponents for improved defense.

Nonetheless, this process demanded a substantial amount of time. Coaches could, at most, review one or two games per day. Locating specific plays required sifting through significant amounts of video, manually fast-forwarding and rewinding, increasing playback speed for quicker navigation, and then slowing it down to scrutinize the desired play in detail. This undertaking demanded significant effort and consumed ample time. Consequently, teams started hiring video coordinators to assume responsibility for these tasks.

One of such new hires was Garrick Barr, later the founder and CEO of Synergy Sports Technology. His journey began when he transitioned from a college basketball coach to an assistant coach and video coordinator for the NBA's Phoenix Suns. Initially, Barr collaborated with engineers to create a digital post-production editing system, streamlining the laborious process of collecting, cataloging, and editing extensive game film used for training. Although this technology served its purpose, it still required teams to compile, edit, and tag the footage manually.

Barr's next endeavor was to devise technology capable of instantly generating advanced analytics, building upon the existing work in video logging and tagging. In 1998, he, along with friend Scott Mossman, established Quantified Scouting Service (QSS), a company that produced computer-generated reports, accessible to teams through dial-up technology, which unveiled the offensive tendencies of every player. In 2004, Barr rebranded the company as Synergy Sports Technology [164] and brought in computer scientist Nils Lahr. Together, they developed groundbreaking application software and streaming media technology that enabled rapid video-data synchronization and upload.

During the 2005–2006 season, they started recording college games featuring the top NBA draft prospects. By the following season, Synergy had acquired college clients and was logging a substantial portion of college games. Data and video for every college game became accessible within 12 to 24 hours after the game's end. For the 2007–2008 season, Synergy entered into a multiyear licensing agreement with the National Basketball Association, providing NBA.com, NBA.tv, and the NBA's television partners access to Synergy's products and services.

NBA and college coaching staffs quickly acknowledged the potential of these innovations in video-based play tracking and promptly incorporated them into their team strategies. In the now iconic 2010 university national championship game (the NCAA final) between the universities of Duke and Butler, a pivotal moment unfolded just before the famous missed shot by Butler's Gordon Hayward at final seconds of the game. With 13.6 seconds remaining and Duke holding a one-point lead, Butler inbounded the ball to Hayward near the top of the key, with Kyle Singler tasked to defend him.

Traditionally, defenders tend to force right-handed players to move left. However, Duke had thoroughly studied Butler's playing style, revealing that allowing Hayward to go left was a risk they could not take. Synergy Sports Technology data showed that Hayward favored going left nearly 70% of the time. Incorporating this data in real time, Singler adjusted his defense strategy by shading Hayward to the right as he made his move. Hayward attempted to go left, but Singler doggedly stayed on his left side, redirecting him towards the right and the baseline. Waiting there was the towering 7'1" center, Brian Zoubek. Under pressure, Hayward had to take an off-balance jumper over Zoubek's extended arm, which ultimately missed. Zoubek secured the rebound, ultimately leading to Duke's victory in the national championship. This story showcased how advanced statistics and video analysis played a pivotal role in Duke's fourth national championship win [124].

In 2009, the NBA introduced a new data storage and analysis tool called Stats Cube exclusively for its franchises. For the 2009–2010 season, StatsCube became accessible to all teams, offering real-time statistical analysis, calculations, customizable notifications, and personalized reports. It contained a comprehensive database of every game event, such as points, rebounds, assists, steals, blocks, turnovers, missed shots, fouls, and substitutions, dating back to the 1996–1997 season when courtside play-by-play data tracking began. This database recorded the exact game moment and the players on the court when each event occurred. Notably,

StatsCube allowed teams to quickly and flexibly analyze this data for insights. The system is now openly available to everyone under the name "nba.com/stats."

4.1.4 The Optical Tracking Era

In Game 4 of the 2009 Finals, the NBA conducted a demonstration of a then in-progress system known as SportVU. SportVU is an automated tracking technology created in 2005 and later acquired by STATS LLC in December 2008. Using a system of six HD cameras arranged in an oval in the rafters of the arena to track player and ball movement, it can obtain precise positioning data for the ball, players, and referees during basketball games, providing insights into defensive and offensive alignments, shot trajectories, dribbles, passes, and player locations in X and Y coordinates. Connected to obtain multiple viewing angles, these cameras capture data at a rate of 25 times per second, recording the two-dimensional positions of players on the court and the three-dimensional location of the ball. Specialized software directs the cameras on what to monitor, utilizing computer vision technology to track individual players based on their jersey numbers and colors. The player-tracking data is transmitted from the cameras to dedicated computer workstations within each arena. From there, it is uploaded to SportVU servers, where it is integrated into an Oracle database and synchronized with play-by-play data generated by human scorekeepers positioned courtside. Teams can access reports based on this combined data in as little as 60 seconds after an event occurs on the court. Approximately 1,000,000 data entries are added to databases per game.

Figures 1 and 2 present each two instances of SportVU images created during a specific play in a match. In these images, players are represented by small circles in blue and green, each labeled with their respective numbers. This technology accurately tracks and visually represents the ball's actual location on the court. These visuals illustrate the spatial arrangement of players on the court at two distinct moments during an offensive phase executed by the blue team and the green team, respectively. The shaded region in pink (Fig. 1) and green (Fig. 2) indicates the convex hull formed by the five defending players of the team in green and in blue, respectively. This data is continuously captured and displayed in real time, facilitating instant access and analysis.

The SportVU system calculates the real-time positions of both players and the ball, enabling the determination of distances between players and the ball itself. Figure 3 illustrates these distances between two pairs of defending-attacking players. In this example, we identify the attacking players numbered 0 and 11 from the blue team and determine the closest players from the opposing green team, along with the distances to these nearest defenders. Similarly, when player number 33 from the blue team passes the ball to teammate number 8, the figure illustrates the distance between the passer and the ball at that specific moment.

SportVU eliminates the need for manual game tracking and provides quantifiable insights into player positions and defensive proximity. This groundbreaking

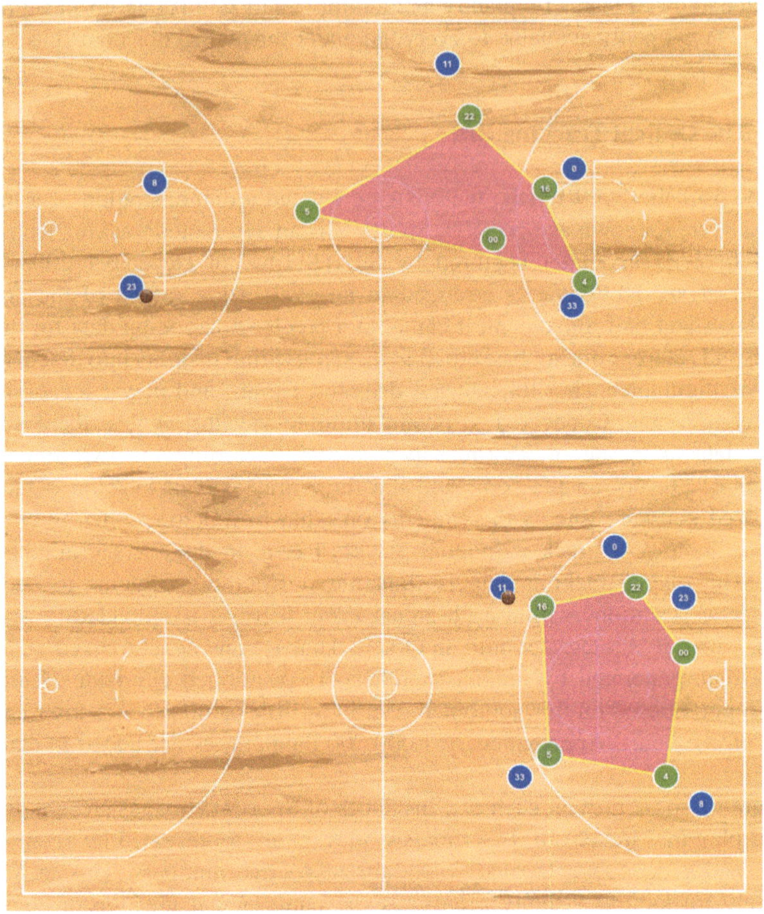

Fig. 1 Two time instances of the visuals generated by the SportVU technology during in-game action of an NBA game. Team players are represented by blue and green circles. The pink area represents the convex hull of the five players in defensive phase (players in green)

technology tracks players' movements during basketball games, providing valuable insights into their speed, agility, and distance covered. It also offers precise data on players' shooting accuracy from different spots on the court and helps assess individual and team defense effectiveness. While the data often confirms coaches' observations, it is particularly valuable for scouting opponents, allowing teams to understand their rivals thoroughly.

For example, the Golden State Warriors used SportVU data to identify areas on the court where their opponents were strong or weak shooters. This information influenced their defensive strategy, aiming to disrupt what their opponents did well. In one case, it pointed out that the rival Clippers' star Blake Griffin primarily takes his shots around or near the basket, where he exhibits excellent shooting accuracy.

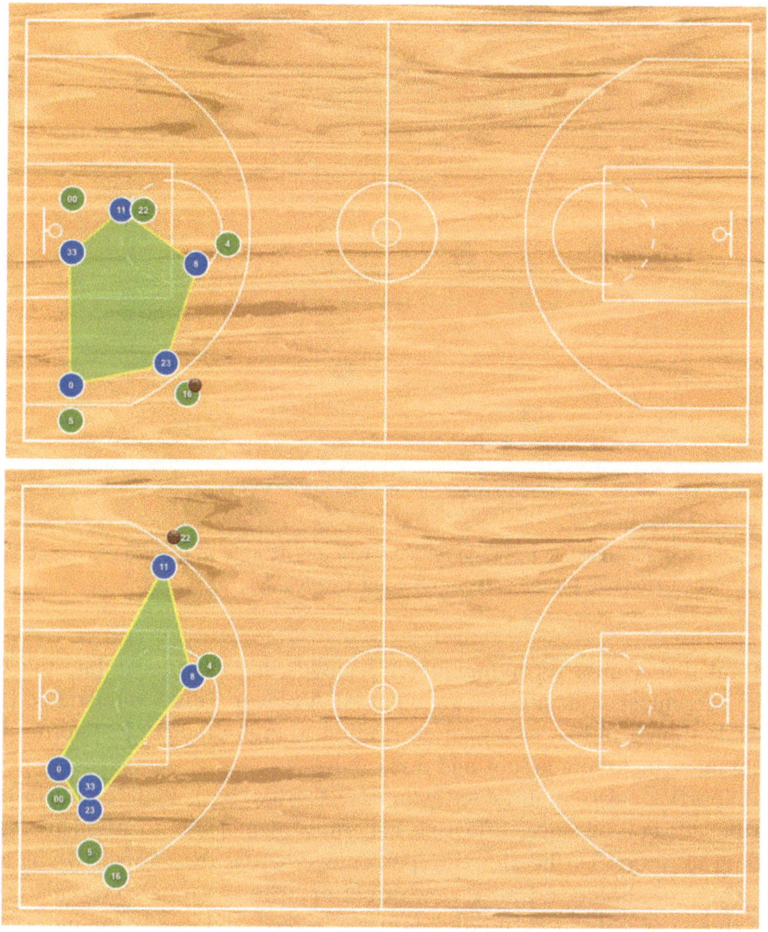

Fig. 2 Two time instances of the visuals generated by the SportVU technology during in-game action of an NBA game. Team players are represented by blue and green circles. The green area represents the convex hull of the five players in defensive phase (players in blue)

However, he also frequently attempts long two-point shots from the area around the foul stripe, a region where his shooting performance is relatively subpar. This information was used to implement defensive strategies aimed at encouraging the player to take more shots in the area where he had a lower shooting efficiency. In another specific example from a Clippers-Warriors game, it was revealed that Warriors' guard Steph Curry had a higher shooting success rate when defended by Clippers' guard Chris Paul compared to other defenders, showcasing the system's detailed insights into player performance. In another example, Celtics' All-Star forward Paul Pierce averaged 4.5 assists in the 2011–2012 season, a strong figure for a scoring wing. However, this stat alone did not capture the whole picture. SportVU

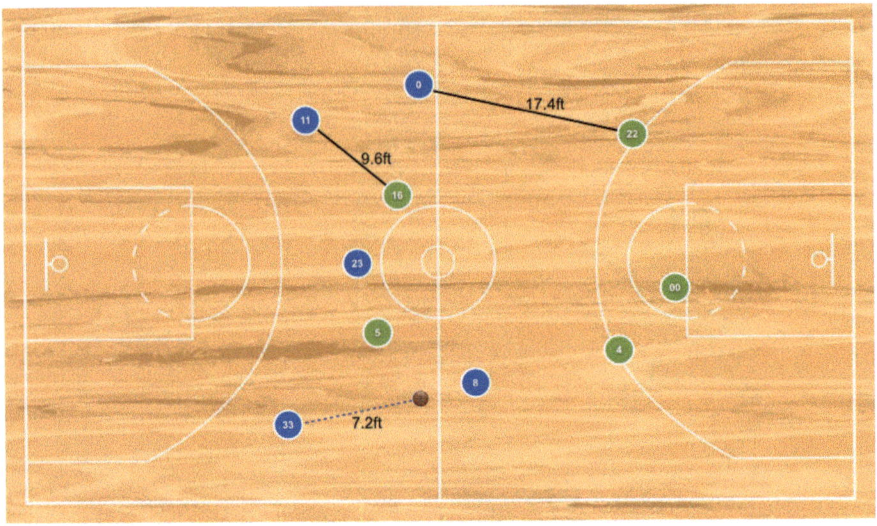

Fig. 3 Illustration of SportVU technology: calculating proximity to the nearest opponent player and the distance to the ball for the player executing a pass

data revealed that after receiving Pierce's passes, his teammates had the highest shooting percentage compared to any other NBA player. This demonstrated Pierce's impeccable timing in his passes, often leading to high-efficiency opportunities for his teammates, including layups and open shots.

The SportVU also highlighted some facts that basketball enthusiasts have long believed. For instance, effective ball movement leads to offensive success. According to SportVU, the shooting percentage across the entire NBA is notably higher when the shooter refrains from dribbling. The system also revealed that the 3-point shot is one of the most efficient ways to score points and the importance of taking uncontested shots, resulting in a sustained rise of the number of 3-point shot attempts and a more dynamic style of play in the NBA.

Teams leverage SportVU for player evaluation, conditioning assessment, and fatigue monitoring throughout the season. Fans can access some of the data on the NBA website, including shooting performance and player matchups. Additionally, even the trajectory of the ball as it travels towards the basket can be readily accessed. At that time, the NBA also envisioned using SportVU to automate certain calls currently handled by human referees and scorekeepers. Then, for the 2010–2011 season, four NBA teams (the Dallas Mavericks, Houston Rockets, Oklahoma City Thunder, and San Antonio Spurs) signed contracts to use SportVU technology in their arenas. That number increased to 10 teams in the following season, and for the 2013–2014 NBA season, the SportVU system was deployed in the arenas of all NBA franchises. This system was in operation between the 2013–2014 and 2016–2017 seasons, when it was replaced by another video tracking technology provider: Second Spectrum.

4.1.5 The Artificial Intelligence and Machine Learning Era

Second Spectrum was founded by Rajiv Maheswaran, Yu-Han Chang, and Jeff Su, all with backgrounds in computer science and data analysis. Maheswaran and Chang (former computer science professors at the University of Southern California) had previously led a USC research group focused on helping people make better decisions using data and machine learning. The company originated from their interest in NBA tracking data, which they first encountered at the MIT Sloan Sports Analytics Conference in 2011. Following Maheswaran's attendance at that conference, he recognized the potential application of their methods in the realm of sports analytics. Subsequently, Maheswaran and Chang embarked on this endeavor, and the following year they presented the research paper in [117]. In their award-winning paper, the authors utilized SportVU data to track the basketball's height trajectory following a missed shot. They meticulously followed the ball's descent from the rim, monitoring its position at various points below ten, nine, and eight feet above the floor. Their goal was to uncover valuable insights regarding offensive rebounding, examining aspects beyond the mere outcome of who secured the rebound.

To achieve this, they analyzed approximately 11,000 field goal attempts. Their aim was to delve deeper into factors such as the influence of shot location on offensive rebounding rates, the likelihood of a shot leading to an offensive rebound, and optimal player positioning for maximizing offensive board opportunities. Intriguingly, their findings revealed that a shot from beyond the three-point line had an equivalent chance of resulting in an offensive rebound as one taken from just seven feet away. Both of these scenarios were more favorable for teammate recovery than shots attempted from 11 feet, highlighting the importance of shot selection. This suggests that if you consider yourself a midrange shooter, your proficiency in this area must be exceptionally high, or alternatively, you should take a step back and attempt a three-pointer.

One of the key conclusions drawn by Maheswaran and his colleagues is that teams can improve their offensive rebounding percentages and potentially enhance their effective field goal percentages by prioritizing shot selection and avoiding areas where offensive rebounds are less likely to occur. These valuable findings and others that followed (see, for instance, [118]) convinced the authors about the great potential and applicability of their approach for enhanced NBA data analytics, leading to the establishment of Second Spectrum in February 2013.

At that juncture, the NBA had already implemented advanced optical tracking technologies across all its courts, courtesy of the SportVU agreement. Nonetheless, the challenge lay not in having the technology or viewing the imagery but in efficiently analyzing the vast volume of underlying data. Second Spectrum's significant contribution stemmed from their adoption of a scientific approach known as "spatiotemporal pattern recognition." While they were not pioneering optical tracking in sports, they were innovators at effectively collecting, recognizing, and leveraging the copious data produced by this tracking method. Their proficiency lay

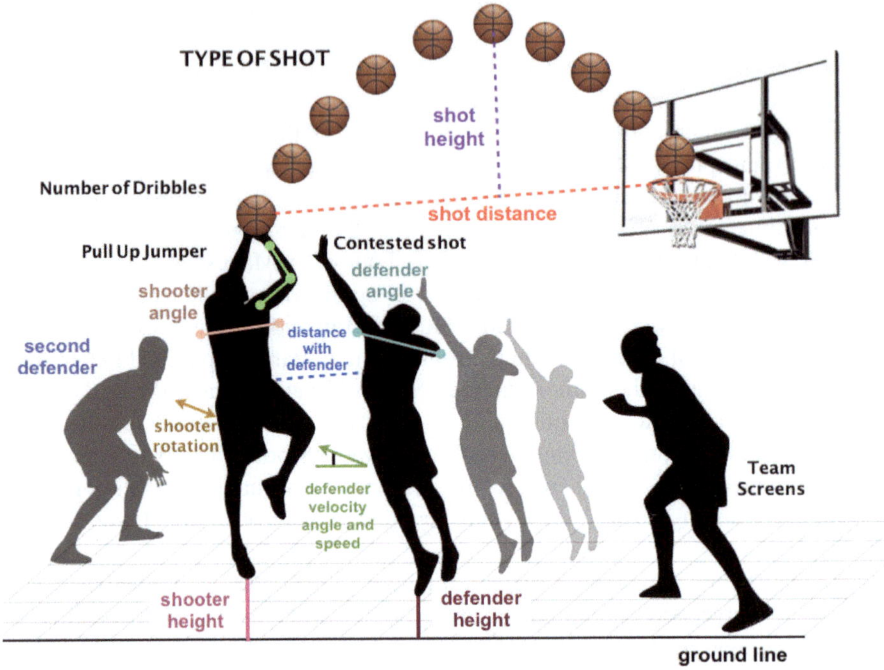

Fig. 4 Pairing optical player tracking and Second Spectrum technology enables obtaining the relevant parameters of a play and the comprehensive data on all participating players with a precision and level of detail unimaginable only a few years ago

in mastering the data rather than the technology itself, setting them apart within the sports analytics industry.

Second Spectrum's innovative approach vastly improved the level of detail available to NBA decision-makers and analysts, surpassing previous optical tracking data solutions. The pairing of player tracking technology alongside Second Spectrum's advanced sports analytics system has ushered in a new era of precision in calculating player and play parameters, surpassing previously unimaginable levels of detail. As depicted in Fig. 4, this system is capable of meticulously tracking the trajectory and positioning of a shooter during an offensive phase, obtaining precise values of variables such as the height of the player in the jump, his angle of rotation (if any), his speed when running or jumping, and the position of his upper arm, lower arm, and wrist. Furthermore, it captures data on the shooting angle of the ball, its linear distance from the rim, shot height, and the trajectory leading to the rim. Likewise, it can compute the same variables of the defending players, whether one or several, as well as the distances of the shooter from the defenders. This information becomes invaluable in determining the level of difficulty inherent in a given shot.

The system can also track other possible players who can intervene in the play, such as possible teammates of the shooter who might set screens for his progress.

Finally, the system determines the parameters of the play, such as whether it is a jump shot, the number of dribbles in the play (if any), and other possible factors. This breakthrough technology accomplishes all of this seamlessly and without human intervention.

In 2017, Second Spectrum secured an agreement with the NBA to become the official optical tracking data provider, replacing SportVU, after demonstrating their superiority in a side-by-side comparison with other tracking companies.

4.1.6 Recent Trends and Future Developments

Cutting-edge technology has become pervasive in basketball and is continually advancing. While certain technological components, like rafter camera tracking technology, have already become integral parts of the sport's infrastructure over the past decade, others are just beginning to emerge in the public's awareness.

One of these thrilling technologies is the use of wearables and other sensor devices for player tracking. In 2013 eight NBA franchises (including Dallas Mavericks, Houston Rockets, San Antonio Spurs, New York Knicks, and four unnamed teams) explored a new technology called OptimEye to monitor their players' movements. These compact devices, roughly the size of old-fashioned beepers, are worn inside the players' jerseys, positioned between the shoulder blades. The OptimEye records every aspect of a player's activity, providing precise measurements of variables such as distance, velocity, changes in direction, acceleration, deceleration, jumps, heart rate, and more.

Teams are employing these monitors primarily for evaluating player exertion levels during workouts. At present, the primary focus is on injury prevention. For example, the New York Knicks used the OptimEye to monitor the health of All-Star point-guard Jason Kidd health during his recovery from an injury. However, these devices offer a wealth of unique information that allows teams to assess the efficiency of their players' movements. For instance, if any team wish to determine how quickly a player accelerates when moving to his right compared to his left, they can simply attach an OptimEye unit to him and analyze the data generated after practice. With this kind of wearable technology, teams can now gain insights into how each player's body responds to their movements.

Another significant innovative step took place during a Women's NBA (WNBA) game in August 2021. This innovation consisted of the pioneering integration of in-game wearable technology with optical tracking, which might have the potential for a profound and long-lasting impact not only on basketball but also on many other team sports. During the Commissioner's Cup tournament's match between the Seattle Storm and the Connecticut Sun, players were equipped with wearable tracking sensors developed by KINEXON, a company providing tracking technology for their practice facilities. Embedding ultra-wideband radio technology, the sensors were discreetly positioned in the waistband of each player's attire and gathered data pertaining to their movements and performance at a remarkable rate of 25 times per second. This technology (not allowed during NBA games) enabled the tracking of critical metrics such as acceleration, changes in direction, speed, and agility while being almost unnoticed by the players. The full integration of KINEXON wearable devices and video tracking technology is predicted to generate 50 million data points from every single game of basketball.

Another illuminating example of the utilization of exciting new technology was described by Nick Nurse, who served as the head coach of the Toronto Raptors during their championship-winning 2018–2019 season. In the midst of that season, Nurse faced a challenge when his star point guard, Kyle Lowry, was struggling with his shooting. Nurse, known for his analytics-driven approach, sought quantifiable answers to the problem. Instead of traditional statistics like box scores or shooting charts, Nurse turned to the Noahlytics Data Service, a proprietary program developed by Noah Basketball. The system employs motion tracking cameras positioned on or above the backboard to follow the ball's trajectory into the basket and determine the shooter's position on the court. This technology provides in-depth data on a shot, including details about the ball's arc, depth in the basket, and alignment on the cylinder.

Nurse and the Raptors were early users of this system in the NBA. To address Lowry's shooting slump, Nurse and his team analyzed data from practice sessions during the 10-game period (NBA does not permit in-game use of this technology). They compared these data with thousands of prior shots that Lowry had taken from similar positions. The key finding was that Lowry's shot arc had dropped to 41 degrees during the slump, while his successful shots typically had arcs of 46 or 47 degrees. Armed with this information, they implemented a solution. Using Noah's audio feedback option, which provides real-time verbal analysis of shots, Lowry began focusing on maintaining the optimal 46- to 47-degree arc angle during practice. This adjustment ultimately improved his shooting performance. This successful outcome encouraged Nurse and his coaching staff to integrate such detailed shot analysis into their player development program as a regular practice.

The significance of this Noah technology is that while anyone can record successful and unsuccessful shots, there is a need to understand what causes the misses. Noah provides much more intricate responses to this question than were ever achievable previously, transforming what used to be a binary variable (successful/missed shot) into a dataset with multiple distinctive data points. With nearly 300 million shots tracked for a 15-year span, this technology captures some interesting insights. For instance, it discovered that the typical jump shooter is more than twice as prone to miss short shoots than to miss long shots, a tendency even more pronounced in many NBA players who are nearly four times as likely. Additionally, they observed that players shooting from the right corner tend to miss more frequently to the left due to a subconscious fear of hitting the backboard, and vice versa. Players are using this technology to identify woes in their shooting mechanics, refine their technique, and attain a higher level of shooting consistency, a critical component of reaching elite status as a shooter.

In March 2023, the NBA and Sony Sports unveiled a multiyear partnership with Hawk-Eye Innovations, marking a significant shift in how the league monitors in-game activities [49]. Hawk-Eye's cutting-edge optical tracking technology is tailored to capture real-time player and ball movements in three dimensions. This move by the NBA to adopt Hawk-Eye technology follows its successful application in various other sports like tennis and soccer, where it expedited officiating decisions, including offside calls during the 2022 Men's Football World Cup in Qatar.

While the NBA had been using tracking cameras for several years (firstly with SportVU and then with Second Spectrum), the existing systems employed only six cameras situated in the arena rafters and primarily tracked player torsos. In contrast, Hawk-Eye's 14-camera setup offers far greater dynamism, as it can meticulously track players' arms and legs from various perspectives, providing a wealth of insights and nuances. This technology supersedes the "center of mass" tracking system used by Second Spectrum, which estimated player locations using a single point.

Sony's Hawk-Eye has previously made its mark in American professional basketball during the Commissioner's Cup final in WNBA basketball in 2021. Utilizing 14 tracking cameras strategically placed around Phoenix's Footprint Center, Hawk-Eye monitored 17 distinct points on each player and referee, as well as the ball. The introduction of pose tracking through a collection of skeletal points introduces new officiating capabilities aimed at enhancing decision-making speed and accuracy while also enabling the NBA and its teams to analyze athletic movement in innovative ways. Sportradar, the NBA's exclusive data provider, will also collaborate with Hawk-Eye to generate tracking data sets, metrics, advanced statistics, and insights. The NBA has revealed that the leaguewide implementation of Hawk-Eye's optical tracking technology is scheduled to start for the upcoming 2023–2024 season. Additionally, the NBA has expanded its partnership with Second Spectrum, designating them as the official team analytics provider. Second Spectrum's augmentation technology is expected to create automated, enhanced graphics based on optical on-court data, incorporating advanced team and player statistics directly into the broadcast.

4.2 How Sports Analytics Shaped the Current NBA Gameplay

The advent of sports analytics has profoundly transformed the NBA landscape, ushering in a veritable revolution. While these transformations had already begun to evolve before the widespread adoption of data analytics across all facets of the NBA, the enthusiasm for statistics, cutting-edge video tracking technologies, and the integration of artificial intelligence and machine learning have catalyzed a revolutionary shift in the very essence of basketball. This transformation is characterized by two primary factors: the rise of small ball and the prominence of the 3-point shot.

4.2.1 Rise of Small Ball

For several decades, the NBA predominantly favored big men. Players like George Mikan, Wilt Chamberlain, Bill Russell, Kareem Abdul-Jabbar, Hakeem Olajuwon, and Shaquille O'Neal dominated the different periods of the league. Between 1950 and 2009, centers accounted for 51% of MVP winners and 33% of top-three draft picks. For many decades, the NBA was mostly a league of giants.

However, the 2010s marked a significant shift as "small ball" gained prominence. Small ball is a playing style prioritizing speed, agility, and increased scoring over height, physical strength, and low post offense/defense. To this aim, it plays without traditional centers and emphasizes the use of smaller players who are often skilled at three-point shooting. This style is associated with concepts like pace and space, which involve speeding up the offense and spreading out the defense with additional shooters on the court. The advantage of small ball lies in having a faster, more agile power forward who can outmaneuver their counterpart and potentially have a better three-point shooting percentage. This not only increases points from three-point plays but also creates space on the court for teammates to score around the basket. Small ball was best exemplified by the Golden State Warriors' use of small ball to a greater extent than any previous team. In the 2014–2015 NBA Finals, they replaced the big man Andrew Bogut in the starting lineup with Andre Iguodala, who went on to earn the Finals MVP title. This successful lineup shift encouraged them to keep small ball lineups on the court featuring 6-foot-6 Draymond Green as a center and sharpshooters Stephen Curry and Klay Thompson, leading them to historic success during the 2014–2019 period, when they reached five consecutive NBA finals, securing three NBA championships and setting the NBA record for wins.

The remarkable achievements of the Warriors' small ball lineups reshaped the NBA landscape, having led some analysts to envision small ball as the future of basketball. This approach also abandons traditional lineups in favor of a more "positionless" style of play that enables teams to go small and adapt to various situations on the court.

4.2.2 Prominence of 3-Point Shooting

The most prominent contemporary trend in basketball is the surge in three-point shooting. This uptick can be attributed primarily to sports analytics. Essentially, it all revolves around the concept of expected points per shot. AI-driven scientific research showed that it makes little sense to attempt a midrange jump shot that yields two points when you can opt for a "similarly challenging" three-point shot that nets one more point. Data analytics evidenced that to match the value of a league-average 36% three-point shooter, a player would need to shoot at an impressive 54% accuracy from midrange to justify opting for a long two-pointer. And finding such a player is not an easy task. In the 2019–2020 NBA season, only one player achieved

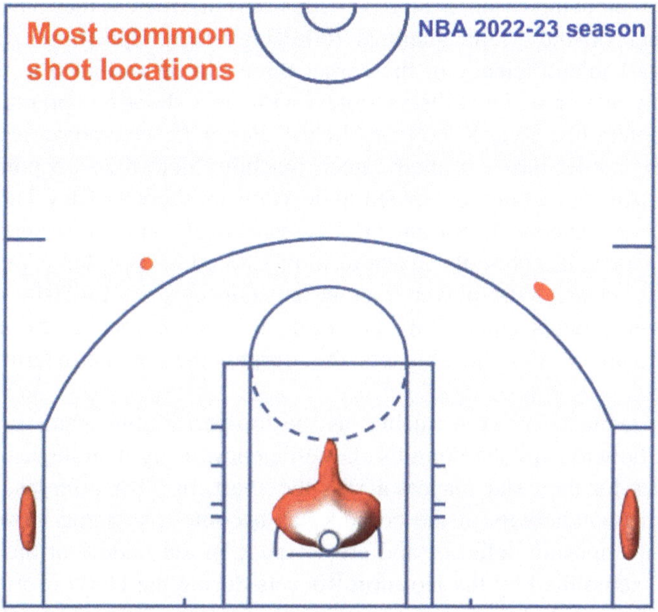

Fig. 5 Diagram showing the most common players' locations (in red) for the NBA 2022–2023 season. Positions near the rim and corners 3-pointers are by far the most common locations, whereas the frequency of midrange jumpers has dramatically decreased

a shooting percentage of at least 54% on midrange jumpers from 15 to 19 feet, with a minimum of two attempts per game.

As shown in Fig. 5, apart from shots taken at the rim, the corner three has emerged as the most frequently attempted field goal in the NBA 2022–2023 season. In fact, corner threes have become not only a prominent element of nearly every NBA team's offensive playbook but also a symbol of the data-driven revolution that has transformed the sport in the twenty-first century. Numbers are clear: despite representing only a small portion of available shooting areas on the court, corner 3-pointers now account for a staggering 19.6% of all jump shot attempts in the league. And the temporal evolution is even more impressive. In 1997–1998, these shoots represented just 3% of all field goal attempts, while now they represent more than 10%.

In terms of expected points per shot during the 2022–2023 season, the average corner three-point shot has delivered 1.17 points per try, while an uncontested shot from the same spot has impressively generated 1.29 points. Due to the limited space available in that area, corner threes are predominantly taken as catch-and-shoot shots rather than off-the-dribble attempts. According to Second Spectrum tracking, over 91% of corner threes in this season qualify as catch-and-shoot attempts. This distinction is significant because NBA shooters have historically converted 36.7%

of catch-and-shoot threes since the 2013–2014 season, whereas they have only made 32.1% of their off-the-dribble attempts, following Second Spectrum data.

The remarkable efficiency of the corner three can be attributed to the fact that it is typically attempted by skilled shooters with clear shooting opportunities. Elite 3-point shooters like Kyle Korver or Michael Porter Jr. have converted more than 50% of their corner threes in their career, reaching more than 1.5 points per shot on average. Another example: Golden State Warriors shooters Klay Thompson and Stephen Curry converted 48.8% and 60.7%, respectively, of their uncontested corner threes. For Curry, it represents an astonishing 1.83 points per shot. Compared with the fact that, in the NBA player tracking era (since the 2013–2014 season), the average layup attempt equals 1.17 points on average, corner threes are not even equivalent to layups. For elite shooters, they surpass them by far in terms of scoring efficiency.

And the tendency is even raising, as the top-performing teams in the league understand how to capitalize on this shot efficiency, using it strategically to create opportunities for their star players across the court. In NBA offenses, positioning one or even two shooters in the corners has become a strategic move to stretch the width of opposing defenses and create space in the middle of the court. This tactic was exemplified by the Houston Rockets during the Daryl Morey (a former statistical consultant) tenure as general manager, when he picked several corner three shooters for his roster to provide superstar James Harden with more freedom and space in his play.

4.2.3 Defensive Switchability

Another remarkable AI-driven NBA gameplay shift is the search for defensive switchability, the ability to switch defenders on screens instead of relying on traditional defensive strategies. Teams are increasingly moving away from traditional centers due to their limitations in defending pick-and-rolls. The trend is evident in the decline of players 6-foot-10 or taller playing significant minutes, from 26.6% in the 2000s to 17.7% in the 2020s.

While small ball offers defensive versatility, it can struggle with rebounding. Some teams aim for the ideal small ball lineup by having taller centers who can both rebound and defend on the perimeter. In this sense, tall two-way wings are highly valued for their ability to create shots, play lockdown defense, and switch positions defensively. This concept extends to "3-and-D" wings who excel in high-level defense and three-point shooting, making them valuable complements to superstar players.

4.2.4 Other Relevant Uses of Sports Analytics in the NBA

There are many things needed to build an NBA championship-winning team in addition to those outlined above. Nowadays, artificial intelligence, machine learning, and data analytics are being intensively used today for several purposes. Some other relevant uses of this technology in current NBA are:

- *Resting players*: NBA teams are now giving much greater attention to monitoring player fatigue, driven by several key factors. One obvious reason is the recognition that resting key players during a relatively unimportant regular-season game is smarter than having them enter the playoffs without the energy needed to perform at their best. The information provided by optical tracking and data analytics systems is filtered and analyzed through AI and ML and closely monitored by the coaches and training staff to develop suitable strategies for resting players during less relevant games for optimal performance when it really counts.
- *Injury prevention*: Another crucial aspect is injury prevention. Data from thousands of injuries suggests that players are more prone to getting hurt when they play while fatigued. In a recent analytics conference, NBA commissioner Adam Silver revealed that teams now have players wear different devices during practice sessions to assess factors like performance and fatigue. They even collect saliva samples, which contain indicators of fatigue. Furthermore, teams diligently monitor and quantitatively assess the nutritional intake of players, along with their conditioning programs, gym sessions, and travel itineraries, all aimed at optimizing the players' health and overall well-being.
- *Player selection*: For ordinary companies, the hiring process typically unfolds relatively quickly, with outcomes becoming evident in a quite short time frame. If a hire does not work out, the company can simply part ways with the employee and find a replacement. However, NBA teams, which largely build their rosters through the draft, face a different situation. A poor choice in the first round can set a team back for years. Therefore, teams are eager to gain any advantage they can in evaluating draft picks, and the use of advanced AI and ML techniques plays a pivotal role in this process. Advanced metrics became instrumental in various aspects of the sports industry, including trade negotiations and contract discussions. Complex models such as Kevin Pelton's Wins Above Replacement Player (WARP) projections and FiveThirtyEight's Career-Arc Regression Model Estimator with Local Optimization (CARMELO) ratings emerged as indispensable tools for evaluating prospects, analyzing drafts, and assessing the value of draft positions. With these computational tools, team analysts are able to analyze huge amounts of data collected from college and high school games, quantify performance in matchups against specific player types, and employ a wide array of analytics fields in their scouting and drafting efforts. The breadth of analytics used in areas like college scouting and international drafting is truly remarkable, reflecting the importance of data-driven decision-making in building successful NBA teams.
- *Game strategies and analytics*: The impact of analytics on basketball is most evident in the significant increase in three-point shot attempts, as discussed in Sect. 4.2.2. In 2012, teams averaged approximately 18 three-point attempts per game. By 2017, this number had surged to 27. This trend is driven by a

combination of common sense and mathematical analysis, supported by data. Essentially, the data demonstrated that the potential rewards of taking three-point shots outweighed the associated risks. On average, teams that prioritize three-point shooting tend to score more points throughout a game. The Golden State Warriors, multiple league champions during the last decade, serve as a prime illustration of this strategic approach. Other game strategies such as small ball (discussed in Sect. 4.2.1), defensive switchability (discussed in Sect. 4.2.3), and many others developed during the last few years are a direct consequence of this new sports analytics era.
- *Player matchups*: Teams extensively analyze data sets concerning defenders from opposing teams. They assess these defenders' performance against different offensive strategies, such as long-range shooting, midrange jump shots, and driving to the basket, identifying areas where they excel and where they struggle. This data guides teams in matchups, helping them pair a player proficient in a specific area against a defender who is less effective in countering that skill. Conversely, teams aim to assign their defenders to specific players, especially in crucial moments, when statistical analysis suggests that a particular defender has a better chance of preventing the opposing team from scoring. This explains why certain players, whose statistical box scores typically appear modest in comparison to their counterparts, are consistently featured in the starting lineup for games, even on championship-caliber teams. A prime illustration of such an undervalued player is former NBA player Shane Battier, who consistently secured a spot in the starting five across various teams due to his capacity to make invaluable contributions through intangible means, particularly on the defensive end.

5 Conclusions

In this chapter, we offer a brief overview of the artificial intelligence, machine learning, and data analytics technologies and their interplay to shape the current landscape of sports analytics at large. The chapter also analyzes their historical evolution and some remarkable applications in various relevant aspects of sports such as sport training, player health and injury prevention, performance analysis, game strategy optimization, business analytics, fan engagement, and game analysis. Some challenges and limitations of sports analytics are also briefly portrayed. Additionally, the chapter delves into a detailed case study involving the NBA, discussing the long journey followed in the NBA in the search for advanced analytics and a better understanding of the professional basketball game and showcasing illustrative examples of the application AI, ML, and data analytics for the NBA in action. Finally, some recent trends and future developments on the way in the field are also outlined.

An obvious strong limitation of this chapter is that it does not describe all currently existing developments in the field. Limitations on chapter length have forced us to restrict our discussion to only a few relevant aspects, while many others have had to be omitted to keep the chapter at a reasonable length. On the other hand, given the vast scope of this field, a comprehensive exploration would likely require several entire volumes. However, we hope we have provided enough interesting discussion points to still make the chapter engaging and informative for our readers.

The AI, ML, and DA technologies have not yet reached their highest level of maturity and are constantly evolving, with a rapid pace of development. It is widely agreed within the field that this progression is irreversible, and these technologies are firmly entrenched in the world of sports. As they continue to advance and gain broader acceptance, we can expect to witness even more groundbreaking applications in the realm of sports in the coming years.

Acknowledgments The authors would like to thank the financial support from the European Union's Horizon 2020 research and innovation programme, Marie Sklodowska-Curie action, RISE program, of the project PDE-GIR with grant agreement reference number 778035 and also from the Agencia Estatal de Investigación (AEI) of the Spanish Ministry of Science and Innovation, for the grant of the Computer Science National Program with reference number #PID2021-127073OB-I00 of the MCIN/AEI/10.13039/501100011033/FEDER, EU).

References

1. Adjabi, I., Ouahabi, A., Benzaoui, A., Taleb-Ahmed, A.: Past present and future of face recognition: A review. Electronics, 9(8), 1188 (2020).
2. Agarwal A.: AI adoption by human resource management: A study of its antecedents and impact on human resources system effectiveness. Foresight, 25(1), 67–81 (2022).
3. Aggarwal, C.C: *Neural Networks and Deep Learning. A Textbook*. Springer (2018).
4. Aguilar, J., Garcés-Jimenez, A., Moreno, M..R., Garcia, R.: A systematic literature review on the use of artificial intelligence in energy self-management in smart buildings. Renewable and Sustainable Energy Reviews, 151,111530 (2021).
5. Akbari, M., Do, T.N.A.: A systematic review of machine learning in logistics and supply chain management: current trends and future directions. Benchmarking: An International Journal, 28(10), 2977–3005 (2021).
6. Al-Garadi, M.A., Mohamed, A., Al-Ali, K., Du, X., Ali, I., Guizani, M.: A survey of machine and deep learning methods for internet of things (IoT) security. IEEE Commun. Surv. Tutorials, 22(3), 1646–1685, (2020).
7. Anantrasirichai, N., Bull, D. Artificial intelligence in the creative industries: a review. Artificial Intelligence Review, 55, 589–656 (2022).
8. Archenaa, J., Anita, E.A.M.: A survey of big data analytics in healthcare and government. Procedia Computer Science, 50, 408–413 (2015).
9. Ariyaratne, M., Silva, R.: Meta-heuristics meet sports: A systematic review from the viewpoint of nature inspired algorithms. Int. J. Comput. Sci. Sport, 21, 49–92 (2022).
10. Atwater, C., Baker, R.E., Atwater, T.: *Applied Sport Business Analytics*. Human Kinetics (2022).
11. Bacic, B., Hume, P.A.: Computational intelligence for qualitative coaching diagnostics: Automated assessment of tennis swings to improve performance and safety. Big Data, 6(4), 291–304 (2018).
12. Bachute, M.R., Subhedar, J.M.: Autonomous driving architectures: insights of machine learning and deep learning algorithms. Machine Learning with Applications, 6, Paper ID 100164 (2021).

13. Batmaz, Z., Yurekli, A., Bilge, A., Kaleli, C.: A review on deep learning for recommender systems: challenges and remedies. Artif. Intell. Rev. 52, 1–37 (2019).
14. Baviskar, D., Ahirrao, S., Potdar, V., Kotecha, K.: Efficient automated processing of the unstructured documents using artificial intelligence: A systematic literature review and future directions. IEEE Access, 9, 72894–72936 (2021).
15. Basketball Reference web site: [online] https://www.basketball-reference.com/ (last accessed on Sept 29 2023).
16. Bauer, P., Anzer, G.: Data-driven detection of counter-pressing in professional football. Data Mining and Knowledge Discovery, 35(5), 2009–2049 (2021).
17. Beal, R., Norman, T.J., Ramchurn, S.D.: Artificial intelligence for team sports: a survey. The Knowledge Engineering Review, 34, e28 (2019).
18. Biswas, C., Chakraborti, A., Majumder, S.: Recent advancements in artificial intelligence and machine learning in sustainable energy management. In: Sustainable Energy Solutions with Artificial Intelligence, Blockchain Technology, and Internet of Things, Chapter 2, CRC Press (2023).
19. Black, S., Keshavarz, S., Souvenir, R.: Evaluation of image inpainting for classification and retrieval. IEEE winter conference on applications of computer vision (WACV), 1049–1058 (2020).
20. Bonidia, R.P., Brancher, J.D., Busto, R.M.: Data mining in sports: A systematic review. IEEE Latin America Transactions, 16(1), 232–239 (2018).
21. Bonidia, R.P., Rodrigues, L.A.L., Avila-Santos, A.P., Sanches, D.S., Brancher, J.D.: Computational intelligence in sports: A systematic literature review. Advances in Human-Computer Interaction, 2018, Article ID 3426178 (2018).
22. Borji, A., Cheng, M., Hou, Q., Li, J.: Salient object detection: a survey. Comput. Vis. Media 5, 117–150 (2019).
23. Bouley, C.: Using Data Science to Analyse Player Performance in Football. Towards Data Science [online] https://towardsdatascience.com/using-data-science-to-analyse-player-performance-in-football-4dd04e579cfa (last accessed on Sept 28th 2023)
24. Briot, J.P., Hadjeres, G., Pachet, F.D.: *Deep Learning Techniques for Music Generation*. Springer, Cham (2020).
25. Broby, D.: The use of predictive analytics in finance. Journal of Finance and Data Science, 8, 145–161 (2022).
26. Brock, H., Ohgi, Y., Lee, J.: Learning to judge like a human: Convolutional networks for classification of ski jumping errors. Proc. ACM Int. Symp. Wearable Computers, F130534, 106–113 (2017).
27. Bull, D., Zhang, F.: *Intelligent Image and Video Compression: Communicating Pictures, 2nd edn.* Elsevier, New York (2021).
28. Bunker, R., Susnjak. T.: The application of machine learning techniques for predicting match results in team sport: A review. Journal of Artificial Intelligence Research, 73, 1285–322 (2022).
29. Buzko, I., Dyachenko, Y., Petrova, M., Nenkov, N., Tuleninova, D., Koeva, K.: Artificial intelligence technologies in human resource development. Computer Modelling and New Technologies, 20(2), 26–29 (2016).
30. Chaffey, D. Patron, M.: From web analytics to digital marketing optimization: Increasing the commercial value of digital analytics. Journal of Direct, Data and Digital Marketing Practice, 14 (1), 30–45 (2012).
31. Chan, C., Ginosar, S., Zhou, T., Efros, A.: Everybody dance now. Proc. of IEEE/CVF international conference on computer vision (ICCV), IEEE, 5932–5941 (2019).
32. Chen, H., Chiang, R.H., Storey, V.C.: Business intelligence and analytics: from big data to big impact. MIS Q., 4 (36), 1165–1188 (2012).
33. Cheng, Z., Yang, Q., Sheng, B.: Deep colorization. Proc. of IEEE international conference on computer vision (ICCV), IEEE (2015).
34. Chouchene, A., Carvalho, A., Lima, T.M., Charrua-Santos, F., Osorio, G.J., Barhoumi, W.: Artificial intelligence for product quality inspection toward smart industries: Quality Control of Vehicle Non-Conformities. In: 9th International Conference on Industrial Technology and Management (ICITM), Oxford, UK, 127–131 (2020).

35. Chowdhary, K.R.: Natural language processing. In: Fundamentals of Artificial Intelligence. Springer (2020).
36. Colombo, E., Mercorio, F., Mezzanzanica, M.: AI meets labour market: Exploring the link between automation and skills. Information Economics and Policy, 47(4), 27–37 (2019).
37. Coombs, C.R., Hislop, D., Taneva, S.K., Barnard, S.: The strategic impacts of intelligent automation for knowledge and service work: An interdisciplinary review. The Journal of Strategic Information Systems, 29(4), 101600 (2020).
38. Das R., Morris, T.H.: Machine learning and cyber security. Proc. of Int. Conf. Comput. Electr. Commun. ICCECE 2017 (2018).
39. Davenport, T., Guha, A., Grewal, D., Bressgott, T.: How artificial intelligence will change the future of marketing. Journal of the Academy of Marketing Science, 48, 24–42 (2020).
40. Davies, J., Klinger, J., Mateos-Garcia, J., Stathoulopoulos, K.: The art in the artificial AI and the creative industries. Creat. Ind. Policy Evid Centre, 1–38 (2020).
41. Della Villa, F., Buckthorpe, M., Grassi, A., Nabiuzzi, A., Tosarelli, F., Zaffagnini, S., Della Villa, S.: Systematic video analysis of ACL injuries in professional male football (soccer): injury mechanisms, situational patterns and biomechanics study on 134 consecutive cases. British Journal of Sports Medicine, 54(23), 1423–1432 (2020).
42. Dixon, M.F., Halperin, I., Bilokon, P.: *Machine Learning in Finance*. Springer (2020).
43. Douglass, E.: *Introduction to Machine Learning (fourth edition)*. Adaptive Computation and Machine Learning series, MIT Press (2020).
44. Drewek-Ossowicka, A., Pietrolaj, M., Ruminski, J.: A survey of neural networks usage for intrusion detection systems. J. Ambient Intell. Humaniz. Comput., 0123456789 (2020).
45. Ellahham, S., Ellahham, N., Simsekler, M.C.E.: Application of artificial intelligence in the health care safety context: opportunities and challenges. American Journal of Medical Quality, 35(4), 341–348 (2020).
46. Elgamml, M.M., Abas, F.S., Goh, H.A.: Semantic analysis in soccer videos using support vector machine. International Journal of Pattern Recognition and Artificial Intelligence, 34(9), Article ID 2055018 (2020).
47. Ellapan, V., Rajkumar, R.: Event detection in sports video based on audio-visual and support vector machine. Case-study: football. International Journal of Internet Technology and Secured Transactions, 9(1/2), 26–36 (2019).
48. Entezari, A., Aslani, A., Zahedi, R., Noorollahi, Y.: Artificial intelligence and machine learning in energy systems: A bibliographic perspective, Energy Strategy Reviews, 45, 101017 (2023).
49. ESPN NBA news: NBA to use Hawk-Eye tracking system to follow players, ball [online] https://www.espn.com/nba/story/_/id/35818363/nba-use-hawk-eye-tracking-system-follow-players-ball (last accessed on Sept 29 2023).
50. Fan, Z., Yan, Z., Wen, S.: Deep learning and artificial intelligence in sustainability: A review of SDGs, renewable energy, and environmental health. Sustainability. 15(18), 13493 (2023).
51. Farrokhi, A., Farahbakhsh, R., Rezazadeh, J., Minerva, R.: Application of internet of things and artificial intelligence for smart fitness: A survey. Computer Networks, 189, Article 107859 (2021).
52. Fenyes, D., Nemeth, B., Gaspar, P.: A predictive control for autonomous vehicles using big data analysis. IFAC-PapersOnLine, 52(5), 191–196 (2019).
53. Fister, I., Brest, J., Iglesias, A., Fister, I.: Framework for planning the training sessions in triathlon. Proceedings of the Genetic and Evolutionary Computation Conference (GECCO'18) Companion, 1829–1834 (2018).
54. Fister, I, Fister Jr., I., Fister, D.: *Computational Intelligence in Sports*. Springer Adaptation, Learning, and Optimization Series, 22 (2019).
55. Fister, I., Fister, D., Iglesias, A., Galvez, A., Rauter, S., Fister, I.: Population-based metaheuristics for planning interval training sessions in mountain biking. In: Advances in Swarm Intelligence. ICSI 2019. Lecture Notes in Computer Science,11655, 70–79 (2019).
56. Fister, I., Iglesias, A., Deb, S., Fister, D., Fister, I.: Development of a framework for modeling preference times in triathlon. Neural Computing and Applications 32, 10833–10846 (2020).

57. Fister, I., Iglesias, A., Galvez, A., Deb, S., Fister, D.: On deploying the artificial sport trainer into practice. Proc. of 8th International Conference on Soft Computing and Machine Intelligence (ISCMI), Cario, Egypt, IEEE, 21–26 (2021).
58. Fister, I., Iglesias, A., Osaba, E., Mlakar, U., Brest, J., Fister, I.: Adaptation of sport training plans by swarm intelligence. In: Recent Advances in Soft Computing, MENDEL 2017. Springer Advances in Intelligent Systems and Computing, 837, 56–67 (2019).
59. Fister, I., Ljubic, K., Suganthan, P.N., Perc, M., Fister, I.: Computational intelligence in sports: challenges and opportunities within a new research domain. Applied Mathematics and Computation, 262, 178–186 (2015).
60. Fister, Jr., I., Salcedo-Sanz, S., Iglesias, A., Fister, D., Gálvez, A., Fister, I.: New perspectives in the development of the artificial sport trainer. Applied Sciences 11 (23), 11452 (2021).
61. FiveThirtyEight web site: [online] https://fivethirtyeight.com (last accessed on Sept 29 2023).
62. Foster, D.: Generative deep learning: teaching machines to paint, write, compose, and play. O'Reilly Media Inc. (2019).
63. Frangoudes, F., Matsangidou, M., Schiza, E.C., Neokleous, K., Pattichis, C.S.: Assessing human motion during exercise using machine learning: A literature review. IEEE Access, 10, 86874–86903 (2022).
64. Geluvaraj, B., Satwik, P.M., Ashok Kumar, T.A.: The future of cybersecurity: major role of artificial intelligence, machine learning, and deep learning in cyberspace. International Conference on Computer Networks and Communication Technologies. Lecture Notes on Data Engineering and Communications Technologies, 15. Springer, Singapore (2019).
65. Giuliodori, P.: An artificial neural network-based prediction model for underdog teams in NBA matches. Proc. of the 4th Workshop on Machine Learning and Data Mining for Sports Analytics, European Conference on Machine Learning and Principles and Practice of Knowledge Discovery in Databases (ECML PKDD 2017), Skopje, Macedonia, 73–82 (2017).
66. Goodfellow, I., Bengio, Y., Courville, A.: *Deep Learning*. Adaptive Computation and Machine Learning series, MIT Press (2016).
67. Grigorescu, S., Trasnea, B., Cocias, T., Macesanu, G.: A survey of deep learning techniques for autonomous driving. Journal of Field Robotics, 37(3), 362–386 (2020).
68. Gupta, A.R.B., Agrawal, J.: A comprehensive survey on various machine learning methods used for intrusion detection system. Proc. of IEEE 9th Int. Conf. Commun. Syst. Netw. Technol. CSNT 2020, 282–289 (2020).
69. Halbouni, A. Gunawan, T.S., Habaebi, M.H., Halbouni, H., Kartiwi and M., Ahmad, R.: Machine learning and deep learning approaches for cyberSecurity: A review. IEEE Access, 10, 19572–19585 (2022).
70. Halilaj, E., Rajagopal, A., Fiterau, M., Hicks, J.L., Hastie, T.J., Delp, S.L.: Machine learning in human movement biomechanics: Best practices common pitfalls and new opportunities. J. Biomech., 81, 1–11 (2018).
71. Han, X., Laga, H., Bennamoun, M.: Image-based 3D object reconstruction: state-of-the-art and trends in the deep learning era. IEEE Transactions on Pattern Analysis and Machine Intelligence, 43(5), 1578–1604 (2021).
72. Harrison, C.K., Bukstein, S.: *Sport Business Analytics. Using Data to Increase Revenue and Improve Operational Efficiency*. CRC Press (2017).
73. He, L., Ren, Y., Cheng, X.: Decision support system for effective action recognition of track and field sports using ant colony optimization. Soft Computing, 1–11 (2023).
74. Henrique, B.M., Sobreiro, V.A., Kimura, H.: Literature review: machine learning techniques applied to financial market prediction. Expert Syst Appl, 124 226–251 (2019).
75. Hobson, H. Scientific Basketball for Coaches, Players, Officials, Spectators and Sportswriters, Second Edition. Prentice-Hall, Inc., New York (1955).
76. Hu, L., Saito, S., Wei, L., Nagano, K., Seo, J., Fursund, J., Sadeghi, I., Sun, C., Chen, Y.C., Li, H.: Avatar digitization from a single image for real-time rendering. ACM Transactions on Graphics, 36(6), 1–4 (2017).
77. Huang M. H., Rust R. T. (2021). A strategic framework for artificial intelligence in marketing. Journal of the Academy of Marketing Science, 49(1), 30–50

78. Iglesias, A., Gálvez, A., Suárez, P.: The role of artificial intelligence and machine learning for the fight against COVID-19. In: Hammouch, Z., Lahby, M., Baleanu, D. (Eds.): Mathematical Modeling and Intelligent Control for Combating Pandemics. *Springer Optimization and Its Applications*, vol. 203, 111–128 (2023).
79. Iglesias, A., Shin, J., Patel, B., Joshi, A.: *Proceedings of World Conference on Information Systems for Business Management: ISBM* 2023, Vols. 1–2. Springer Lecture Notes in Networks and Systems, Vols. 833, 834 (2024).
80. Inan, T., Cavas L.: Estimation of market values of football players through artificial neural network: a model study from the Turkish Super League. Applied Artificial Intelligence, 35(13), 1022–1042 (2021).
81. Jarvinen, J., Karjaluoto, H, The use of Web analytics for digital marketing performance measurement. Industrial Marketing Management, 50, 117–127 (2015).
82. Jurafsky, D., Martin, J.H.: *Speech and Language Processing: An Introduction to Natural Language Processing, Speech Recognition, and Computational Linguistics, 2nd edition*. Prentice-Hall, Upper Saddle River, NJ (2008).
83. Justesen, N., Bontrager, P., Togelius, J., Risi, S.: Deep learning for video game playing. IEEE Transactions on Games, 12(1), 1–20 (2020).
84. Kabudi, T., Pappas, I., Olsen, D. H.: AI-enabled adaptive learning systems: A systematic mapping of the literature. Computers and Education: Artificial Intelligence, 2, 100017 (2021).
85. Kakani, V., Nguyen, V.H., Kumar, B. P., Kim, H., Pasupuleti, V.R.: A critical review on computer vision and artificial intelligence in food industry. Journal of Agriculture and Food Research, 2, 100033 (2020).
86. Kaminskas, M., Ricci, F.: Contextual music information retrieval and recommendation: state of the art and challenges. Comput. Sci. Rev., 6(2), 89–119 (2012).
87. Kang, Y., Cai, Z., Tan, C.W., Huang, Q., Liu, H.: Natural language processing (NLP) in management research: A literature review. Journal of Management Analytics, 7(2), 139–172 (2020).
88. Kelleher, J.D., Mac Namee, N., D'Arcy, A.: *Fundamentals of Machine Learning for Predictive Data Analytics, Algorithms, Worked Examples, and Case Studies (2nd Edition)*. MIT Press (2020).
89. Khan, A.A., Laghari, A.A., Awan, S.A.: Machine learning in computer vision: A review. EAI Endorsed Scal Inf Syst., 8(32), 1–11 (2021).
90. Khurana, D., Koli, A., Khatter, K. et al. Natural language processing: state of the art, current trends and challenges. Multimed. Tools Appl., 82, 3713–3744 (2023).
91. Kiruthiga, A.S., Arunkumar, S., Thirisha, R., Lilian, J.F.: Role of machine learning in renewable energy. In: Sustainable Energy Solutions with Artificial Intelligence, Blockchain Technology, and Internet of Things, Chapter 3, CRC Press (2023).
92. Klein, C., Luig, P., Henke, T., Bloch, H., Platen, P.: Nine typical injury patterns in German professional male football (soccer): a systematic visual video analysis of 345 match injuries. British Journal of Sports Medicine, 55(7), 390–396 (2021).
93. Knight, R.F.: An Analysis of Basketball Offensive Efficiency Rating. M.Sc. Thesis, Kansas State University, Manhattan, Kansas (1967).
94. Kotaro, Y., Nakada, Y.: Fast implementation for computational method of optimum attacking play in rugby sevens. Modeling, Simulation and Optimization, Springer Nature, 97–109 (2022).
95. Kowsari, K.,Meimandi K.J., Heidarysafa, M., Mendu, S., Barnes, L., Brown, D.: Text classification algorithms: a survey. Information 10(4), 150 (2019).
96. Krstic, D., Vuckovic, T., Dakic, D., Ristic, S., Stefanovic, D.: The application and impact of artificial intelligence on sports performance improvement: A systematic literature review. Proc. of 4th International Conference on Communications, Information, Electronic and Energy Systems (CIEES), IEEE, 1–8 (2023).
97. Kruse, R., Mostaghim, S., Borgelt, C., Braune, C., Steinbrecher, M.: *Computational Intelligence. A Methodological Introduction (3rd Edition)*. Springer Texts in Computer Science, Springer (2022).

98. Kumyaito, N., Yupapin, P., Tamee, K.: Planning a sports training program using adaptive particle swarm optimization with emphasis on physiological constraints. BMC Research Notes, 11, Article 9 (2018).
99. Kuutti, S., Fallah, S., Katsaros, K., Dianati, M., Mccullough, F., Mouzakitis, A.: A survey of the state-of-the-art localization techniques and their potentials for autonomous vehicle applications. IEEE Internet of Things Journal, 5(2), 829–846 (2018).
100. Langaroudi, M.K., Yamaghani, M.R.: Sports result prediction based on machine learning and computational intelligence approaches: A survey. Journal of Advances in Computer Engineering and Technology, 5(1), 27–36 (2019).
101. Lee, G.J., Jung, J.J., Camacho, D.: Exploiting weighted association rule mining for indicating synergic formation tactics in soccer teams. Concurrency and Computation: Practice and Experience, 35(16), e6221 (2021).
102. Leung, C.K., Joseph, K.W.: Sports data mining: predicting results for the college football games. Proc. of 18th International Conference on Knowledge-Based and Intelligent Information and Engineering Systems, Procedia Computer Science, 35, 710–719 (2014).
103. Lewis, M.: *Moneyball: The Art of Winning an Unfair Game*. W.W. Norton & Company (2003).
104. Li, B.H., Hou, B.C., Yu, W:T., Lu, X.B., Yang, C.W.: Applications of artificial intelligence in intelligent manufacturing: a review. Frontiers Inf. Technol. Electronic Eng. 18, 86–96 (2017).
105. Li, S., Jang, S., Sung, Y.: Automatic melody composition using enhanced GAN. Mathematics, 7, 883 (2019).
106. Lim, S.M., Oh, H.C., Kim, J., Lee, J., Park, J.: LSTM-guided coaching assistant for table tennis practice. Sensors, 18(12), 4112 (2018).
107. Liu, N., Liu, L., Sun, Z.: Football game video analysis method with deep learning. Computational Intelligence and Neuroscience, vol. 2022, Article ID 3284156 (2022).
108. Liu, L., Ouyang, W., Wang, X., Fieguth, P., Chen, J., Liu, X., Pietikainen, M.: Deep learning for generic object detection: a survey. Int. J. Comput. Vis., 128, 261–318 (2020).
109. Lopez de Prado, M.: *Advances in Financial Machine Learning*. Wiley (2018).
110. Lopez-Gomez, J.A.,, Romero, F.P., Angulo, E.: A feature-weighting approach using metaheuristic algorithms to evaluate the performance of handball goalkeepers. IEEE Access, 10, 30556–30572 (2022).
111. López-Serrano, C,, Zakynthinaki, M., Mon-López, D., Molina, J.J.: Contextualizing evaluation of performance in volleyball: introducing contextual individual contribution coefficients to assess technical actions. Perceptual and Motor Skills, 130(6) 2663–2684 (2023).
112. Lu, H., Li, Y.: *Artificial Intelligence and Computer Vision*. Springer Studies in Computational Intelligence, 672 (2017).
113. Ma, S., Zhang, X., Jia, C., Zhao, Z., Wang, S., Wang, S.: Image and video compression with neural networks: a review. IEEE Trans. Circuits Syst. Video Technology, 30(6), 1683–1698 (2020).
114. Madhavan, V.: Predicting NBA Game Outcomes with Hidden Markov Models, Berkeley University, (2016).
115. Mahadevkar, S. V., Khemani, B., Patil, S., Kotecha, K., Vora, D.R., Abraham, A., Gabralla, L.A.: A review on machine learning styles in computer vision –techniques and future directions. IEEE Access, 10, 107293–107329, (2022).
116. Mahanjan, P.: *Artificial Intelligence in Healthcare: AI, Machine Learning, and Deep and Intelligent Medicine Simplified for Everyone (3rd edition)*. MedMantra (2021).
117. Maheswaran, R., Chang, Y.H., Henehan, A., Danesis, S.: Deconstructing the rebound with optical tracking data. *Proceedings of 6th MIT Sloan Sports Analytics Conference*, Boston, MA, USA (2012).
118. Maheswaran, R., Chang, Y.H., Su, J., Kwok, S., Levy, T., Wexler, A., Hollingsworth, N.: The three dimensions of rebounding. *Proceedings of 8th MIT Sloan Sports Analytics Conference*, Boston, MA, USA (2014).
119. Makino, K., Kitano, Y., Nishizaki, H.: Classification of swing motion of tennis using a recurrent-based neural network. Proc. 12th Int. Conf. Human Syst. Interact. (HSI), 237–242 (2019).

120. Manesh M.F., Pellegrini, M.M., Marzi G., Dabic, M.: Knowledge management in the fourth industrial revolution: Mapping the literature and scoping future avenues. IEEE Transactions on Engineering Management, 68(1), 289–300 (2021).
121. Maszczyka, A., Golasa, A., Stanulaa, A., Pietraszewskia, P., Roczniok a, R., Zajaca, A.: Application of neural and regression models in sports results prediction. Elsevier, Social and Behavioral Sciences 117 (2014).
122. Mauro, A., Pacifico, M.: Data-driven transformation: how to drive substantial business value with data analytics. Financial Times Publishing (2024).
123. Min, B., Kim, J., Choe, C., Ian, R.: A compound framework for sports prediction: the case study of football. Knowledge-Based Systems, 21(7), 551–562 (2008).
124. Moore, C.J.: Meet Synergy Sports Technology, the Company Revolutionizing College Basketball. Bleacher Report (2013).
125. Murphy, K.P.: *Probabilistic Machine Learning. An Introduction*. Adaptive Computation and Machine Learning series, MIT Press (2022).
126. Murphy, K.P.: *Probabilistic Machine Learning. Advanced Topics*. Adaptive Computation and Machine Learning series, MIT Press (2023).
127. Mutijarsa, K., Ichwan, M., Utami, D.B.: Heart rate prediction based on cycling cadence using feedforward neural network. Proc. of International Conference on Computer, Control, Informatics and Its Applications (IC3INA), IEEE, 72–76 (2016).
128. Nalbant, K.G., Aydin, S.: Literature review on the relationship between artificial intelligence technologies with digital sports marketing and sports management. Indonesian Journal of Sport Management, 2(2), 135–143 (2022).
129. Naranjo, J.E., Serradilla, F., Nashashibi, F.: Speed control optimization for autonomous vehicles with metaheuristics. Electronics, 9 (4) 1–15 (2020).
130. NBAstuffer Reference web site: [online] https://www.nbastuffer.com/ (last accessed on Sept 29 2023).
131. Nikitina, M.A.: Development of a personalized diet using structural optimization. Society 5.0: Cyber-Solutions for Human-Centric Technologies, Springer Nature Switzerland, 43–52 (2023).
132. Oliveira, R.S., Leicht, A.S., Bishop, D., Barbero-Alvarez, J.C., Nakamura, F.Y.: Seasonal changes in physical performance and heart rate variability in high level futsal players. Int. J. Sports Med., 34, 424–430 (2013).
133. Ohanian, T.: How artificial intelligence and machine learning may eventually change content creation methodologies. SMPTE Motion Imaging Journal 128(1), 33–40 (2019).
134. Otter, D.W., Medina, J.R., Kalita, J.K. A survey of the usages of deep learning for natural language processing. IEEE Transactions on Neural Networks and Learning Systems, 32(2), 604–624 (2020).
135. Owramipur, F., Eskandarian, P., Mozneb, F.S.: Football result prediction with Bayesian network in Spanish League–Barcelona Team. International Journal of Computer Theory and Engineering, 5(5), 812–815 (2013).
136. Paden, B., Cap, M., Yong, S.Z., Yershov, D., Frazzoli, E.: A survey of motion planning and control techniques for self-driving urban vehicles. IEEE Transactions on Intelligent Vehicles, 1(1), 33–55, (2016).
137. Pappalardo, L., Cintia, P., Ferragina, P., Massucco, E., Pedreschi, D., Giannotti, F.: PlayeRank: data-driven performance evaluation and player ranking in soccer via a machine learning approach. ACM Trans. Intell. Syst. Technol. 10(5), Article 59 (2019).
138. Patricio, D.I., Rieder, R.: Computer vision and artificial intelligence in precision agriculture for grain crops: A systematic review. Computers and Electronics in Agriculture, 153, 69–81 (2018).
139. Petrunin, Y.Y.: Analysis of the football performance: from classical methods to neural network. Journal of Journal of Human Activity Theory, 2, (2011).
140. Provost, F. Fawcet, T.: Data Science for Business: What You Need to Know About Data Mining and Data-Analytic Thinking. Athenaeum Uitgeverij (2013).

141. Rajsp, A., Fister Jr., I.: A systematic literature review of intelligent data analysis methods for smart sport training. Applied Sciences, 10(9), 3013 (2020).
142. Rana, M., Mittal, V.: Wearable sensors for real-time kinematics analysis in sports: A review. IEEE Sensors J., 21(2), 1187–1207 (2021).
143. Ratten, V., Dickson, G.: *Statistical Modelling and Sports Business Analytics*. Routledge, London (2020).
144. Reddy, P.S., Ghodke, P.K., Reddi, K., Akiti, N.: Recent developments of artificial intelligence for renewable energy. In: Sustainable Energy Solutions with Artificial Intelligence, Blockchain Technology, and Internet of Things, Chapter 1, CRC Press (2023).
145. Rodriguez-Gallego, L., López-Contreras, G., Garcia, G.A.D., Cuenca-Fernández, F.: Assessment of feedback devices for performance monitoring in master's swimmers. International Journal of Performance Analysis in Sport, 22(5), 701–714 (2022).
146. Rotshtein, P., Posner, M.,Rakityanskaya, A.B.: Football predictions based on a fuzzy model with genetic and neural tuning. Cybernetics and Systems Analysis, 41(4), 619–630 (2015).
147. Russell, S., Norvig, P.: *Artificial Intelligence: A Modern Approach (3rd Edition)*. Pearson (2009).
148. Sagar, B.S.: Providing cyber security using artificial intelligence—A survey. Proc. of 3rd Int. Conf. Comput. Methodol. Commun., 717–720 (2019).
149. Sakabe, H., Nakada, Y.: Computational method for determining optimal dribbling routes in basketball. Proc. of IEEE Eighth International Conference on Multimedia Big Data (BigMM), 107–08 (2022).
150. Sakabe, H., Nakada, Y.: Enhanced method for computing optimal dribbling routes using tracking data in basketball. Proc. of IEEE Ninth Multimedia Big Data (BigMM), IEEE, 11–18 (2023).
151. Saidali, J., Rahich, H., Tabaa, Y., Medouri, A.: The combination between big data and marketing strategies to gain valuable business insights for better production success. Procedia Manufacturing, 32, 1017–1023 (2019).
152. Salih, A., Zeebaree, S.T., Ameen, S., Alkhyyat, A. Shukur, H. M.: A survey on the role of artificial intelligence, machine learning and deep learning for cybersecurity attack detection. Proc. of 7th International Engineering Conference Research & Innovation amid Global Pandemic (IEC), 61–66 (2021).
153. Explore the FC Bayern journey with SAP [online] https://www.sap.com/about/customer-stories/fcbayern.html (last accessed on Sept 29 2023).
154. Saura, J.R.: Using data sciences in digital marketing: framework, methods, and performance metrics. Journal of Innovation & Knowledge, 6 (2), 92–102 (2021).
155. Sautoy, M.D.: *The Creativity Code: Art and Innovation in the Age of AI*. Harvard University Press (2019).
156. Schwarting, W., Alonso-Mora, J., Rus, D.: Planning and decision-making for autonomous vehicles. Annual Review of Control, Robotics, and Autonomous Systems, 1(1), 187–210 (2018).
157. Sen, A., Hossain, S.M.M., Uddin, R., Deb, K., Jo, KH. (2022). Sequence recognition of indoor tennis actions using transfer learning and long short-term memory. In: Frontiers of Computer Vision. IW-FCV 2022. Communications in Computer and Information Science, 1578, 312–324 (2022).
158. Serner, A., Mosler, A.B., Tol, J.L., Bahr, R., Weir, A.: Mechanisms of acute adductor longus injuries in male football players: a systematic visual video analysis. British Journal of Sports Medicine, 53(3), 158–164 (2019).
159. Singh, A., Wiktorsson, M., Hauge, J.B.: Trends In machine learning to solve problems In logistics. Procedia CIRP, 103, 67–72 (2021).
160. Stockl, M., Morgan, S.: Visualization and analysis of spatial characteristics of attacks in field hockey. International Journal of Performance Analysis in Sport, 13(1), 160–78 (2013).
161. Stone, J.V.: *Intelligence Engines: A Tutorial Introduction to the Mathematics of Deep Learning*. Sebtel Press (2009).

162. Sundaram, S., Zeid, A.: Artificial intelligence-based smart quality inspection for manufacturing. Micromachines, 14, 570 (2023).
163. Sutton, R.S., Barto, A.G.: *Reinforcement Learning, An Introduction (second edition)*. Adaptive Computation and Machine Learning series, MIT Press (2018).
164. Synergy Sports Technology web site: [online] https://synergysports.com/ (last accessed on Sept 29 2023).
165. Szeliski, R.: *Computer Vision: Algorithms and Applications*. New York, NY, USA, Springer (2010).
166. Thakkar, A. Lohiya, R.: A review of the advancement in intrusion detection datasets. Procedia Comput. Sci., 167, 636–645 (2020).
167. Taulli, T.: *Artificial Intelligence Basics: A Non-Technical Introduction (1st Edition)*. Apress (2019).
168. Taylor, B.: *Thinking Basketball*. CreateSpace Independent Publishing Platform (2016).
169. The Guardian newspaper: Are super-nerds really ruining US sports? [online] https://www.theguardian.com/sport/2018/aug/16/sports-nerds-analytics-data (last accessed on Sept 29th 2023).
170. Tilp, M., Schrapf, N.: Analysis of tactical defensive behavior in team handball by means of artificial neural networks. IFAC-Papers On-Line 48-1 (2015).
171. Tomczak, J.M.: *Deep Generative Modeling*. Springer (2022).
172. Troilo, M., Urban, T.L., Mondello, M., Sutton, W.A.: Business analytics, revenue management and sport: evidence from the field. International Journal of Revenue Management, 11(4), 277–296 (2020).
173. Ugail, H., Aggarwal, R., Iglesias, A., Suarez, P., Maqsood, M., Aadil, F., Campuzano, A., Gleghorn, S., Mehmood, I., Taif, K.: An optimisation model for designing social distancing enhanced physical spaces. 2020 International Conference on Internet of Things and Intelligent Applications (ITIA), Zhenjiang, China. IEEE CS Press, 1–5 (2020).
174. Ugail, H., Mehmood, I., Aggarwal, R., Iglesias, A., Suarez, P., Maqsood, M., Aadil, F., Campuzano, A., Tajouri, L., Gleghorn, S., Taif, K., Reeve, C., Wan, T., Redman, R.: On rearranging physical spaces for enhancing social distancing measures to combat the COVID-19 infection rates. 2020 International Conference on Internet of Things and Intelligent Applications (ITIA), Zhenjiang, China. IEEE CS Press, 1–5 (2020).
175. Ugail, H., Aggarwal, R., Iglesias, A., Howard, N., Campuzano, A., Suarez, P., Maqsood, M., Aadil, F., Mehmood, I., Gleghorn, S., Taif, K., Seifedine, K., Muhammad, K.: Social distancing enhanced automated optimal design of physical spaces in the wake of the COVID-19 pandemic. Sustainable Cities and Society, 68, Paper ID 102791 (2021).
176. Ugurlu, M. Dogru, I. A.: A survey on deep learning based intrusion detection system. Proc. of 4th Int. Conf. Comput. Sci. Eng., 223–228 (2019).
177. Van Bulck, D., Vande Weghe, A., Goossens, D.: Result-based talent identification in road cycling: discovering the next Eddy Merckx. Annals of Operations Research, 325, 539–556 (2023).
178. Vaz, M.S., Ribeiro, Y.S., Pinheiro, E.S., Del Vecchio, F.B., Psychophysiological profile and prediction equations for technical performance of football players, Revista Brasileira de Ciencias do Esporte, 41(2), 215–221 (2019).
179. Wakelam, E., Steuber, V., Wakelam, J.: The collection, analysis and exploitation of footballer attributes: A systematic review. J. Sport. Anal., 8, 31–67 (2022).
180. Wang, G., Gunasekaran, A., Ngai, E.W.T., Papadopoulos, T.: Big data analytics in logistics and supply chain management: Certain investigations for research and applications. International Journal of Production Economics, 176, 98–110 (2016).
181. Wang, W., Gang, J.: Application of convolutional neural network in natural language processing. Proc. of International Conference on Information Systems and Computer Aided Education (ICISCAE 2018), IEEE, 64–70, (2018).
182. Wang, Y., Zheng, P., Xu, X., Yang, H., Zou, J.: Production planning for cloud-based additive manufacturing–A computer vision-based approach., Robot. Comput. Integr. Manuf., 58, 145–157 (2019).

183. Wang, X., Ellul J., Azzopardi, G.: Elderly fall detection systems: A literature survey. Frontiers in Robotics and AI, 7, 71 (2020).
184. Wang, Z.: Semantic analysis based on fusion of audio/visual features for soccer video. Procedia Computer Science, 183, 563–571 (2021).
185. Wiafe, I., Koranteng, F.N., Obeng, E.N., Assyne, N., Wiafe, A., Gulliver, S.R.: Artificial intelligence for cybersecurity: A systematic mapping of literature. IEEE Access, 8, 146598–146612 (2020).
186. Wilson, J.: *Inverting the Pyramid—The History of Football Tactics*. Nation Books (2018).
187. Xiong, S., Li, X.: Intelligent strategy of internet of things computing in badminton sports activities. Wireless Communications and Mobile Computing, 1–9 (2022).
188. Xu, T., Zhou, Y., Zhu, J.: New advances and challenges of fall detection systems: a survey. Appl. Sci. 8, 418 (2018).
189. Yan, W.Q.: *Computational Methods for Deep Learning. Theory, Algorithms, and Implementations*. Springer Texts in Computer Science, Springer (2023).
190. Yang, T.Y., Swartz, T.: A two-stage Bayesian model for predicting winners in Major League Baseball. Journal of Data Science, 2(1), 61–73 (2004).
191. Young, T., Hazarika, D., Poria, S., Cambria, E.: Recent trends in deep learning based natural language processing. IEEE Computational Intelligence Magazine, 13(3), 55–75 (2018).
192. Yurtsever, E., Lambert, J., Carballo, A., Takeda, K.: A survey of autonomous driving: Common practices and emerging technologies. IEEE Access, 8, 58443–58469 (2020).
193. Zaib, A., Ahmad, M.T.: Research on biomechanical analysis of football player using information technology in sports field. Journal of Sport Psychology, 31(3), 21–30 (2022).
194. Zhilenkov, A., Chernyi, S., Emelianov, V.: Application of artificial intelligence technologies to assess the quality of structures. Energies, 14, 8040 (2021).
195. Zhang, H. Lin,K.Y., Chen, W., Genyuan, L.: Using machine learning techniques to improve Intrusion detection accuracy. Proc. 2nd IEEE Int. Con;f. Knowl. Innov. Invent. 2019 ICKII, 308–310 (2019).
196. Zhang, Y., Zhang, Y.: Sports training system based on convolutional neural networks and data mining. Computational Intelligence and Neuroscience, 2021, Article ID 1331759 (2021).
197. Zhao, Q.J., Cao, P., Tu, D.W.: Toward intelligent manufacturing: Label characters marking and recognition method for steel products with machine vision. Adv. Manuf., 2(1), 3–12 (2014).
198. Zhou, H., Li., Y.: Design of intelligent analysis system of basketball skilled movement based on data mining technology. Proc. of IEEE 5th Eurasia Conference on IOT, Communication and Engineering (ECICE), IEEE, 457–459 (2023).
199. Zhou, L. Zhang, L., Konz, N.: Computer vision techniques in manufacturing. IEEE Transactions on Systems, Man, and Cybernetics: Systems, 53(1), 105–117 (2023).
200. Zhu, R.: Application of business analysis in sports business. Proc. of the 2020 2nd International Conference on Economic Management and Cultural Industry (ICEMCl 2020), 129–132 (2020).

An Ecological Dynamics Approach to the Use of Artificial Intelligence and Machine Learning to Analyze Performance in Football

Sofia Ferreira, Daniel Carrilho, and Duarte Araújo

Abstract Technological advances in the measurement of players' activities have revealed new ways for performance analysis in sports. From wearable devices to optical tracking systems, positional data, events data, and biological parameters are becoming more accessible and thus collected in bigger amounts, which opens opportunities for the use of artificial intelligence (AI) techniques in sports, such as machine learning (ML). This chapter aims to discuss AI and ML applications to football, especially from a performance analysis perspective. We also present how ecological dynamics theoretical approach can be used as guidance when studying performance. We finalize by presenting a case study where a new spatial-temporal indicator called density zone is described.

Keywords Artificial intelligence · Machine learning · Football · Performance · Density zones

1 Introduction

In a digital era, technology is embedded in our lives, and its usage is extended in different domains. Although with application in many areas like finance, logistics, or medicine, data analysis in sports has been following its own path and growth [6]. In this book chapter, we discuss theoretical and applied perspectives for the use of AI in sports, particularly football.

Collective sports can be analyzed at different levels: the individual player performance, the team as a collective unit, and the match with both teams. So, depending on the aim of our analysis, different performance questions can be made, and different metrics are required. What is considered a performance indicator for

S. Ferreira (✉) · D. Carrilho · D. Araújo
CIPER, Faculdade de Motricidade Humana, Universidade de Lisboa, Cruz Quebrada-Dafundo, Portugal
e-mail: a.sofia2@edu.ulisboa.pt; carrilho.daniel@edu.lisboa.pt; daraujo@fmh.ulisboa.pt

M. J. Blondin et al. (eds.), *Artificial Intelligence, Optimization, and Data Sciences in Sports*, Springer Optimization and Its Applications 218,
https://doi.org/10.1007/978-3-031-76047-1_6

doctors or physiotherapists does not represent necessarily a performance indicator for coaches.

This chapter is specifically dedicated to football, the sport with more than five billion football fans around the world [30]. In elite football, performance is routinely analyzed [59, 68, 76], and due to its complexity, team performance can be evaluated from different perspectives. Definitely, it is important to understand physical demands during a competition or a season, analyzing which skills are needed to achieve the best performance [60], making sure players are able to participate in different competitions during the week in his/her best shape [13, 43, 51] or even analyzing how injuries can be managed or its risk [7]. But this is just a part of the full picture. It is also relevant, for instance, to understand how players are interacting with each other during a match [42] or how many goals or opportunities the team is able to do during the match [44, 62]. These are different questions already being asked by sport scientists and practitioners. All of these questions are based on improving performance, with the final aim of improving team success.

In short, football has evolved over the years, together with the development of technological and computerized systems, enabling a more in-depth understanding of performance. Recent reviews of performance analysis in soccer (e.g., [57, 58]) showed that match analysis is slowly, but progressively, considering the dynamic, interactive properties in complex systems that explain performance [26]. Aligned with this idea, Araújo [4] proposed that analysts need to move beyond documenting "performance statistics" to study the emergent behaviors among players, in key areas of the field, which underpin success in soccer. Motion analysis methods help to identify players' performances in a continuum and transitions in their course of action during competitive performance [71] in futsal. These studies develop our understanding of how players interact with the performance environment to create and use information to support successful decision-making and actions.

Considering these aspects requires an extended knowledge in each specific subdomain, but could all of them be supported by the same dataset? Could AI be used to help understand better each phenomenon and have a more global performance picture? We argue that data can be collected, treated, and analyzed to help physiologists, coaches, doctors, and players to have contextualized knowledge about team and players performance and make their professional decisions.

This chapter starts by briefly describing big data evolution in sports. Secondly, it discusses how key performance indicators are relevant, suggesting literature examples of how AI is being used. Thirdly, it presents ecological dynamics approach as the theory that can guide AI applications in sports. Fourthly, it presents an innovative proposal of analyzing collective behavior in football called density zones based on positional data usage.

2 Big Data in Sports

Sport sciences are actively following the technological development, with particular emphasis on global navigation satellite systems (GNSS) and optical tracking systems, which allow practitioners to have access to detailed player positional data. These systems collect and save data with higher frequencies than before, e.g., 25 Hz, and higher resolution, generating big amounts of data that are typically complemented with other data sources such as biological (e.g., heartbeat), sleep, clinical information, or athletes' periodic well-being reports. This large variety of data, in many cases unstructured and typically managed by different types of profiles inside an organization, originates new challenges and demands in data management.

The "value" of big data refers to how an organization can extract usable information from the complex sport environment. Since only a small part of the data is available to be analyzed, it is crucial to implement a management process that contemplates different data processing stages and clearly identifies which data sources are needed when analyzing sport performance. For example, when players are wearing GPS systems during a training session, the entire session is being recorded, but probably only a small part of that dataset will be actually used by physiologists to summarize that training session demands in a report. Or, when accessing tracking data from an optical tracking system, hundreds of events are available, but probably the technical team wants to select specific parts of the match, depending on the coaches' demands. More than the data that can be obtained, it is important to ask what data the sport organization needs. To answer such question about sport performance, a theory about performance is needed. Such theory would guide toward the most relevant variables to be measured, instead of simply collecting data because it is technologically available. Sport scientists are concerned with using data not only to maximize one's ability to predict meaningful outcomes but also to develop and further elaborate theories that provide meaningful explanations for observed relationships in performance, practice, and development [2, 6]. However new explanations from big data and artificial intelligence seem to be rare in practice. It may be due to the complex interactions between players, as it happens in team sports, and the lack of theory informing the use of methods. For example, there are many inconsistencies on how match events are identified. This is where the ecological dynamics approach to sport performance comes in.

The dynamical interactions of an athlete with a performance environment help her detect information from different perceptual modalities and guide decision-making and actions [17]. The ecological dynamics approach stresses the primacy of establishing functional individual-environment relations in understanding sport behavior. Studying the couplings formed between a performer and performance environment helps us to understand complex aspects of human behavior in the actual world such as moving around, selecting paths, or deciding with whom to cooperate and compete. Action is an ecologically flexible process (self-organized, emergent). Sport actions that emerge in a performance environment are holistic entities that are different from the components that make up the system (e.g., the field, the ball, performer's legs) and cannot be predicted solely from the characteristics of those components. Consequently, many sorts of solutions to a sport problem can emerge given the many ways its elements can interact under the same constraints. But instead of being a random process or, in the other extreme, a process that is internally programmed in advance, performers are perceptively attuned to affordances (action possibilities offered by the environment, [32]) from the context that guide self-organizing action toward achieving (or not) a task goal [27]. The dynamic process implied in the perception of affordances provides the basis by which performers can (self) regulate their behavior prospectively.

Ecological dynamics construes that goal-directed behaviors in sports emanate from the hard-assembled (physical) and soft-assembled (informational) links between performers and their performance environments [41]. This idea implies that ecophysical variables [5, 6] provide the most appropriate starting point to capture sport behavior. These are variables that express the fit between the environment and the performer's adaptations. As mentioned before, environmental properties may directly inform what an individual can and cannot do [32]. For example, Carrilho et al. [19] captured team synergic properties, using spatiotemporal data from one match of the 2018 FIFA World Cup in Russia, to analyze the ecophysical variable "player-ball-goal angles" in a cluster phase analysis. Relative phase analysis (RPA) has been used in team sports to measure synchronization (e.g., [69]). RPA is based on the difference of the oscillation between two phases, expressed as a measure of their relative phase angle, defined by their angular frequency (ω) and initial phase (ϕ_0). Two oscillators in a time series can be locked in an in-phase mode of synchronization if their angle difference holds near 0° or in an anti-phase if the angle is near 180°. Based on this idea, cluster phase analysis (CPA) also measures synchronization, but it is based on the relationship among the oscillatory movements of a group of elements, using the Kuramoto order parameter [28]. Therefore, differently from RPA, CPA measures synchrony among the oscillating phases of the elements within a group. In the study of Carrilho et al. [19], they analyzed player-ball-goal angles (PBGA) in CPA to directly measure player-environment relationships and capture the properties of team synergic behavior [2]. The ball, the goals, and the players were identified as key ecological variables of the football match, as they inform and constrain the behavior of players. Therefore, by using the PBGA they capture big data that informs about the link between players and performance environment, providing ecological explanatory value to CPA. The synchronization measures obtained from CPA were a direct result of changes in the PBGA, which continually captured the displacement of the players on the pitch, in relation to the goal and the ball.

An important challenge for researchers and practitioners using big data and artificial intelligence is to capture ecophysical variables in their work to enable understanding of how sport performance might be predicated on perception-action couplings during continuous, emergent, performer-environment interactions.

From a resource-based perspective, greater access to big data and the knowledge and skills for their analysis can lead to sustainable performance advantages [72]. Collecting and managing big data facilitates the development of analytical tools like the ones used to produce reporting. Big data and analytics need to develop together, by creating bridges to foster validated sources (also called "golden sources"), timely updated and structured to be analyzed in a "performance-driven" way.

3 Artificial Intelligence in Sport Performance Analysis: The Importance of Relevant Key Performance Indicators

AI and ML are part of the digital revolution and are bringing novel perspectives also to sport performance analysis. Hughes et al. [37] indicate that the practical value of performance analysis is that well-chosen performance indicators highlight successful or unsuccessful influences on performance output. Key performance indicators (KPIs) are typically used to help stakeholders take better decisions or, at least, provide insight information that can be wisely used by experts to better understand a phenomenon [50]. But how should a KPI be built? Are KPIs well defined in sports? Or if not, how can they be improved? Sport experts collect data from their athletes and use these data to better understand their training and competition performance. But data was mainly collected by different means, by different people in different competitions. Scouting in football is a good example of that. When asking a scout how they decide on a relevant player during a certain match, probably their performance decision criteria will be different from that of another scout, considering the performance by different indicators and with high-level inferences [14, 15]. If by chance both are collecting data from the same match, their results may not be directly comparable. Consequently, data availability, methodological application, and task expertise need to work together to generate robust key performance indicators and avoid random interpretations that are difficult or impossible to assess. Therefore, expertise is essential for the relevant use of data, and data can be used to generate more task expertise. When well implemented, it works as a bidirectional interaction with key performance indicators in the center.

Nowadays, there are databases available with historical player statistics such Wyscout, Stats Perform, or Statsbomb. But there are some concerns about using data, namely, (1) data quality and cost, (2) the analyst has knowledge about how a model should be developed, and (3) which technology should be used to perform that development. Moreover, there are already different scouting apps available in the market, which do not necessarily implicate any model or computation knowledge to use it. Again, these applications are only available because technological development existed, but they are not completely customized to what the analyst needs. The quality of the data, the information about how the model was built, and the associated user costs are important factors to take into account when considering its usage. Nevertheless, generally speaking, either building an "in-house" model or using a third-party app, key performance indicators should act as a performance assessment driver for a specific phenomenon explanation. But how can AI contribute to that? Nowadays, data is everywhere, and having algorithms capable of dealing with this big amount of data is essential. AI can provide such algorithms. This implies that sports and data expertise can connect, giving the possibility of creating relevant information for sports professionals.

In football, both physical and tactical indicators are needed for a better team and player management. Decades ago, coaches would need to write on paper any annotation they would like to do about their players. But considering a team of

11 players, playing at the exact same time, how could this be physically possible? Perhaps looking for one player at a time could work, but how contextualized is this? And how dynamically can it be analyzed? Measuring maximum running speed during a workout training is not exactly the same as measuring the contextual information about how that type of running happens during a match or who was next to him/her when it happens [54]. Then, video started to be used as a match analysis tool. With video, analysts can watch the match several times, reviewing what happens, cutting and highlighting what they want. It can represent a huge difference compared to human observational analysis only. Even so, when video started to be used, it was also dependent on the subjective analysis of analysts. How to measure, for instance, how far one player was from another one? How to define what was a fast attack from a non-fast attack? Would this change if two different analysts were checking the same match? To answer this type of questions, new technologies started to be developed, giving the possibility to measure distance based on video images [67]. That was a great achievement. It means that analysts could actually measure distances and more accurately calculate player/match values, not without error but with much fewer erratic interpretations. However, the first ways of doing it were not very efficient. It required many hours tackling all the points so the system could translate it in data coordinates from where variables could then be calculated. If considering during a competition where each match occurs with 2 or 3 days apart, it required a lot of effort and time consuming just to have all matches data points, which means, although possible, the solution was not exactly able to retrieve in a timely manner. Technology and methodologies evolved, and AI turned more actively applied to sports. Different companies (e.g., Stats Perform) started to propose optical tracking systems capable of recognizing players and the ball and, on top of that, retrieving a full dataset after each match ends. These data providers also developed apps capable of retrieving real-time data, allowing the technical team to interpret predefined calculated measures during the match. In parallel, other companies (e.g., STATSports, KINEXON Sports, Catapult, WIMU) developed wearable systems capable of measuring time, distance, or velocity. What previously required hours and hours of measurement through human observation now only takes a few minutes or a couple of hours to be available.

So now that data is available, were sports scientists and technical teams ready to ask questions about this big amount of data? Were sports practitioners able to follow this technological progress? How can they best take advantage of the data? These questions can be answered perhaps by proposing interesting key performance indicators, making sure that metrics are more accurately calculated from a trustworthy source and are available in a timely manner. A key performance indicator should help sports practitioners to know better their phenomena, despite being physical or more tactical questions. To do so, we need to start from the beginning, knowing first how trustworthy a source is, what variables do I want to calculate to get my answers, and how do I retrieve them from the data so I can analyze my key performance indicators dynamically. AI proposes promising methodologies to be used in this subject.

Nowadays, there are mainly two datasets being used: match events datasets and tracking datasets. Regarding tactical performance analysis, match events data and match positional tracking data are the commonly used data sources [21, 33, 77]. Match events data is a type of data source that identifies match actions like passing, dribbling, or shooting while the match is occurring, being already performed by automatic AI software. When events are synchronized with positional data, it is possible to analyze detailed and specific match performance. For example, a commonly used KPI in sports is the goal scoring probability, also known as expected goals (xG) [56]. xG measures quantitatively the probability that a shot turns into a goal based on various factors such as the distance and angle of the shot to the goal, the number of defenders between the ball and the goal, or the position of the goalkeeper. However, in sports with a low scoring nature such as football, calculating the probability that an action, such as a shot, results in a goal might be challenging. This happens because the goal/shot ratio is low and goals can be scored in a variety of ways, from different locations, and by any player on the pitch.

The literature shows that most of the studies related to AI and ML in football are investigating pattern recognition, performance prediction, and injury prevention [6]. Prediction is being studied from different perspectives. Barron et al. [10], for example, investigated performance indicators through a neural network algorithm, trying to identify predictive indicators of a player's career. Jaspers et al. [39] studied the ML capacity of predicting perceived exertion rate (RPE) for a training session also recurring to neural networks. From another perspective but still related with forecasting, Pappalardo and Cintia [47] assessed the quantification between players' performance and success, presenting a ranking by team. The authors analyzed ten million game events like tackles, dribbles, or passes and quantified the relation between those technical performance metrics and the team's success, measured by the team's final ranking. ML algorithm was used to understand how technical metrics could explain game outcome. They continued their investigation later [48] where "PlayeRanking" was suggested as a new generation of data-driven framework to evaluate football players.

Regarding pattern recognition, there are studies recognizing activity patterns such as walking or jumping through ML algorithms [9, 25] or tactical patterns of team activities through the identification of football actions like ball possession, quick attack, and set pieces [45].

A big part of these studies recurred to ML techniques, considering hundreds of input features which, as mentioned previously, are usually obtained from match events frequencies, e.g., number of goals, or from biological parameters collection, e.g., heart rate and the number of accelerations and decelerations [55]. Based on algorithm results, the most important features are then pointed as performance indicators [1, 8, 11, 34]. However, there is still a lack of studies suggesting variables capable of measuring not only individual aspects but also the relation between the player and the environment. We argue that ecological dynamics theoretical approach may be a helpful guide.

4 An Ecological Dynamics Approach to Sport Performance Analysis

Due to the complexity of a football match, a robust development of KPIs needs a theoretical rationale to be useful to practitioners when choosing metrics and making training session decisions. The ecological dynamics theoretical approach to sport behavior [3, 6] argues that athletes' actions are explained at the level of individual-environment system. This means that the environment should be perceived in behavioral terms, or as Gibson [32] argued, the environment is perceived in terms of opportunities for action called "affordances." Affordances indicate how an athlete can act within a specific performance environment guided by the detection of ecological information. This information is not based on the physical (neutral) properties of the environment (e.g., the ball is 3 m away) but instead is based on the individual ability of an athlete to act in such environment. For example, a defender in football perceives an opportunity to slide tackle not by calculating the distance between himself/herself and the ball or by calculating the speed of the attacker with the ball. Instead, they do so by detecting optical, acoustical, or haptical information that shows how to tackle according to their slide tackle ability in that moment.

Football players tend to move in coordinated patterns having as objective to invade their opponent's space and score a goal. This coordination occurs at different levels: each player organizes himself in a neuromusculoskeletal way so it can perform a pattern of movement [73] as much as the different players coordinate between them both in attacking or defending situations. Players need to detect information about the match, so their actions are adapted in a proper way to the environmental demands. A player who is playing their first match in a final competition probably won't have the same feeling from another player who is ending his career and is doing their last match. Playing with a stadium full of adepts probably will be also different from playing in a stadium completely empty. Playing with different opponents can also represent different psychological pressures to players, consequently affecting team behavior. So, contextual environment matters, and ecological dynamics emphasizes the importance of studying behavior at the performer-environment system level rather than considering the individual as an isolated part. In football, team behavior coordination has been studied in different match occasions such as synchronization of player-ball-goal during a match [19] or analyzing players' interpersonal linkages [18]. Capturing player-environment interactions demands technology capable of collecting data and methodological ways of relating these data with sports phenomena. That is how AI can be used wisely, since it typically comprehends the existence of huge amounts of data and proposes algorithms to manage and to analyze them. Additionally, but not less important, AI, and more specifically ML usage, demands comprehensive phenomena knowledge to better assess algorithms and cleverly select variable inputs which will impact ML model results. The ecological dynamics approach is an excellent theory guidance for this, guarantying that ML model application is adapted to understanding football phenomena and requirements. The theoretical guidance of ecological dynamics to the use of ML in sports, such as football, means that the features used in the models to capture behavior should measure the athlete-environment system, since it is this interplay that guides behavior. In football this system is the match, and it includes 22 players, the ball, and the whole context where the match is being played. Due to their ability to deal with complex data, ML algorithms are good tools to study sport performance.

There are already some studies guided by the ecological dynamics' theoretical framework for the analysis of performance in sports. Passos et al. [49] studied interpersonal dynamics in rugby and used an artificial neural network algorithm to test the presence of dynamical systems properties, such as phase transitions and symmetry-breaking processes. They found that interpersonal interactions in rugby express dynamical properties, and by means of artificial neural network, they could classify players' interactions in the specific rugby actions of tackle or try. In another study, Couceiro et al. [23] used a sample of five classification algorithms to investigate players' individual cinematics profile in golf. The authors concluded that ML methods could capture the intra- and interpersonal variability of golf performance, identifying each player whenever a put was performed. In football, the ARCANE project [24] described how an ecological dynamics framework could be implemented to organize data collection and their analysis to provide information that could be used to design practice tasks or evaluate athletic performance.

From another perspective, Ramos et al. [53] recurred to ML techniques to study the different dynamic relations that can occur in a relational space like a football pitch, i.e., (1) interactions considered in a geographical space, (2) time structure (order of sequence or simultaneity of interactions), and (3) co-existing cooperation and competition interactions [53]. Ecological dynamics offered guidance on how to measure these relations by means of the dynamics of complex networks. Ramos et al. [52] argued that the usage of a flexible time structure of temporal networks could be an alternative approach for capturing interactions during a game, especially in contrast with the traditional social network analysis method, which is represented by a single cumulative snapshot of frequencies (e.g., passing) only. This alternative approach, called hypernetworks, can capture the interactions of sets of players that may include an arbitrary number of teammates and opponents. Ramos et al. [52] represented a multilevel team behavior dynamic including micro (interactions between players), meso (dynamics of a given critical event), and macro (interactions between sets of players) levels. After analyzing matches from the English Premier League, the authors verified that at the micro level, the most frequent player configuration is 1 vs. 1 (one attacker vs. one defender), and at the meso level, the interaction dynamics near the goal depends on significant changes in the players' speed and direction. Importantly, these variables captured the players' relationship with the performance environment and not the players' variables per se.

The ecological dynamics approach can offer guidance on how to interpret the meaning of large and complex volumes of data, theoretically identifying key data sources and constraining how to detect patterns in intraindividual and interindividual behaviors, even in real-time performance. The studies previously mentioned all have in common variables that capture the relation between the athletes and their context, which, depending on each investigation, can be the coordination between two players or, in bigger scales, the coordination between all players from one team and the pitch where they play. These studies helped us to understand how ecological dynamics can guide the creation of performance indicators. There are several studies using AI and ML in football, but most of them are based on methodology application only, and they are not guided by any theory that explains performance. Taking advantage of AI and ML techniques to apply to football data would offer the possibility to have access to more situated data as guided by an ecological dynamics approach, as we suggest next.

4.1 *Artificial Intelligence to Characterize Successful Line Breaking Events Through Density Zone Analysis: An Exploratory Approach*

Analyzing subgroups of players like dyads, e.g., 1 vs. 1, is commonly mentioned in the literature [78]. More recently, team-based variables are becoming more explored, going further with the analysis not only by player or dyad but also by other collective team structures. A team's center, for example, has been used in various ways to evaluate intra- and inter-team coordination in team sports, representing in a single variable the relative positioning of both teams in forward-backward and side-to-side movement displacements [2]. The effective playing space (or surface area) is also being used to assess spatial distributions and capture specific measures of team coordination [2]. These team-based variables aim to capture team players' coordination rather than looking into a single player's behavior. Following this concept, we propose a new collective variable called density zones (DZ) and demonstrate how ML algorithms can leverage performance analysis by the creation of such variables.

DZ are zones delimitated by groups of players who, by their proximity level, are associated to a zone. These groups are not predefined by any position type or match lineup, but they are simply created and dissolved dynamically according to the moment of the match being able to analyze their structure across time. This is different from creating subgroups of players by default (e.g., 2 vs. 2 or 3 vs. 3). Since players are constantly changing/adjusting positions, DZ are emerging or being dissolved based on each player's distance to each other, calculated based on positional coordinates x, y. This collective and spatial-temporal variable includes both teams so that the attacking structure from one team and the defending opponent structure can be analyzed together. In summary, density zones propose to capture how player grouping (team and opponent) emerges during the match, forming different structures across time. Our aim includes verifying the characteristics of those, e.g., number or size, depending on the match phase where they are. For the purpose of this case study, we selected a match of the Women's World Championship 2023 Australia and New Zealand and calculated DZ recurring to

ML clustering technique called hierarchical agglomerative clustering algorithm. All calculations were performed using Python 3.9.12 through Spyder IDE version 5.1.5, and the demonstrative code is available at https://github.com/sofiafe/clustering_dynamiczones for consultation.

Clustering algorithms are well known for their pattern recognition capabilities [61]. This is an unsupervised learning technique and a connectivity-based clustering model which groups data points. It considers points that are closer to each other to be more similar or related than those located farther apart. It is also called agglomerative since it groups data points from a bottom-up approach model, where having an observation set, I, sets each of the observations, i, j, k, etc. $\in I$, to be a singleton cluster. Then, agglomerate the closest pair of clusters and redefine the inter-cluster dissimilarities with respect to the newly created cluster. If n is the cardinality of observation set I, then this agglomerative hierarchical clustering algorithm completes $n - 1$ agglomerative steps. All distances are measured by the Euclidean distance (1) between each data point using positional coordinates,y.

$$d_{ij} = d\left(\{X_i\}, \{X_j\}\right) = \left\|X_i - X_j\right\|^2 \tag{1}$$

Ward's method is then used to decide the cluster division. It is the only agglomerative clustering methods that is based on a classical sum-of-squares criterion, producing groups that minimize within-group dispersion at each binary fusion while looking for clusters in multivariate Euclidean space [46].

A hierarchical algorithm yields a dendrogram (illustrated by Fig. 1) representing the nested grouping of patterns and similarity levels at which groupings change [38]. The clustering process is performed by merging the most similar patterns in the cluster set to form a higher one [16].

When computing the algorithm for every frame, hierarchical clustering will group players into subgroups based on a player's distance (Fig. 1). Dendrograms give a visual assessment about the best number of clusters to be considered, given by the location of the largest vertical difference between nodes, and in the middle pass a horizontal line. The number of vertical lines intersecting it is the optimal

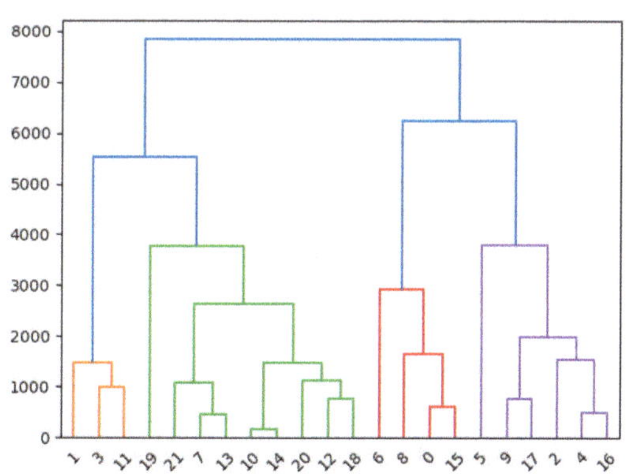

Fig. 1 The dendrogram illustrates, in arbitrary units and variables in both axes, how each cluster is composed by drawing a U-shaped link between a non-singleton cluster and its children. The top of the U-link indicates which clusters were merged. The length of the two legs represents the distance between the child clusters [66]

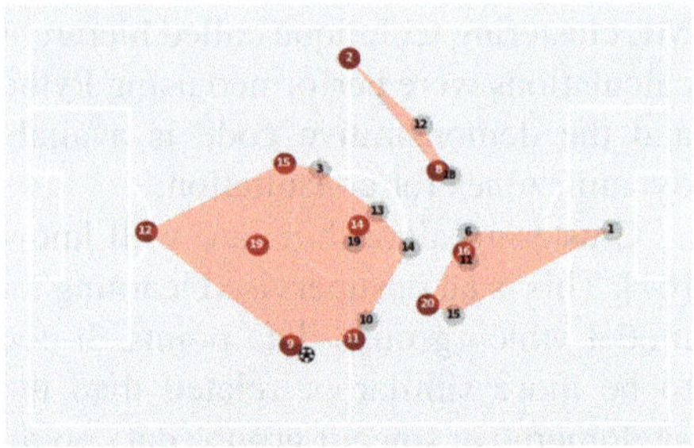

Fig. 2 Illustration of DZ for a specific match frame. Each red/gray point represents a player, and each red shadow represents a DZ (cluster). In this specific frame, we can verify that the player carrying the ball belongs to the biggest DZ

number of clusters. These groups are then illustrated by positioning players and groups in an image (Fig. 2). DZ are constrained by the match situation in each instant. Along players movement, DZ will emerge differently in size, position in the field or number of players inside. Then, DZ is a collective variable characterizing the match environment constrained by the action that players are performing (e.g., attacking vs. defending situations).

Having DZ as a collective variable leads us to think how we can use them to analyze performance. Considering that performance analysis relies on what the analysis objective is, we decided to explore the link of DZ to performance through the analysis of successful line breaking events actions. Line breaking events are football match events where the attacking team successfully advances into the opponents' field and gets closer to the opponents' goal, passing through opposing players. It can be considered as a successful progression for the attacking team. In this section we demonstrate how DZ characteristics behave during a line breaking events situation.

4.2 *Exploratory Approach to Density Zones*

Ten successful line breaking events ($n = 10$) were randomly selected from an events dataset belonging to a match from the Women's World Championship 2023 Australia and New Zealand. Following FIFA's definition, a line breaking event is defined by a ball progression that cuts out a number of opposition players in a penetrative action [29]. A unit line must consist of at least two players.

Using positional tracking data (25 Hz), players' positioning was synchronized with each line breaking event previously selected. Each line breaking event was checked using match video, and an initial timeframe and a final timeframe were defined for an easier analysis. The initial timeframe is defined when the player in possession of the ball is in the moment of doing the passing, which will result in the line breaking event (Fig. 3), while the final timeframe is defined when the new player receives the ball, concretizing the line breaking event successfully (Fig. 4). Hierarchical clustering algorithm run is performed in both frames. Three main points are analyzed: (1) if in the initial frame the player with the ball is inside

Fig. 3 The initial frame when the line breaking event started

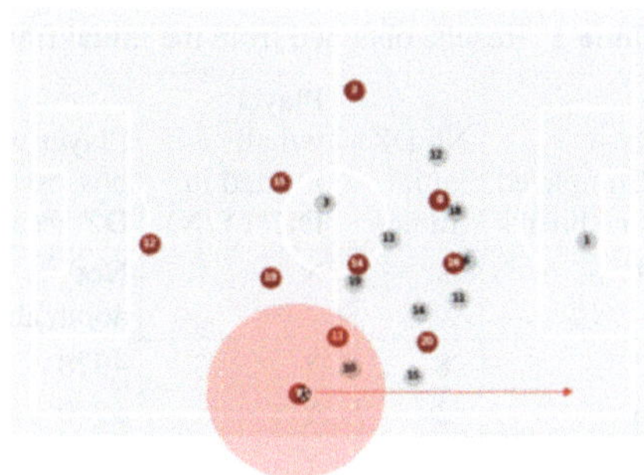

Fig. 4 The final frame when the new player receives the ball, concretizing the line breaking successfully

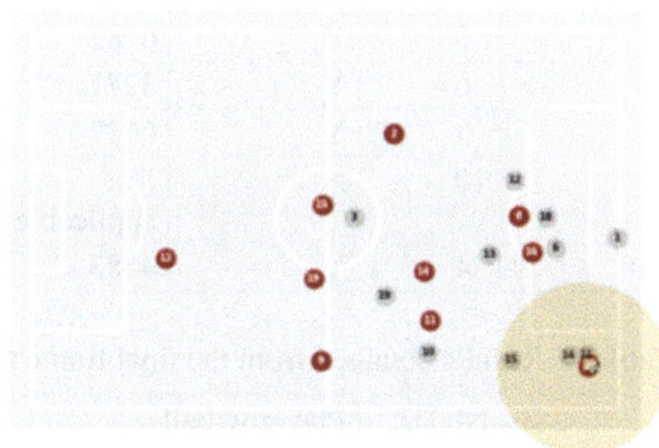

or outside a DZ; (2) if she was located inside a DZ, we considered how many teammates vs. opponents constitute that DZ; and (3) we considered how DZ size behaves during a successful line breaking event.

For point 1, after DZ was defined, we identified if the player carrying the ball belonged to a cluster or was a singleton (cluster formed by one player only). Point 2 occurs as a sequence of point 1, and it is only considered when a player is not identified as a singleton. If a player belongs to a DZ cluster, we identified how many teammates and opponents were part of that cluster and calculate the ratio between them. Considering DZ_i as a DZ defined in frame i, team possession status is calculated by the sum of teammates belonging to DZ_i, divided by the sum of opponents belonging to DZ_i (2). When Team in Possession $\text{status}_i \geq 1$, the attacking team is in numeric superiority; when Team in Possession $\text{status}_i = 0$, the team is in numeric equality; and when Team in Possession $\text{status}_i < 0$, the team is in numeric inferiority.

$$\text{Team in Possession status}_i = \frac{\sum \text{teammates players}_i}{\sum \text{opponent players}_i} \tag{2}$$

Point 3 regards to DZ size, and it is defined by DZ polygons area calculation (m^2).

When running the ML clustering model, the results showed that in 80% of the cases, the player in possession of the ball belonged to a DZ when a line breaking event started (Table 1), which means that in only 20% of the cases, the player

Table 1 Results obtained from the initial frame for every line breaking event

Completed line breaks	Nb DZ initial frame	Player w/ball located in DZ? (Y/N)	Player w/ball possession DZ area (m^2)	Team in possession status	Home team vs opponent team	Completed line break occurred in the same DZ? (Y/N)
1	8	N	Not applicable	Not applicable	1 vs 0	N
2	8	Y	4978	Superiority	3 vs 2	N
3	3	Y	5270	Inferiority	3 vs 5	Y
4	6	Y	5707	Equality	3 vs 3	Y
5	5	Y	4346	Equality	4 vs 4	N
6	3	Y	6782	Inferiority	3 vs 4	N
7	6	Y	3271	Superiority	2 vs 1	N
8	3	Y	6627	Superiority	3 vs 2	N
9	10	N	Not applicable	Not applicable	1 vs 0	N
10	4	Y	4383	Equality	5 vs 5	Y

Table 2 Results obtained from the final frame for every line breaking event

Completed line breaks	Nb DZ final frame	Player w/ball located in DZ? (Y/N)	New player w/ball possession DZ area (m^2)	Team in possession status	Home team vs opponent team
1	3	Y	12814	Equality	2 vs 2
2	3	Y	9941	Inferiority	3 vs 8
3	3	Y	4339	Inferiority	3 vs 5
4	3	Y	8280	Equality	5 vs 5
5	4	Y	11257	Inferiority	5 vs 8
6	3	Y	11641	Inferiority	5 vs 7
7	3	Y	11945	Inferiority	4 vs 7
8	6	Y	6457	Inferiority	3 vs 4
9	8	Y	3682	Superiority	2 vs 1
10	4	Y	4676	Equality	5 vs 5

was completely isolated from teammates or opponents. Regarding the structure, we verified that in the initial frame, in 60% of the cases, the DZ where the player with ball was located, the team attacking was in superiority or in equal number of the opponent team, while in the final frame, in 90% of the cases, the team in possession of the ball was in inferiority or in equal number than the opponent team. When considering the DZ size, we verified that in the initial frame, the DZ were around 40% smaller compared with the final frame (Table 2). In addition, we verified that successful line breaking events occurred 70% of the time from one DZ to another, which indicates a prevalence of inter-DZ dislocation rather than intra-DZ dislocation.

Despite the limited sample size, this case study shows how DZ can analyze player structures recurring to ML techniques in a player-environment relation to characterize match situations such as successful line breaking events. We believe that there is a lot of potential in this symbiotic idea of guiding AI and ML through a theory, enriching football performance analysis development. Using AI and ML through an ecological dynamics approach means studying the athlete-environment relation rather than only individual performance. Collective variables like DZ can help practitioners understand how players interact during a match, adding insights about, for instance, ball possession behaviors. Linking this type of variables to successful actions can be a way to improve performance analysis. We propose that more development studies around this subject should be tested to better assess this novel approach.

5 Conclusion and Challenges

This chapter contextualized big data in sports and illustrated some of the current AI and ML applications in football. Most of the studies are based on outcome prediction and pattern recognition subjects. Although the number of studies published increased in the recent years, there is still a lack of theory guidance when finding an explanatory cause for performance. This happens, for example, during feature selection process. Typically, these models use numerous variables without any variable reduction technique or including any contextually relevant aspects. This sometimes creates naïve solutions or difficulties in establishing an ecologically meaningful relationship between performance and input features. To fill this gap, this chapter discusses the contribution of the ecological dynamics theoretical framework in guiding football performance analysis. This theory relates each individual or team with the specific performance environment where they are playing, analyzing ecological interactions more than individual context-free variables. Aligning ecological dynamics with AI and ML methodologies opens the possibility to a better understanding of complex environments like football, enhancing the value of the analysis for the stakeholders like athletes, coaches, and managers. We argued that ecological dynamics theory can be used to guide performance analysis, demonstrating different investigations, where this theory was already applied with success.

As an example of theory guidance application, we proposed a new collective indicator called density zones. These structures emerge and dissolve during match timeframes and can be linked with performance analysis when goal oriented. Our case study showed how these collective structures can be used to explore line breaking events performance, understanding DZ's characteristics in different match context situations. Preliminary results seem to show some trends in how a successful line breaking event can be characterized in terms of the spatial-temporal behavior of players. Further studies on this subject need to be developed including more matches from different teams and how density zones behave in other match situations like ball possession or a counter-attack occasion.

In summary, we believe that aligning ecological dynamics with more sophisticated AI techniques can be considered as a future challenge in football performance analysis. New key performance indicators can be suggested where context is intrinsically embedded and a timeframe analysis is highlighted, bringing a new perspective of analysis for sports practitioners.

Bibliography

1. Almulla, J. & Alam, T., (2020). Machine Learning Models Reveal Key Performance Metrics of Football Players to Win Matches in Qatar Stars League. IEEE Access, 8. https://doi.org/10.1109/ACCESS.2020.3038601
2. Araújo, D., Davids, K., (2016). Team Synergies in Sport: Theory and Measures. Frontiers in Psychology, 7. https://doi.org/10.3389/fpsyg.2016.01449
3. Araújo, D., Davids, K. & Hristovski, R. (2006). The ecological dynamics of decision in sport. Psychology of Sport and Exercise, 7(6), 653–676
4. Araújo, D. (2017). Research topics in Soccer. In Passos, P., Araújo, D., & Volossovitch, A. (Eds) Performance Analysis in Team sports. London: Routledge.
5. Araújo, D., Davids, K., & Renshaw, I. (2020). Cognition, emotion and action in sport: An ecological dynamics perspective. In G. Tenenbaum, R. C. Eklund, & N. Boiangin (Eds.), Handbook of sport psychology: Social perspectives, cognition, and applications (4th ed., pp. 535–555). John Wiley & Sons, Inc. https://doi.org/10.1002/9781119568124.ch25
6. Araújo, D., Couceiro, M., Seifert, L., Sarmento, H. & Davids, K., (2021). Artificial Intelligence in Sport Performance Analysis. Routledge
7. Argibay-González, J.; Vázquez-Estévez, C.; Gutiérrez-Santiago, A.; Paramés-González, A., Reguera-López-De-la-Osa, X. & Prieto-Lage, I., (2022). Analysis of injury patterns in men's football between the english league and the spanish league. International Journal of Environmental Research and Public Health, 19, 18.
8. Bai, L., Gedik, R. & Egilmez, G., (2023). What does it take to win or lose a soccer game? A machine learning approach to understand the impact of game and team statistics, Journal of the Operational Research Society, 74:7, 1690–1711, https://doi.org/10.1080/01605682.2022.2110001
9. Barra, S., Carta, S.M., Giuliani, A., Pisu, A., Podda, A.S. & Riboni, D., (2023). FootApp: An AI-powered system for football match annotation. Multimed Tools Appl 82, 5547–5567. https://doi.org/10.1007/s11042-022-13359-0
10. Barron D., Ball G., Robins M. & Sunderland C., (2018). Artificial neural networks and player recruitment in professional soccer. PLoS ONE 13(10): e0205818. https://doi.org/10.1371/journal.pone.0205818
11. Barron, D., Ball, G., Robins, M. & Sunderland, C., (2020). Identifying playing talent in professional football using artificial neural networks, Journal of Sports Sciences, 38:11–12, 1211–1220. https://doi.org/10.1080/02640414.2019.1708036
12. Bačić B, (2016). Predicting golf ball trajectories from swing plane: An artificial neural networks approach. Expert Systems with Applications, 65, 423–438, https://doi.org/10.1016/j.eswa.2016.07.014
13. Beato, M.; Youngs, A. & Costin, A., (2024). The analysis of physical performance during official competitions in professional English football: do positions, game locations, and results influence players' game demands? Journal of Strength and Conditioning Research, 38, 5, e226–e234.
14. Bergkam, T., Frencken, W., Niessen, A., Meijer, R. & Hartigh, R., (2022). How soccer scouts identify talented players. European Journal of Sport Science, 22, 7, 994–1004. https://doi.org/10.1080/17461391.2021.1916081

15. Bergkam, T., Niessen, A., Hartigh, R., Frencken, W. & Meijer, R., (2019). Methodological issues in soccer talent identification research. Sports Medicine, 49, 1317–1335. https://doi.org/10.1007/s40279-019-01113-w
16. Bouguettaya, A., Yu, Q., Liu, X., Zhou, X. & Song, A., (2015). Efficient agglomerative hierarchical clustering. Expert Systems with Applications, 42, 2785–2797
17. Button, C., Seifert, L., Chow, J., Araujo, D. & Davids, Keith. (2021). Dynamics of Skill Acquisition: An Ecological Dynamics Approach. Human Kinetics. https://doi.org/10.5040/9781718214125.
18. Caldeira, N., Lopes, R., Fernandes, D. & Araújo, D. (2023). From optical tracking to tactical performance via Voronoi diagrams: team formation and players' roles constrain interpersonal linkages in high-level football. Sensors, 23, 1.
19. Carrilho, D., Couceiro, M., Brito, J., Figueiredo, P., Lopes, R. & Araújo, D., (2020). Using Optical Tracking System Data to Measure Team Synergic Behavior: Synchronization of Player-Ball-Goal-Angles in a Football Match. Sensors, 20. https://doi.org/10.3390/s20174990.
20. Catapult. https://www.catapult.com. Accessed June 2024
21. Chawla S., Estephan J., Gudmundsson J. & Horton M. (2017). Classification of Passes in Football Matches Using Spatiotemporal Data. ACM Trans Spatial Algorithms Syst. 3:1–30.
22. Clemente, F., Couceiro, M., Martins F., Mendes, R. & Figueiredo, A., (2013). Measuring collective behavior in football teams: inspecting the impact of each half of the match on ball possession. International Journal of Performance Analysis in Sport, 13, 3, 678–689.
23. Couceiro, M., Dias, G., Mendes, R. & Araújo, D. (2013). Accuracy of pattern detection methods in the performance of gold putting. Journal of Motor Behavior, 45, 1.
24. Couceiro, M., Dias, G., Araújo, D. & Davids, K. (2016). The ARCANE Project: how an ecological Dynamics Framework can enhance performance assessment and prediction in football. Sports Medicine, 48(12), 1781–1786
25. Cuperman, R., Jansen, K. & Ciszewski, M. (2022). An End-to-End Deep Learning Pipeline for Football Activity Recognition Based on Wearable Acceleration Sensors. Sensors. 22(4). https://doi.org/10.3390/s22041347
26. Davids, K., Araújo, D. & Shuttleworth, R. (2005). Applications of Dynamical Systems Theory to Football. In T. Reilly, J. Cabri & D. Araújo (Eds.), Science and Football V, pp. 537–550. London: Routledge, Taylor & Francis.
27. Davids, K., & Araújo, D. (2010). The Concept of "Organismic Asymmetry" in Sport Science. Journal of Sports Science and Medicine, 13, 663–640. https://doi.org/10.1016/j.jsams.2010.05.002
28. Duarte, R., Araújo, D., Correia, V., Davids, K., Marques, P. & Richardson, M., (2013). Competing together: Assessing the Dynamics of team-team and player-team synchrony in professional association football. Human Movement Science, 32(4), 555–556.
29. FIFA Training Center (2021) https://www.fifatrainingcentre.com/en/resources-tools/football-language/in-possession/penetrative-action/line-break/index.php. Accessed 11th March 2024.
30. FIFA. https://inside.fifa.com/about-fifa/organisation. Accessed June 2024
31. Gibson, J. J. (1967). "James J. Gibson and the psychology of perception," in The Century Psychology Series. A History of Psychology in Autobiography, Vol. 5, eds E. G. Boring and G. Lindzey (East Norwalk, CT: Appleton-Century-Crofts), 127–143. https://doi.org/10.1037/11579-000
32. Gibson, J. J. (1979). The ecological approach to visual perception. Houghton, Mifflin and Company.
33. Goes F.R., Kempe M., Meerhoff L.A. & Lemmink K.A.P.M. (2019). Not Every Pass Can Be an Assist: A Data-Driven Model to Measure Pass Effectiveness in Professional Soccer Matches. Big Data, 7:57–70
34. Guan, S. & Wang, X. (2022). Optimization analysis of football match prediction model based on neural network. Neural Comput & Applic 34, 2525–254. https://doi.org/10.1007/s00521-021-05930-x

35. Hu, K. (2023, February 2). ChatGPT sets record for fastest-growing user base – analyst note. https://www.reuters.com/technology/chatgpt-sets-record-fastest-growing-user-base-analyst-note-2023-02-01/
36. Hubáček, O., Šourek, G. & Železný, F., (2019). Learning to predict soccer results from relational data with gradient boosted trees. Mach Learn 108, 29–47. https://doi.org/10.1007/s10994-018-5704-6
37. Hughes, M., Franks, I., Dancs, H., (2020). Essentials of Performance Analysis in Sport Third edition. Routledge.
38. Jain, A., Murty, M. & Flynn, P., (1999). Data clustering: A review. ACM Computing Surveys, 31(3), 264–323.
39. Jaspers, A., De Beék, T., Brink, M., Frencken, W., Staes, F., Davids, J. & Helsen, W. (2018). Relationships between the External and Internal training load in professional soccer: what can we learn from Machine Learning? International Journal of Sports Physiology and Performance, 12(5), 625–630.
40. Kinexon Sports. https://kinexon-sports.com. Accessed June 2024
41. Kugler, P. & Turvey, M., (1987). Information, Natural Law, and the Self-Assembly of Rhythmic Movement. Routledge.
42. Low, B., Coutinho, D., Gonçalves, B., Rein, R., Memmert, D. & Sampaio, J., (2020). A systematic review of collective tactical behaviors in football using positional data. Sports Medicine, 50, 2, 343–385.
43. Maikiniemi, J.; Savolainen, E., Finni, T. & Ihalainen, J., (2023). Position specific physical demands in different phases of competitive matches in national level women's football. Biology of Sport, 40, 3, 629–637.
44. Mitrotasios, M., Gonzalez-Rodenas, J., Armatas V. & Aranda, R., (2019). The creation of goal scoring opportunities in professional soccer. Tactical differences between Spanish La Liga, English Premier League, German Bundesliga and Italian Serie A. International Journal of Performance Analysis in Sport, 19, 3, 452–465.
45. Montoliu, R., Martín-Félez, R., Torres-Sospedra, J., & Martínez-Usó, A., (2015). Team activity recognition in Association Football using a Bag-of-Words-based method. Human Movement Science, 41, 165–178.
46. Murtagh, F. & Legendre, P., (2014). Ward's Hierarchical Agglomerative Clustering Method: Which Algorithms Implement Ward's Criterion? Journal of Classification, 31, 274–295
47. Pappalardo, L. & Cintia, P., (2018). Quantifying the relation between performance and success in soccer. Advances in Complex Systems, 21. https://doi.org/10.1142/S021952591750014X
48. Pappalardo, L., Cintia, P., Ferragina, P., Massucco, E., Pedreschi, D. & Giannotti, F., (2019). PlayeRank: Data-driven Performance Evaluation and Player Ranking in Soccer via a Machine Learning Approach. ACM Transactions on Intelligent Systems and Technology. 10. 1–27. https://doi.org/10.1145/3343172.
49. Passos, P., Araújo D., Davids, K., Gouveira, L. & Serpa, D. (2006). Interpersonal Dynamics in sport: the role of artificial neural networks and 3-D analysis
50. Passos, P., Araújo, D. & Vollossovitch, A. (2017). Performance Analysis in Team Sports. Routledge.
51. Pillay, L., Burgess, D., van Rensburg, D., Kerkhoffs, G. & Goutterbarge, V., (2022). The congested international match calendar in football: views of 1055 professional male players. BMC Sports Science Medicine and Rehabilitation, 14, 1.
52. Ramos, J., Lopes, R., Marques, P. & Araújo, D. (2017b). Hypernetworks reveal compound variables that capture cooperative and competitive interactions in a soccer match. Frontiers in Psychology, 8. https://doi.org/10.3389/fpsyg.2017.01379
53. Ramos, J., Lopes, R. & Araújo, D. (2017a). What's next in complex networks? Capturing the concept of attacking play in invasive team sports
54. Rein, R. & Memmert, D., (2016). Big data and tactical analysis in elite soccer: future challenges and opportunities for sports science. SpringerPlus, 5, 1410.
55. Rico-Gonzalez, M., Pino-Ortega, J., Mendez, A., Clemente, F.M., Baca, A. (2023). Machine learning application in soccer: a systematic review. Biology of Sport, 40(1), 249–263

56. Rowlinson, A. (2020). Football Shot Quality: Visualizing the Quality of Soccer/Football Shots, Master's thesis, Aalto University School of Business
57. Sarmento, H., Clemente, F. M., Afonso, J., Araújo, D., Fachada, M., Nobre, P., & Davids, K. (2022). Match Analysis in Team Ball Sports: An Umbrella Review of Systematic Reviews and Meta – Analyses. Sports Medicine – Open, 8(66), 1–22.
58. Sarmento, H., Clemente, F. M., Araújo, D., Davids, K., McRobert, A., & Figueiredo, A. (2018). What performance analysts need to know about research trends in association football (2012-2016): A systematic review. Sports Medicine, 48(4), 799–836.
59. Sarmento, H., Marcelino, R., Anguera, M.T., Campaniço, J., Matos, N. & Leitão, J.C., (2014). Match analysis in football: a systematic review, Journal of Sports Science, 32(20), 1831–1843.
60. Sarajarvi, J., Freitas, R., Elovaara, M. & Volossovitch, A., (2023). Skill-related studies from youth to high-performance football: a scoping review. German Journal of Exercise and Sport Research.
61. Saxena, A., Prasad, M., Gupta, A., Bharill, N., Patel, O., Tiwari, A., Joo Er, M., Ding, W. & Lin, C. (2017). A review of clustering techniques and developments. Neurocomputing, 267, 664–681.
62. Scanlan, M., Harms, C., Wilkie, J. & Ma'ayah, F., (2020). The creation of goal scoring opportunities at the 2015 women's world cup. International Journal of Sports Science & Coaching, 15, 5–6, 803–808.
63. Stats Perform. https://www.statsperform.com. Accessed June 2024
64. Statsbomb. https://statsbomb.com. Accessed June 2024
65. Stats Sports. https://statsports.com. Accessed June 2024
66. SciPy v1.12.0 Manual (2024). https://docs.scipy.org/doc/scipy/reference/generated/scipy.cluster.hierarchy.dendrogram.html (access performed on 10th March 2024)
67. Serrano, J., Shahidian, S. & Fernandes, O., (2014). Validation of a manual position tracking software (TACTO) to quantify the football movements. Journal of Sport Sciences, 4(6A), 28–34
68. Travassos, B., Araújo D., Correia, V. & Esteves, P., (2010). Eco-dynamics approach to the study of team sports performance. Open Sport Sci J, 3, 56–57.
69. Travassos B., Araújo D., Duarte R. & McGarry T. (2012). Spatiotemporal coordination behaviors in futsal (indoor football) are guided by informational game constraints. Hum Mov Sci, 31(4):932–45. https://doi.org/10.1016/j.humov.2011.10.004.
70. Vilar L., Araújo D., Davids K. & Travassos B., (2012). Constraints on competitive performance of attacker-defender dyads in team sports. J Sports Sci., 30(5):459–69. https://doi.org/10.1080/02640414.2011.627942.
71. Vilar, L., Araújo, D., Davids, K., Correia, V., & Esteves, P. (2013). Spatial-temporal constraints on decision-making during shooting performance in the team sport of futsal. Journal of Sports Sciences 31 (8), 840–846.
72. Wamba, S. F., Gunasekaran, A., Akter, S., Ren, S. J. F., Dubey, R., & Childe, S. J. (2017). Big data analytics and firm performance: Effects of dynamic capabilities. Journal of Business Research, 70, 356–365.
73. Warren, W., (2006). The dynamics of perception and action. Psychological Review, 113, 2, 358–389.
74. Wimu. https://www.hudl.com/en_gb/products/wimu. Accessed June 2024
75. Wyscout. https://www.hudl.com/en_gb/products/wyscout. Accessed June 2024
76. Hughes, M., Franks, I., Hughes, M., Franks, I.M., & Dancs, H. (Eds.). (2019). Essentials of Performance Analysis in Sport: Third edition (3rd ed.). Routledge. https://doi.org/10.4324/9780429340130
77. Szczepański, Ł., & McHale, I.G. (2016). Beyond completion rate: evaluating the passing ability of footballers. Journal of the Royal Statistical Society: Series A (Statistics in Society), 179.
78. Vilar L., Araújo D., Davids K. & Travassos B., (2012). Constraints on competitive performance of attacker-defender dyads in team sports. J Sports Sci., 30(5):459–69. https://doi.org/10.1080/02640414.2011.627942.

A Supervised Learning Approach for Evaluating Football Performances

Stefania Corsaro, Giuseppina Dello Ioio, Vincenzo Di Sauro, and Zelda Marino

Abstract Over the past decades, the global sports market has experienced significant growth, with a compound annual growth rate (CAGR) of approximately 5% per year. The market value increased from \$486.61 billion in 2022 to \$512.14 billion in 2023, providing profitable opportunities for bookmakers and investors. In this context, the use of data is becoming increasingly significant, focusing on acquiring information about players and sporting events to predict outcomes and optimize team strategies. Therefore, the ability to accurately predict match outcomes is of growing research interest. This study addresses the crucial challenge of predicting the number of goals scored in football matches of the main Italian league (Serie A) based on data from previous seasons to investigate which approach can significantly improve the accuracy of predictions and contribute to the understanding of the determining factors behind the results of football matches. The research employs specialized regression models, such as machine learning methods, generalized linear models and some of their extensions designed to handle count-type response variables and account for overdispersion and excess zeros commonly found in goal-scoring patterns. By utilizing these models, we aim to improve prediction accuracy and facilitate informed decision-making for sports fans and betting professionals.

Keywords Machine learning · Football events forecast

S. Corsaro · Z. Marino
Department of Management and Quantitative Studies, University of Naples "Parthenope", Napoli, Italy
e-mail: stefania.corsaro@uniparthenope.it; zelda.marino@uniparthenope.it

G. Dello Ioio (✉) · V. Di Sauro
Department of Economics and Legal Studies, University of Naples "Parthenope", Napoli, Italy
e-mail: giuseppina.delloioio@studenti.uniparthenope.it;
vincenzo.disauro001@studenti.uniparthenope.it

M. J. Blondin et al. (eds.), *Artificial Intelligence, Optimization, and Data Sciences in Sports*, Springer Optimization and Its Applications 218,
https://doi.org/10.1007/978-3-031-76047-1_7

1 Introduction

Over the past few decades, sports have played a significant role in finance and economics worldwide, involving several opportunities to profit, mainly for bookmakers and investors. The global sports market grew from 486.61 billion dollars in 2022 to 512.14 billion dollars in 2023 at a compound annual growth rate (CAGR) of 5.2%. At the same time, the sports betting industry is constantly growing, involving an increasing number of fans betting on sporting events [1].

Professional sports are followed by millions of people globally, providing an audience that is captive to advertisement and marketing opportunities, particularly in the United States and Europe. This trend is particularly notable in the United States and Europe, where sports are widely popular. The most important sports in the United States are the National Football League (NFL), the National Basketball Association (NBA), the National Hockey League (NHL) and the Major League Baseball (MLB). The NFL includes the Dallas Cowboys, the team with the highest franchise value of any sports team worldwide. In Europe, the football sport dominates the professional sports market. The so-called Big Five leagues represent most of the European professional sports market. This category includes the Premier League in England, La Liga in Spain, the Bundesliga in Germany, Serie A in Italy and Ligue 1 in France. Among these, the Premier League is financially the most successful sports league worldwide.

In this framework, it is important to be able to give value to the soccer clubs. The income for clubs is generated from broadcast agreements and match day ticket sales, media, brand and product licensing and merchandise and concessions sales. But what is crucial is the identification of the value of the team players that depends on performance, role, age and behaviour. Some financial instruments have been developed to book the mark-to-market value of the team players in the balance sheet. The platform Kickoffers allows to invest in *KickBond*, i.e. bonds with a variable interest rate, determined by the team's performance and the performance of its player.

Moreover, the increasing demand for sports betting puts licensed sportsbooks at risk due to imbalances in their books. These imbalances often arise when, for example, licensed sportsbooks, which are legally allowed to do business in their licensed state, receive wagers that disproportionately favour a local team. This problem occurs mainly in the United States, where licensed sportsbook operators can only service customers within the state where they are licensed. In order to hedge the risk of significant losses, ErisX, a digital asset exchange, has proposed to list sporting event-based futures contracts in an effort to meet the hedging needs of commercial operators [2].

Therefore, there is evidence of the need to develop a model that is able to predict the performance of a player or a team, which could be the underlying of a derivative contract or even the results of a match and that can indicate to the bettor the most probable outcome of a match and the reliability of the results.

When we talk about the performance of a football team, we can refer to different aspects. As a performance measure, it is possible to consider the victory/loss/draw in a match or the number of points obtained. Again, it is possible to refer to the number of goals scored or once a goal threshold has been set, and it is possible to identify whether the team is above or below the threshold. This work aims to predict the number of goals scored in a match since, in this way, it is possible to include different aspects of the performance measure. By predicting the number of goals scored in a match, we can establish whether the team won, drew or lost, and we can calculate the number of points obtained based on this information. We can also set a goal threshold and determine whether the team is over or under.

The objective is to compare different methods to increase the accuracy of predictions of the number of goals scored to develop effective game strategies and manage risk, both for teams and coaches and for bettors and bookmakers. In addition to a generalized linear model, which is considered the benchmark model of the work since it is the most widely used in the literature, other models are analysed since they take into account other aspects, such as the hurdle model, which overcomes the problem of excess zeros that the distribution of the number of goals presents, and the generalized additive model, which investigates non-linear relationships between the response variable and the explanatory variables. Finally, machine learning techniques are explored, which allow us to manage large amounts of data, simultaneously analysing numerous aspects to obtain more accurate predictions, and can capture complex relationships between data, which are not easy to capture with other models. These techniques can be trained and adapted frequently to account for changes in player conditions and team performance over time. Deep learning is also a powerful technique that can predict the number of goals in football matches by providing a better understanding of the game's dynamics. However, our data may limit the use of neural networks, which require a substantial amount of data to train effectively and achieve optimal performance.

We based our analysis solely on end-of-game statistics, which pertain only to the end of the match. We chose to use this type of statistics because of their greater availability and our interest in analysing and predicting the number of goals scored at the end of the match. As a result, the number of observations is lower than if we had used in-play statistics, which are collected throughout the entire match. However, the statistics collected at the end of the match summarize the team's performance during all minutes.

The aim of this work is to predict the number of goals in football matches of the main Italian league, Serie A, based on previous seasons. Therefore, we assume that the number of goals to predict in a season depends on the data on previous seasons, in accordance with the results of Andrews et al. [13]. It is also assumed that the number of goals scored is influenced by various game variables, such as the offensive or defensive capacity of the team. Although Serie A is not the most followed championship, and so there are few studies on it, it is considered one of the most complex and most popular championships; therefore, it could be interesting to study and explore the aspects linked to the predictions of the results of this championship.

Forecasting using data from previous seasons in Serie A is limited due to the rules that change the championship's structure. Every year, the last three teams operate on a system of promotion and relegation with Serie B, the second professional division of the Italian men's football championship. The bottom three teams in Serie A will be downgraded to Serie B, while the top three teams in Serie B will gain promotion to Serie A. Through this dynamic system, new teams are introduced into the league each year, diversifying the matchups and challenges and refreshing the competitive landscape, as the new teams will bring new strengths and weaknesses and new game strategies. However, this system limits our work because there is no information available to predict the number of goals scored in the first games of the championship for teams newly promoted to Serie A. We will explain how this problem is treated in Sect. 5.1.

After the overview of the current state of the art in Sects. 2, 3 will describe the formulation of the problem and the different methods used in the present work in order to identify the best one for predicting the number of goals. In Sect. 4, the data set and the preliminary analyses on the data will be described; these analyses were carried out in order to be able to develop the previously described goal number prediction models, which will then be analysed and tested through a rigorous comparative analysis in Sect. 5. Finally, Sect. 6 will conclude the work, also providing some ideas for future work.

2 Related Work

Interest in predicting the results of football matches as a research topic began as early as the mid-twentieth century, with the first insights given in 1956 by Moroney in [3], in which the author discussed the use of statistics in football and how to calculate probabilities based on historical data. However, among the first real works on predicting football results, we find the work of Reep and Benjamin in 1968 [4] and the work of Reep et al. of 1971 [5], in which the authors examine the relationship between skill and randomness in ball games, including football. They argue that the skill of players and teams is an important factor affecting match outcomes but that randomness still plays a role in football match outcomes. The authors use a statistical method to model the number of goals using a negative binomial distribution.

It is only in 1974 that Hill in [6] demonstrates that the outcome of a football match can be predicted based on data from previous seasons, despite there being a component of error linked to chance and the lack of some aspects in the data. Based on these works, Maher in [7] proposes a statistical method to model the scores of football matches, using a Poisson distribution for these scores. Afterwards, Dixon and Coles, in their work Dixon and Coles [8], use a model based on the Poisson distribution to predict the number of goals scored in a match and examine how such models can be used to identify inefficiencies in the sports betting market.

A great deal of research has been done on predicting the number of goals, which can be divided into two groups: one that assumes that the number of goals follows a Poisson distribution and one that assumes that it follows a negative binomial distribution. These distributions are frequently used for count variables, especially when we want to know the number of times an event occurs in a fixed period or space. Therefore, they are well suited for the number of goals in a football match. The choice between using the Poisson distribution and the negative binomial distribution to model goal scoring is crucial due to the variability of match data. The Poisson distribution is appropriate for scenarios where the variance of the data is equal to its mean, while the negative binomial distribution is more suitable for situations with overdispersion, where the variance exceeds the mean, thanks to an additional parameter that takes this variability into account. Therefore, the choice between these distributions depends on the nature of the data. In uniform scenarios, the simplicity of Poisson is preferred, while in more variable contexts, the negative binomial distribution is recommended due to its flexibility and accuracy. In our work, we applied both distributions to identify which was the most suitable.

Over the years, there have been many works about predictions of results in football, all based on linear statistical models, in particular on the generalized linear model (GLM) [9–11].

We aim to overcome the GLM model by applying its extensions to include aspects that often arise in football event data but are not captured by the simple GLM. In detail, we apply the hurdle model to overcome the problem of the excess of zeros in the distribution of the number of goals. We also apply the generalized additive model (GAM) to investigate non-linear relationships between the response and explanatory variables. However, we apply the GLM to the same data as it is the most commonly used model in the literature and will serve as the benchmark for our work.

The era of big data has brought an unprecedented impact on the development of the sports industry, shifting the interest to machine learning techniques for predicting the results of football matches. In 2011, in [12], the authors aimed to develop a system capable of outperforming bookmakers' odds in predicting football match outcomes. Their objective was to identify key features that play a significant role in predicting football matches and to calculate the probabilities associated with these features to maximize potential profits from betting. To achieve this goal, the authors employed different machine learning algorithms. In [13] authors also aim to predict the results of a match based on data from previous seasons. The study employs various models, such as logistic regression, SVM and XGBoost. Among these models, logistic regression demonstrated the best results and quickest predictions using the given attributes. In [14] authors apply machine learning concepts to the football match result prediction field. Three classifiers, LR, GBDT and RF, were examined and fine-tuned using a training set. The authors used the R^2-score and accuracy as the primary evaluation metrics for each classifier. Among all the three classifiers, random forest achieved the highest accuracy and the best R^2-score.

Considering the strength of these techniques for forecasting problems, we also want to include machine learning techniques in our work, in particular random forest. The aim is to enhance the use of this technique compared to the existing literature by implementing a sliding window approach; this enables the model to be trained by updating the available information periodically, resulting in more precise predictions.

To ensure a comprehensive analysis, we aim to compare the linear models mentioned above with machine learning techniques, such as random forest, to identify the best method for solving this problem and obtaining accurate predictions.

In Sect. 5.1, the results obtained by applying the benchmark model, the GLM, will be compared with the extensions proposed in our work, hurdle model and GAM, and with the random forest technique.

3 Modelling

Since the number of goals in a match is a random counting variable, adequate statistical models are fundamental in the analysis to understand the factors that influence events' incidence and formulate meaningful predictions. Therefore, the following sections present the methods that will be applied to identify the best predictions on the number of goals.

Section 3.1 presents the problem to be analysed, i.e. the prediction of the number of goals in a football match. Section 3.2 presents the advantages of machine learning techniques for this type of problem and the technique we have chosen, the random forest. Section 3.3 describes the statistical models used in our analysis. In particular, we present the generalized linear model (GLM), which we assume to be our benchmark model since it is widely used in the literature. Our analysis, however, goes beyond this model to include aspects that often arise in football event data but are not captured by the simple GLM. In detail, we apply the hurdle model to overcome the problem of the excess of zeros in the distribution of the number of goals. We also apply the generalized additive model (GAM) to investigate non-linear relationships between the response and explanatory variables.

3.1 Problem Statement

This work aims to predict the number of goals scored by a team in a Serie A match; as explained in Sect. 1, this measure can be used as an indicator of team performance. This is a count random variable, and it is our response variable.

Suppose we have a set of observed games for some teams. Let T be the number of observed games and $\mathcal{K}$ be the set of all teams. Then, $\boldsymbol{x}_{t,k} \in \mathbb{R}^q$ is the vector of features of the kth team at the tth match. We assume that the number of goals of team $k \in \mathcal{K}$ at time t, $y_{t,k}$ depends on certain features, that is, a function f_w exists such that:

$$y_{t,k} = f_{\boldsymbol{w}}(\boldsymbol{x}_{t,k}) \quad 1 \le t \le T \quad k \in \mathcal{K}.$$

Function f depends on parameters $\boldsymbol{w}$ and is unknown, and an approximation of it is sought. The set of parameters $\boldsymbol{w}$ is estimated by solving:

$$\underset{\boldsymbol{w}}{\operatorname{argmin}} \sum_{t=1}^{T} \sum_{k \in \mathcal{K}} L(y_{t,k}, f_w(\boldsymbol{x}_{t,k})), \tag{1}$$

where L is a loss function.

Once the set of parameters $\boldsymbol{w}$ has been estimated, the forecast value is obtained as:

$$y_{T+1,k} = \hat{f}_w(\boldsymbol{x}_{T+1,k}), \qquad k \in \mathcal{K},$$

where $\hat{f}_w$ is the estimated function.

The features that can influence the performance of a team include aspects related to on-field skills; however, it is important to note that other aspects can also influence a team's performance, such as the psychological aspect of the players, the public's support and the refereeing errors, which can "distort" the results. However, it is impossible to predict such events; therefore, there will always be an error in the forecast. Furthermore, it is not reasonable to think that a team always has the same probability of scoring, since in achieving a goal, not only aspects related to the offensive ability of the team in question come into play, but also aspects related to the defensive ability of the opposing team [7]. Therefore, we assume that the number of goals scored by a team depends on the team considered and the opponent team.

If i is the home team and j is the away team, the number of goals scored in a match is a pair:

$$(y_{t,ij}^{(i)}, y_{t,ji}^{(i)}), \qquad 1 \le t \le T, \quad i, j \in \mathcal{K}, \; i \ne j,$$

where the superscript (i) denotes which is the home team and:

- $y_{t,ij}^{(i)}$ is the number of goals at time t scored by the team i against the team j, when i is the home team
- $y_{t,ji}^{(i)}$ is the number of goals at time t scored by the team j against the team i, when j is the away team

Previous studies on predicting the results of a football match are mainly divided into two groups: one that assumes that the number of goals follows a Poisson distribution and one that assumes that it follows a negative binomial distribution. The difference between the two models is that the negative binomial distribution has an overdispersion parameter, which is not present in the Poisson distribution.

In our work, we applied both distributions, obtaining an overdispersion parameter not significantly different from zero. Thus, the negative binomial distribution reduces to a Poisson distribution. Therefore, we assume:

$$y_{t,ij}^{(i)} \sim Poisson(\mu_{t,ij}^{(i)}),$$

where $\mu_{t,ij}^{(i)}$ depends on the vector of covariates $\boldsymbol{x}_{t,ij}^{(i)} \in \mathbb{R}^q$ such that:

$$\mu_{t,ij}^{(i)} = f(\boldsymbol{x}_{t,ij}^{(i)}).$$

3.2 Machine Learning Techniques

Football matches produce a considerable amount of statistical data about football players, team performance and other factors. This data can be exploited using machine learning techniques to predict various information related to a football match, i.e. the result of the match, a player's performance, a player's injury, etc. These techniques have several advantages; first, they allow managing a large amount of data, simultaneously analysing numerous aspects to obtain accurate forecasts. Furthermore, these techniques can capture complex relationships between data, which are not easy to capture with other models. Another important aspect is that by using these techniques, it is possible to make more accurate forecasts than classic statistical models; it is possible to develop effective gaming strategies and manage risk for both teams and coaches and for bettors and bookmakers.

Machine learning techniques, then, can be trained and adapted frequently to take into account changes in the conditions of the players and the teams' performances over time. Finally, these techniques can help mitigate the problem of overfitting, i.e., the problem where the model is excessively adapted to the training data, so it does not perform well on new data, producing inaccurate predictions.

Since the data set at our disposal, which will be described in Sect. 4, is relatively small, we select the random forest (RF) technique among the different machine learning ones since it is successfully used in these situations. In particular, this technique has a natural tendency to reduce overfitting, a problem that tends to appear more often with small data sets since there is less variation in the data to train the model. Furthermore, it can handle complex and non-linear data, capturing complex relationships between team characteristics, players and other variables related to match results. Also, this technique is robust to noise in the data, another problem that is very common in football data, which often has registration errors, missing data or random variations in performance. RF can address these challenges without significantly compromising performance.

In order to describe the selected methods, we use the standard notation denoting with x the explanatory variable and with y the response variable.

3.2.1 Random Forest

A random forest is a supervised machine learning algorithm introduced in [15] as an extension of the method proposed in [16]. It is an ensemble learning method used to solve regression and classification problems because the underlying principle of this technique is the aggregation of a large number of classification or regression trees. This work will use the random forest technique for a regression problem.

If $\mathcal{S}$ is our training data, the random forest technique generates B different trees (all grown independently from each other) to reduce variance compared to a single decision tree. For this purpose, two different randomization phases are applied; the trees are not applied to the original sample of data, but to random subsamples, and at each node, a random subset of explanatory variables is used to find the best split.

If we denote by N the bootstrap sample size of the training data, m the number of random variables that are selected at each step among the q variables available to identify the best division in a node of the tree and $n_{\min}$ the minimum size of a leaf node in the tree, the random forest procedure can be applied. It is described by Algorithm 1.

Algorithm 1 Random forest

% Given $\mathcal{S}$, B, N, q, m, n_{min}
for $b = 1$ to B **do**
    Draw a bootstrap sample $\mathcal{S}^*$ of size N from $\mathcal{S}$
    $n = 1$
    while $n < n_{min}$ **do**
        % Grow a random forest tree T_b
        Select m variables at random from the q variables
        Pick the best variable and/or split-point among the m
        Split the node into two daughter nodes $n = n \times 2$
    end while
end for
Output the ensemble of trees $\{T_b\}_1^B$

In a random forest regression, each tree produces a specific prediction, and the random forest prediction is the mean of these different predictions:

$$\hat{f}^B(x) = \frac{1}{B}\sum_{b=1}^{B} T_b(x).$$

It is important to correctly establish the number of random trees that the technique must produce from the data set and the number of explanatory variables to use for the split since these settings significantly influence the performance of the model and its behaviour. In particular, a too-low number of trees causes an unstable model that is more subject to the problem of overfitting. In contrast, a high number of trees causes an increase in training times and computational resources without necessarily improving the model's performance. It also makes the model more difficult to interpret. Regarding the number of explanatory variables, overfitting results if this number is too high, as the trees become too specialized in the training data. A too-low value, however, avoids overfitting but can reduce the diversity between trees and the model's ability to capture complex relationships in the data.

3.3 Statistical Models

For a long, statistical models have been the pillar on which research for predicting the results of a football match has been based. Although in recent years, interest has shifted to more complex techniques, such as machine learning techniques, statistical models are still used nowadays. In this section, we present the generalized linear model, one of the main models used over the years for predicting results. Still, we also aim to explore some extensions of this model to capture information regarding aspects often present within football event data sets. In particular, we present hurdle models and generalized additive models. All these models are based on the exponential family of probability distributions.

The probability function in the exponential family has this form:

$$f(y) = c(y, \phi) \exp\left\{\frac{y\theta - a(\theta)}{\phi}\right\},$$

where θ is the canonical parameter, and ϕ is the dispersion parameter; $c(y, \phi)$ and $a(\theta)$ are functions, and their choice determines the actual probability function, such as binomial, normal, or gamma.

3.3.1 Generalized Linear Model

To model the relationship between variables, we used the generalized linear model (GLM), which expands the traditional linear model, providing methods to analyse non-normal data often encountered in sports analytics. In this case, GLM offers a flexible tool to analyse and understand complex relationships.

The probability function *f(y)* associated with the variable *y* is in the exponential family, and it is:

$$f(y) = c(y, \phi) \exp\left\{\frac{y\theta - a(\theta)}{\phi}\right\}, \qquad g(\mu) = x'\boldsymbol{\beta},$$

where $g(\mu)$ is called the link and specifies that a transformation of the mean, μ, is linearly related to explanatory variables contained in x. g is a monotonic, differentiable function (for example, square root or lag). The choice of $g(\mu)$ determines how the mean is related to x.

A GLM model can be constructed using Algorithm 2.

Algorithm 2 Generalized linear model (GLM)

1: Choose $f(y)$ and $a(\theta)$
2: Choose $g(\mu)$
3: Choose explanatory variables x in terms of which $g(\mu)$ is to be modelled
4: Collect observations $y_1, ..., y_n$ on the response y and corresponding values $x_1, ..., x_n$ on the explanatory variables $\boldsymbol{x}$
5: Fit the model by estimating $\boldsymbol{\beta}$ and ϕ (if unknown)
6: Generate predictions of y for different settings of x, given the estimate of $\boldsymbol{\beta}$

After building the model and obtaining predictions, it is necessary to evaluate the fit of the model by examining the deviation of the fitted values from the actual values. The estimated value of $\boldsymbol{\beta}$ is also used to check whether or not certain explanatory variables are important in determining μ.

In this case, the Poisson regression model is:

$$y \sim Poisson(\mu), \qquad g(\mu) = x'\boldsymbol{\beta}.$$

Common choices for $g(\mu)$ are the identity link, $\mu = x'\beta$, and log link, $\ln \mu = x'\boldsymbol{\beta}$. We use a logarithmic link function to connect the mean of the predicted counts to the predictors. Then, the Poisson regression model can be written as:

$$\log(E(y)) = \beta_0 + \beta_1 x_1 + \cdots + \beta_n x_n$$

where $E(y)$ is the expected value of the response variable y, $x_1, ..., x_n$ are the explanatory variables and $\beta_0, ..., \beta_n$ are the estimated coefficients measuring the effect of each explanatory variable on the response variable when the other explanatory variables are held constant.

3.3.2 Hurdle Model

Since the number of goals in a match is a distribution with a high presence of zeros, it is useful to use hurdle models, which combine a binary model for the zero/positive count decision with a truncated-at-zero count data model for positive outcomes [17].

More in detail, hurdle models are a class of statistical models used to analyse data with excess zeros. They are particularly useful for count data where most of the observations are zero, and the remaining observations represent positive counts, as they are models consisting of two components: a binary component and a count component. The model combines the binary component and the count component to provide an overall data distribution, accounts for excess zeros by modelling the probability of a zero and models positive counts for non-zero observations.

The first component (binary component) is responsible for modelling the presence of excess zeros in the data. The model in this part considers two phases:

(a) A binary process that determines whether the observation is a zero or not. This process is usually modelled using logistic regression to estimate the probability of an observation being zero.
(b) A count process that models the actual counts (positive values) for the non-zero observations.

The second component (count component) focuses on modelling the positive counts in the data. It uses a count distribution, such as Poisson or negative binomial, to model the counts for the non-zero observations. This distribution describes the probability of observing a specific count given the non-zero nature of the observation.

The hurdle model is defined as:

$$\log(E(y)) = \log(\boldsymbol{\pi}) + \log(E(y|y > 0))$$

where $\boldsymbol{\pi}$ is the vector of probabilities that $y = 0$ for each observation.

Algorithm 3 explains the procedure for a hurdle model.

Algorithm 3 Hurdle model

1: Choose $f(y)$ and $a(\theta)$
2: Choose $g(\mu)$
3: Choose explanatory variables x in terms of which $g(\mu)$ is to be modelled
4: Choose binary process for modelling excess zeros
5: Collect observations $y_1, ..., y_n$ on the response y and corresponding values $x_1, ..., x_n$ on the explanatory variables $\boldsymbol{x}$
6: Fit the binary component to model the presence of excess zeros by estimating the probability that y is zero (π)
7: Fit the count component (conditional on $y > 0$) by estimating $\boldsymbol{\beta}$
8: Generate predictions of y for different settings of x, combined the binary and count components

3.3.3 Generalized Additive Model

In addition to generalized linear models (GLMs), generalized additive models (GAMs) are also often used for the analysis of count data, as they offer a flexible and robust approach to modelling the nonlinear relationships between the predictors and the response variable; for this reason, this model is also applied in the present work.

GAM models the effect of explanatory variables in terms of non-parametric functional forms. As for the GLMs, the distribution of the response variable is of the exponential family and its mean, μ, is related to the explanatory variables through a link, represented by $g(\mu)$.

A GAM is expressed in the general form as follows:

$$\log(E(y)) = \beta_0 + s_1(x_1) + \cdots + s_n(x_n)$$

where $x_1, ..., x_n$ are the explanatory variables and $s_1, ..., s_n$ are regular functions. Functions s_i, $i = 1, \ldots, n$ are the smoothing functions that will be applied to each predictor variable to capture the non-linear relationships with the response variable. To model these relationships, you can use three different methods, which are:

- Splines: use polynomial curves or chunks of lines to fit the data flexibly
- Local regressions (LOESS): use local weightings to fit the data in the vicinity of each point
- Regular splines: use spline functions with penalties to control for excess flexibility

To fit a GAM to the data, it is necessary to estimate the parameters of the smoothing functions and any other covariates of the model. This process is usually performed using maximum likelihood techniques or minimization of a penalty function. Algorithm 4 explains the procedure for a generalized additive model.

Algorithm 4 Generalized additive model (GAM)

1: Choose $f(y)$ and $a(\theta)$
2: Choose $g(\mu)$
3: Choose explanatory variables x in terms of which $g(\mu)$ is to be modelled
4: Collect observations $y_1, ..., y_n$ on the response y and corresponding values $x_1, ..., x_n$ on the explanatory variables $\boldsymbol{x}$
5: Fit the model by estimating the additive components $s_i(x_i)$ using smoothing techniques
6: Estimate any additional parameters such as $\boldsymbol{\beta}$ and ϕ if needed
7: Generate predictions of y for different settings of $\boldsymbol{x}$, given the estimates of the additive components

4 Analysis and Processing of Data

A fundamental challenge in sports prediction, particularly for football match events, is the collection of a comprehensive and representative data set. Football matches are inherently influenced by myriad factors, making data richness and diversity imperative for reliable outcomes. Consequently, our primary endeavour was assembling a robust data set, leveraging the intricacies and insights furnished by various sources, as delineated in Sect. 4.1. Section 4.2 elucidates the chosen variables for the predictive models, which will be further explored in Sect. 5.

4.1 Data Description

As just said, we construct and analyse a football data set. The matches are related to the top professional division of the Italian men's football championship; the official name is Serie A. It is made up of 20 teams, which compete in two rounds, called the first round and the second round.

It adopts a round-robin format where each team in each of the two groups competes against all the other teams in both home and away matches. Therefore, a single team plays 19 games in each group, for a total of 38 games. For clarity, consider the example of two teams: Juventus and Inter. They play each other twice during the championship, once in the first round and once in the second round, with one game taking place at Juventus' home and one at Inter's home. The order in which the teams meet is random and is based on Berger's algorithm, which determines the "days" by pairing the teams. For each match day, points are assigned, the number of which varies according to the results: the losing team receives no points, and the winning team receives 3 points, while in the event of a draw, both teams receive 1 point. The ranking is gradually updated, and at the end of the championship, the complete ranking is drawn up, in descending order, based on the points won. The first-place winner is the one who wins the championship, while the last three teams operate on a system of promotion and relegation with Serie B, the second professional division of the Italian men's football championship. The bottom three teams in Serie A will be downgraded to Serie B, while the top three teams in Serie B will gain promotion to Serie A.

Through this dynamic system, new teams are introduced into the league each year, diversifying the matchups and challenges and refreshing the competitive landscape, as the new teams will bring new strengths and weaknesses and new game strategies. The starting point is a confidential data set provided by an Italian start-up whose activity concerns the development of advanced systems for analysing the performance of professional athletes. The data set collects player-specific performances within individual football matches for four seasons, from 2016/2017 to 2019/2020. This data set comprises 48 features, with each observation intricately capturing a multitude of metrics representing a player's game contribution. Notably, it includes identifiers such as an identification number for the match, for the team and for the player and personal details such as first name and name and extends to in-game statistics like shots, key passes, interceptions and duels won. These combined attributes show aspects of a player's performance in any match. However, in line with the primary objective of predicting the number of goals scored by teams, we have streamlined the data set to focus on team-specific aggregates, and we definitively selected 39 features.

While the initial data set provided valuable insights, the complexity of machine learning techniques and the drive to leverage more advanced match metrics led to the search for a broader data set. Recognizing the potential benefits of a larger and more comprehensive data set, we expanded our data collection approach to incorporate alternative sources. In our pursuit, we identified data sets that not only mirrored the variables present in our original collection but also introduced more advanced metrics. Furthermore, these new data sets extended our temporal coverage by including seasons starting from 2014/2015. This addition not only enriched our data pool in terms of metrics but also in terms of time span, thereby amplifying the comprehensiveness of our analytical platform. Consequently, we amalgamated these data sets to curate a richer and more nuanced analytical platform that resonates with our primary objective. As a result, we merged some data sets to provide a richer and more detailed analysis in line with our primary objective. Data was collected and combined from two sources: Understat[1] and Football-Data.[2]

The first website offers free descriptive “simple” match statistics such as goals scored, shots taken, the number of corners, fouls committed, yellow and red cards received and the final match result. In addition, the website provides new metrics that delve deeper into the performance characteristics and skill sets of both teams. One such metric is deep completion (DC), which represents the number of passes a team completes within the last 18 meters of the field, excluding crosses. Another metric, the passes allowed per defensive action (PPDA), quantifies the intensity of each team’s pressing; it is calculated by dividing the number of passes allowed in the opponent’s half by the number of defensive actions, including interception attempts, tackles (both won and lost) and fouls committed. This metric indicates how frequently a team attempts to regain possession early in an opponent’s play, offering an objective and comparable measure of their defensive stance and aggressiveness. From Understat, we selected six more advanced metrics.

The second source, Football-data.co.uk, primarily offers bookmaker odds but also includes match statistics. These odds will be integral for later calculations related to the betting strategies we apply. Many of the data set’s variables are discrete and pertain to team performance, such as goal count. There are also binary variables, indicating, for instance, whether a team played at home or away. Four variables were added from Football-Data to provide a strategic perspective and a glimpse into the betting realm. This includes the odds and details of the home and away teams, as well as the HomeFlag variable, which indicates whether the team in the observation played the match at home.

Then, through a series of modifications to the two data sets, they are combined with the initial data set so that each match is divided into two rows, each representing one of the two participating teams. Following the data collection phase, a meticulous examination was conducted to address inconsistencies. Thus, the data set used for the analysis examines data drawn from 6080 soccer matches with 49 variables over eight seasons, from 2014/2015 to 2021/2022. Table 1 presents the descriptive statistics for the 6080 Italian Serie A under-study matches. From the data set, the home team won in approximately 42.9% of the matches, with 1303 wins in home matches. Matches that ended in a draw accounted for 25.2%. Conversely, away teams won in 971 encounters, equivalent to 31.9% of the total away games. Further breakdowns, including goal rates and match outcomes with various goal thresholds, are detailed in the table.

To delve deeper into the distribution of response variable, a comprehensive analysis was undertaken. Figure 1 presents a histogram that delineates the frequency distribution of the dependent variable by both home and away teams. This graphical representation not only highlights the empirical distribution of goals but also contrasts it with a theoretical distribution, providing a comparative perspective.

[1] https://understat.com/

[2] https://football-data.co.uk/

Table 1 Italian Serie A: descriptive statistics

Event	Rate
Home win	42.9%
Away win	31.9%
Draw	25.2%
Goal	55.9%
Over 1.5	77.4%
Over 2.5	54.4%
Over 3.5	32.6%

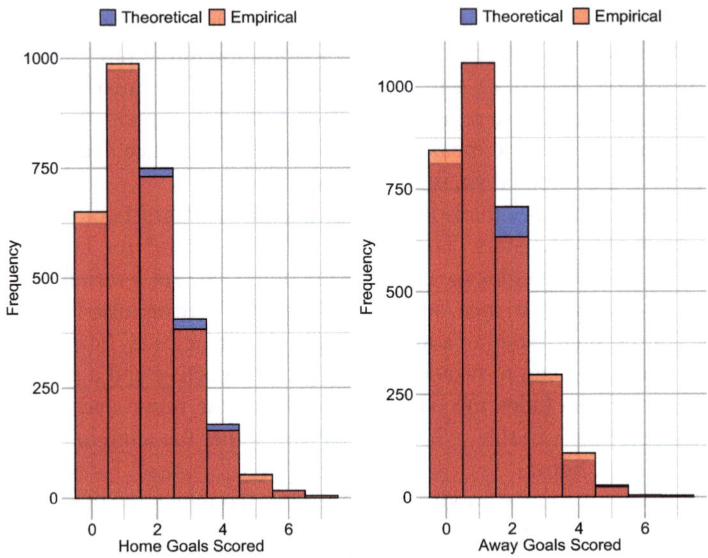

Fig. 1 Comparison of theoretical and empirical distribution

In analysing the distribution of the number of goals scored by the home and away teams, we embarked on a twofold approach. First, we derived observed proportions directly from our data set; for each distinct goal count, we calculated the observed proportion by dividing the number of matches that ended with that specific goal count by the total number of matches in the data set. Second, we assessed expected proportions based on the Poisson distribution; the expected proportions were obtained by modelling the goal counts using a Poisson distribution. The rate parameter λ of this distribution was set to the sample mean of goals from our data set. Specifically, for each goal count k, the expected proportion was computed using the formula:

$$P(Y = k \mid \lambda)$$

where Y is the random variable representing the number of goals. The comparison of the observed and expected proportions provides a clear insight into how well the

Table 2 Observed and expected proportions of the number of goals scored by the home team

Number of goals	0	1	2	3	4	≥5
Observed	0.221	0.331	0.245	0.128	0.050	0.025
Expected	0.222	0.320	0.244	0.133	0.061	0.020

Table 3 Observed and expected proportions of the number of goals scored by the away team

Number of goals	0	1	2	3	4	≥5
Observed	0.288	0.356	0.211	0.099	0.035	0.011
Expected	0.268	0.351	0.237	0.100	0.034	0.010

Table 4 Statistical parameters describing the number of goals scored by the observed team

Min	1st qu.	Median	Mean	3rd qu.	Max	Sd
0.000	0.000	1.000	1.409	2.000	7.000	1.229

Poisson distribution approximates the actual distribution of goals scored. This is true for both the home team, as reported in Table 2, and the away team, as shown in Table 3.

Lastly, Table 4 provides a statistical summary of the goals scored by the observed team. The descriptive statistics capture the distribution's key characteristics, from the minimum and maximum number of goals scored to the mean and standard deviation. Notably, the average number of goals per game stands at approximately 1.409, with teams scoring up to 7 goals in some instances.

4.2 *Data Exploration*

After analysing the target variable, the next step is to identify which variables can most accurately predict it and which variables can be excluded. For this task, feature selection was employed to identify critical attributes while discarding unnecessary, redundant or noisy information [18]. Numerous studies underscore the ability of feature selection to decrease problem dimensionality, improve classification accuracy and prevent overfitting [19–21].

Among the numerous feature selection techniques available, filter methods stand out. These methods, including approaches like correlation analysis, are employed to assess the relevance of individual features and to identify variables to be removed from the initial data set, independent of the learning algorithm.

From the initial set of 49 variables chosen for analysis, we selected numerical attributes, resulting in a subset of 21 distinct features to produce a correlation plot. This streamlined approach was adopted to gain a clearer visual understanding of the interrelationships and potential collinearities among the selected features, as represented in Fig. 2. At this stage, variables representing the bookmaker's odds were omitted; since the process leading to the creation of these odds is unknown, it

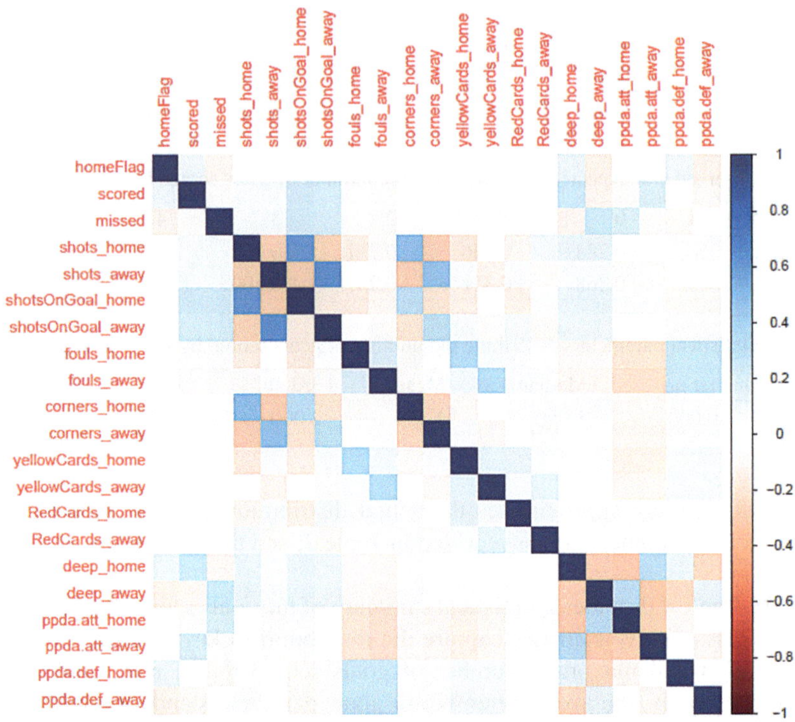

Fig. 2 Matrix correlation between variables

was decided not to include them as covariates in our models. Variables that correlate highly with other variables must be removed [22]. This must be done to prevent the importance of these variables from being overestimated, damaging the prediction of results. In the case where there are two highly correlated variables, one of them becomes redundant, not adding relevant information to the training model.

For this reason, the variables `shots_home` and `shots_away`, which correspond to the number of shots of the teams at home and away, were eliminated, as they have a high correlation (greater than 0.75) with the variables `shotsOnGoal_home` and `shotsOnGoal_away`, which correspond to the number of shots on goal by the home and away teams, respectively. Among the variables shots and the shots on goal, we decided to keep the latter since they are of greater relevance for the aim of predicting the number of goals.

In addition, it can be seen that the variables related to the home team have a greater correlation with each other, as the variables related to the away team also have a greater positive correlation with each other. The variables related to the number of fouls, yellow and red cards do not have a positive relationship with other variables.

Additionally, tools like the boxplot are used to visualize data distribution and identify potential outliers or peculiarities in the features to assess the relationship

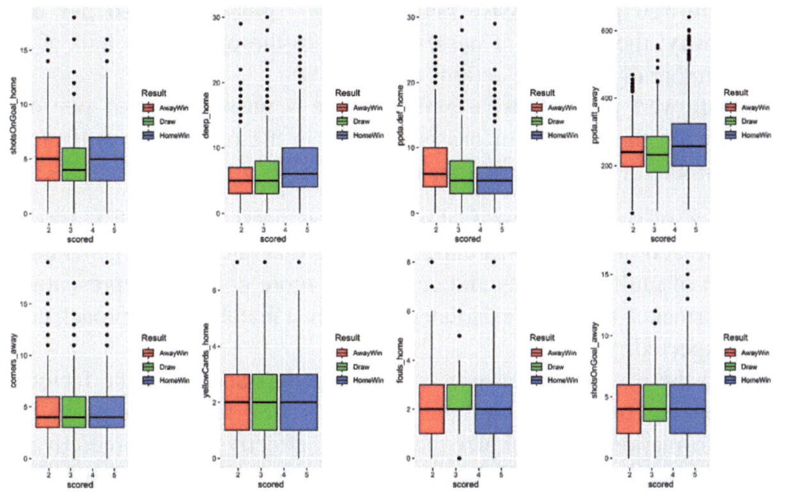

Fig. 3 Boxplot analysis

between all variables and the match events: win, loss and draw. From the visualizations, certain variables have emerged as being particularly useful in predicting the final match results, while other variables are less important for training the forecast model.

Figure 3 presents boxplots of selected variables from our data set. The first four boxplots represent some of the variables that were included in the training phase of the model due to their significant impact on the events under consideration. In contrast, the subsequent four boxplots depict variables that were excluded from the training phase. The exclusion was a result of their demonstrated lack of substantive impact on the events of interest. This visual representation offers a clear contrast between variables that were deemed influential and those that were not, based on our analytical criteria.

At the end of the data exploration process, to select our features, we use the backward elimination approach, a method used in regression analysis and machine learning techniques to select a subset of explanatory variables for the model. In the first step, this method inserts all the variables into the model and removes the least significant one (based on p-value), and it does this until all variables have a p-value below a certain threshold (the significance level α); in this way, the optimal explanatory variables are obtained.

Then, the selected variables for the final model are the following:

- **Home flag**: it is equal to 1 if the team plays at home and 0 if it plays away.
- **Shots on goal**: the shots on target taken by a team.
- **Deep home – deep away**: the number of passes that a team completes in the last 18 metres of the field net of crosses.

- **Passes allowed per defensive action home – passes allowed per defensive action away**: the number of passes granted in the opponent's half of the field on the number of defensive actions taken.
- **Passes allowed per offensive action home – passes allowed per offensive action away**: the number of passes granted in the opponent's half of the field on the number of defensive actions taken.

The relevance of some of these features, such as deep values and passes allowed per defensive action values, was expected since they are metrics generated by the combination of multiple events and can offer a more complete representation of a team's performance than single, isolated events in a match. The approach confirmed what was expected.

Including these variables ensures a deeper understanding of the factors potentially influencing the target attribute. This selection strategy aims to use only pertinent information and maximize the potential effectiveness of predictive models. As part of ongoing research efforts, it would be beneficial to periodically reevaluate these variables to account for changing dynamics or new data trends.

To predict football match outcome events before they occur, it is crucial to feed prediction models with data available before the commencement of each match. Since the data related to match outcomes, such as the number of goals and shots from each team, is collected post-match, it is not directly applicable to training the prediction models, so data needs to be transformed. The chosen approach is to use lags in the available data, considering attributes of previous games, such as the number of goals or shots a team had scored before a given game. Specifically, for each game and its participating teams, lags were computed for the selected variables. So the problem (1) becomes:

$$\underset{\boldsymbol{w}}{\operatorname{argmin}} \sum_{t=lag+1}^{T} \sum_{k \in \mathcal{K}} L(y_{t,k}, f_w(\boldsymbol{x}_{t-1,k}, \boldsymbol{x}_{t-2,k}, \dots, \boldsymbol{x}_{t-lag,k}))$$

where lag is the dimension of the lag. In our analyses, we use $lag = 2$. This choice was grounded in both empirical evidence and domain-specific reasoning. In football matches, the immediate previous game and the one before it often carries significant weight in determining a team's morale and form. By considering attributes from the last two matches, we effectively capture a concise yet potent window of a team's recent performance. Furthermore, from a data-driven perspective, preliminary analyses demonstrated that introducing $lag > 2$ did not significantly improve model accuracy but added computational complexity and reduced the number of observations. This trade-off underscored the value of the most recent two matches as the most predictive of the imminent match outcome, while games before these introduced noise or redundant information.

5 Numerical Experiments

This section provides a comprehensive evaluation of the methods employed in our study, followed by an assessment of the generated predictions based on specific metrics, as described in Sect. 5.1. While the primary focus is on the algorithms' accuracy in predicting football match events, we also delve into their implications in the field of sports betting. This exploration, presented in Sect. 5.2, offers insights into the balance between theoretical modelling capabilities and their tangible, real-world applications. In addition to the scientific results and the financial evaluation through betting strategies, we also compare our predictions with the final ranking of the football teams. Through this comparison, we aim to bridge the gap between our theoretical predictions and real-world results, providing a deeper understanding of the algorithms' performance and reliability.

5.1 Algorithms

Choosing the right method to predict the number of goals scored by a team requires a detailed comparison of different algorithms, each boasting distinct computational and statistical features. This study employs multiple algorithms, listed below, along with their associated R software packages:

- **Generalized linear model** (GLM) – stats package
- **Hurdle model** (HM) – pscl package
- **Generalized additive model** (GAM) – mgcv package
- **Random forest** (RF) – randomForestSRC package

As described in Sect. 3, we applied the generalized linear model as our benchmark model and two extensions (the hurdle model and the generalized additive model) to capture some aspects. Then, among machine learning techniques, we chose the random forest since it works well even with small data sets, like our data set, and brings some advantages.

Our analysis starts with the random forest algorithm. In order to correctly apply this technique and have accurate predictions, we need to optimize the calibration of some parameters, such as the number of trees (B), the number of explanatory variables to use for the split (m) and the split rule (*splitrule*).

A systematic grid search methodology is used to determine the optimal set of parameters; this involves examining a range of parameter values, combining them and optimizing algorithms to ensure maximum accuracy.

The metric used for evaluation is the out-of-bag (OOB) error, which measures the prediction error of random forest. It allows for robust internal validation, providing valuable insights into model performance on new data without requiring a separate validation set. In Algorithm 5, the parameter selection process is reported. The parameters set {B, splitrule, m} that minimizes the OOB error will be used in the random forest procedure.

Algorithm 5 Parameter tuning

```
 1: for B in {500, 1000, 1500} do
 2:     for splitrule in ['mse', 'quantile.regr'] do
 3:         for m in {2, 5, 7, 10} do
 4:             model ← RandomForest(B, splitrule, m)
 5:             oob_error ← calculate_oob_error(model)
 6:         end for
 7:     end for
 8: end for
 9: OOB_min ← min(oob_error)
10: Select parameters for OOB_min
```

Table 5 shows the OOBs obtained with the various parameter configurations for the RF algorithm. The lowest OOB error gives us the optimal combination of parameters to use.

We can see that the combination of the following parameters obtains the lowest OOB error:

- **B:** 1000
- **splitrule:** quantile.regr
- **m:** 2

Specifically, $B = 1000$ is used to achieve a good balance between computational efficiency and prediction performance, while $splitrule = quantile.regr$ suggests a choice of splitting the quantile regression via the "check-loss" function, emphasizing a more robust and resistant regression method that can be particularly effective in the presence of anomalous. $m = 2$ guarantees that only two predictors are considered for each division, preventing excessive complexity and favouring the interpretability of the results.

We also add the parameter:

- **family:** Poisson

The choice is dictated by the fact that we assume that our response variable follows a Poisson distribution.

Table 5 Random forest parameter tuning

	B					
	500		1000		1500	
m	mse	q.regr	mse	q.regr	mse	q.regr
2	1.4700	1.4568	1.4698	1.4025	1.4686	1.4266
5	1.4870	1.4623	1.4866	1.4614	1.4863	1.4608
7	1.4944	1.4639	1.4935	1.4627	1.4938	1.4632
10	1.5036	1.4650	1.5030	1.4650	1.5022	1.4650

Table 6 Choice of training set size

Season	RMSE	OOB
Two	1.2484	1.5671
Three	1.2236	1.4850
Four	1.2235	1.4739
Five	1.2145	1.5185
Six	1.2140	1.4886
Seven	1.2019	1.4225

Once the parameters have been set, to make predictions, it is necessary to train our model on previous data; therefore, the next step is to split the data set into training sets and test sets and determine the optimal size of the two sets. In our study, the test set is derived from the most recent championship in our data set: the 2021/2022 season. The training set can take on different dimensions based on the number of previous seasons we choose to insert; for an optimal choice, we made several attempts with different numbers of previous seasons (from 2 to 7), and we evaluated two different error measures, the root mean square error (RMSE) and out-of-bag (OOB) error rates. Table 6 shows the results obtained.

The RMSE and OOB fluctuate with the length of the season data set. The lowest errors occur for seven seasons; therefore, we used seven previous seasons to train the model.

Now, the model can be trained, and predictions can be made. The first idea is to train the model on the seven previous seasons available and make a prediction on the entire new season; however, this approach has a limitation due to the nature of Serie A. As explained in Sect. 4.1, three teams are downgraded from Serie A every year, allowing three new teams to enter the league. Training the model on previous seasons leads to a need for more information on newly promoted teams, thus rendering prediction-making impossible for matches involving these teams. To counter this issue, we insert the previously predicted day into the training set at each step (day) to maintain the latest information and obtain data for newly promoted teams. This way, we face an information loss only in the early stages of the model.

To achieve this, we employ the sliding window approach, which maintains constant the size of the training set by removing the oldest day as each newly predicted day is added. This approach reduces the risk of overfitting by ensuring that the model is manageable to a specific data segment. Due to its value in capturing temporal consistency, this approach is efficient when there is a need to maintain chronological order between training and testing data, and it adeptly identifies any inherent seasonality or trends within the data.

Using the sliding window approach, the random forest algorithm was trained and applied to generate the predictions. Afterwards, we implement the remaining methods utilizing the sliding window approach with a seven-season training set for each in order to compare the predictions. The generalized linear model, the hurdle model and the generalized additive model were applied. Each method can bring advantages; exploring them in this research grants us a broader perspective, ensuring comprehensive understanding and reducing potential blind spots in the analysis.

Suppose that we have results from n_t previous championship days, and we want to predict the number of goals for team x on each championship day of the new season; n_s is the number of championship days of the new season. The procedure is summarized in Algorithm 6.

Algorithm 6 Model training and prediction with sliding window approach

```
% {teams} is the set containing all the teams of the new season
Given {teams}
for each x in {teams} do
    % ws is the number of days on which we train the model
    given ws
    for i in 1:n_s do
        define L_train using matches from (n_t + 1 − ws) to n_t
        train the model on L_train
        define L_test using match at (n_t + 1)
        predict the model on L_test
        % increments the size of the numbers of previous matches
        n_t = n_t + 1
    end for
end for
```

In our case, we have seven previous seasons for a total of $n_t = 266$ championship days, and we want to predict the number of goals on the eighth season made of $n_s = 38$ championship days. {teams} is the set containing all the Seri A teams of the eighth season.

As mentioned in Sect. 2, using Table 7 we compare the different methods used. In particular, Table 7 provides a detailed breakdown of the predictive performance across several algorithms. For each algorithm, the RMSE on the test set is reported and also the percentage of success in the evaluation of the final score of all the matches in the test set, the percentage of success in the identification of victories at home and away, of draws and the over 2.5 predictions (i.e. when there are more than 2.5 goals in a match).

The random forest (RF) stands out remarkably for its performance, demonstrating exceptional predictive ability with the lowest RMSE value of 1.1909.

This positions it as the most accurate model in terms of prediction errors among all the models evaluated. Furthermore, the total success rate for random forest is the highest at 64.89%. The random forest shows higher percentages and more accurate predictions for all the metrics. Another notable observation is the inherent

Table 7 Forecast results: success percentage

Algorithm	RMSE	%success	%victories HT	%draws	%victories AT	%over2.5
GLM	1.5060	60.08%	60.64%	53.63%	62.95%	63.1%
HURDLE	1.4229	60.38%	62.91%	53.63%	63.30%	64.3%
GAM	1.3438	62.64%	64.49%	54.91%	64.57%	66.6%
RF	1.1909	64.89%	68.53%	56.02%	65.22%	69.8%

challenge all models face when predicting draws compared to other outcomes, such as home or away team victories. This can be attributed to the relative rarity of draws in football matches compared to a victory for one of the teams. However, random forest surpasses others in this regard, achieving a success rate close to 60%. The generalized additive model performs admirably among traditional methods, showcasing an RMSE of 1.3438. This contrasts with 1.5060 for the GLM and 1.4229 for the hurdle model, emphasizing its superior predictive capacity among conventional techniques. The GLM is the least-performing model, with the highest RMSE and the lowest success rates for all events. This underscores the limitations of simpler models in accurately predicting match outcomes, especially when compared to more intricate models encompassing a broader spectrum of influencing factors.

In addition to providing evaluation metrics for our forecasts, Table 8 offers an intriguing comparative insight, comparing the actual points accumulated by the football teams over the season against the points generated by the random forest algorithm predictions.

The correspondence between the actual results and those predicted by the random forest algorithm is considerable; for most teams, such as Milan, Inter and Napoli, there is perfect alignment between the actual and expected results. This agreement accentuates the accuracy and reliability of the algorithm in modelling and predicting team performance throughout the season.

Table 8 Final ranking

Position	Team	Points	Team	Predicted points
1	Milan	86	Milan	86
2	Inter	84	Inter	84
3	Napoli	79	Napoli	79
4	Juventus	70	Juventus	70
5	Lazio	64	Fiorentina	65
6	Roma	63	Roma	63
7	Fiorentina	62	Lazio	61
8	Atalanta	59	Atalanta	59
9	Verona	53	Verona	53
10	Torino	50	Sassuolo	50
11	Sassuolo	50	Torino	47
12	Udinese	47	Udinese	47
13	Bologna	46	Bologna	46
14	Empoli	41	Empoli	44
15	Sampdoria	36	Sampdoria	36
16	Spezia	36	Spezia	36
17	Salernitana	31	Salernitana	31
18	Cagliari	30	Cagliari	30
19	Genoa	28	Genoa	28
20	Venezia	27	Venezia	27

Furthermore, it is noteworthy to observe the consistency with which the random forest algorithm has produced predictions for teams at the top of the table and across the entire league standings. Even for mid-table teams like Verona and bottom-table contenders like Genoa and Venezia, the predictions are spot on, signifying that the model captures the underlying dynamics irrespective of a team's position in the league. This precision, spread across the range, underscores the robustness of the random forest algorithm in comprehensively understanding the myriad of factors that drive team performance.

5.2 Betting

In predictive modelling, particularly in sports, it is essential to evaluate the performance of models in a strictly quantitative way and contextualize their implications in real-world scenarios. It is fascinating to see how predictions from complex mathematical models align with tangible outcomes. These results can be used in different contexts, for example, for tactical planning in sports or for making financial decisions in betting. This comparative analysis helps evaluate the model's reliability and potential applications, especially in industries such as sports betting.

Previously, the performance of various algorithms designed for improved prediction accuracy was presented; if integrated with the bookmaker's odds, these enhanced predictions set the stage for the subsequent betting experiments. More in detail, we use the betting data sets of the latest Serie A season included in our data set, and we apply two different betting strategies: the fixed strategy and a strategy based on the Kelly criterion [23]. Below are the details.

We adopt the most straightforward approach for both betting strategies, according to which a bet is placed on any prediction the model considers worthy. Furthermore, for both betting strategies, we assume we start from a bankroll of 1000 €.

The fixed betting strategy divides the initial bankroll by the number of picks, obtaining the sum to bet, which is constant for each bet. The Kelly criterion strategy places a different amount for each bet. This sum is calculated based on the eighth-Kelly, which establishes betting only one-eighth of the optimal odds calculated with the Kelly criterion. This is because the full-Kelly model is particularly aggressive and often culminates in substantial financial setbacks [23]; the eighth-Kelly, being a more conservative approach, is more in tune with real-world betting scenarios where both the odds provided and those perceived by the bettor influence betting decisions [24, 25].

These strategies are used to determine the profit from betting. Starting from the initial bankroll, for each game, the results predicted by the models are multiplied by the odds provided by the bookmakers; if the prediction tends to place a bet, the amount to bet is determined and subtracted from the existing bankroll. If the bet wins, the earnings and the initial bet are added to the bankroll. The results are shown in Table 9.

Table 9 Profits from betting strategies

Algorithm	Fixed profit	Kelly criterion profit
GLM	1020.67€	1132.12€
HURDLE	1091.78€	1189.11€
GAM	1106.70€	1167.43€
RF	1112.75€	1263.38€

Since the bets are based on our predictions, we expect random forest to produce the highest profits; the table confirms what has been said. A general trend is the higher profit obtained through the strategy based on the Kelly criterion; this highlights the benefits of using a dynamic, data-sensitive betting method over a static one. These empirical insights pave the way for further research to refine the algorithmic models and prediction strategies.

6 Conclusions and Future Work

This research leveraged a machine learning approach and other statistical methods to model and predict the number of goals scored in football matches. The implications in financial contexts, especially within betting strategies and financial derivatives related to sports outcomes, are profound. As sports markets become more financially sophisticated, the potential for derivatives anchored on team or player performances grows. These instruments can act as hedges against risks or investment opportunities in their own right, making the accurate prediction of outcomes all the more critical. Refining and deploying precise models cannot be underrated in an industry where prediction accuracy can have significant financial implications. These models could be the backbone for constructing and pricing such financial derivatives in the sports industry. Our analysis and comparative evaluation highlighted the random forest (RF) algorithm as particularly influential. The precision and robustness of this algorithm make it an indispensable tool for academic investigations. Future research might not only refine data collection processes but also delve into neural network methods or the inclusion of economic and financial variables of the clubs to improve prediction accuracy further. In conclusion, the nexus of machine learning and sports analytics presents a compelling avenue for academic exploration and real-world financial impact.

References

1. Russell, A.M.T., Hing, N., Browne, M., Li, E., Vitartas, P: Who bets on micro events (microbets) in sports?. In: Journal of Gambling Studies. **35(1)**, 205–223 (2019)
2. ErisX, Eris Exch. Submission No. 2020-11E, CFTC regulation 40.2(A) Certification. Notification Regarding the Initial Listing of Eris Exchange RSBIX NFL Futures at 2, /Dec. 14,2020), https://bit.ly/3BznRg2

3. Moroney, M.J.: Facts from figures. In: Penguin Books (1956)
4. Reep, C., Benjamin, B.: Skill and chance in association football. In: Journal of the Royal Statistical Society Series A: General, **131(4)**, 581–585 (1968)
5. Reep, C., Pollard, R., Benjamin, B.: Skill and chance in ball games. In: Journal of the Royal Statistical Society Series A: Statistics in Society, **134(4)**, 623–629 (1971)
6. Hill, I.D.: Association football and statistical inference. In: Journal of the Royal Statistical Society: Series C (Applied Statistics), **23(2)**, 203–208 (1974)
7. Maher, M.J.: Modelling association football scores. In: Statistica Neerlandica, **36(3)**, 109–118 (1982)
8. Dixon, M.J., Coles, S.G.: Modelling association football scores and inefficiencies in the football betting market. In: Journal of the Royal Statistical Society: Series C (Applied Statistics), **46(2)**, 265–280 (1997)
9. Rue, H., Salvesen, Ø.: Focus on Sport: Prediction and Retrospective Analysis of Soccer Matches in a League. Journal of the Royal Statistical Society Series D: The Statistician, **49(3)**, 399–418 (2000)
10. Adam, A.: Generalised Linear Model for Football Matches Prediction. MLSA PKDD/ECML, (2016)
11. Nguyen, Q.: Poisson Modeling and Predicting English Premier League Goal Scoring. The New England Journal of Statistics in Data Science, **1(1)**, 95–101 (2021)
12. Buursma, D.: Predicting sports events from past results: Towards effective betting on football matches. In: Conference Paper, presented at 14th Twente Student Conference on IT, Twente, Holland (Vol. 21), (2011)
13. Andrews, S.K., Narayanan, K.L., Balasubadra, K., Josephine, M.S.: Analysis on sports data match result prediction using machine learning libraries. In: Journal of Physics: Conference Series, **1964(4)**, 042085 (2021)
14. Hu, S., Fu, M.: Football Match Results Predicting by Machine Learning Techniques. In: 2022 International Conference on Data Analytics, Computing and Artificial Intelligence (ICDACAI) (pp. 72–76), IEEE, (2022)
15. Breiman, L.: Random forests. In: Machine Learning, **45**, 5–32 (2001)
16. Ho, T.K.: The random subspace method for constructing decision forests. In: IEEE Transactions on Pattern Analysis and Machine Intelligence, **20(8)**, 832–844 (1998)
17. Mullahy, J.: Specification and testing of some modified count data models. In: Journal of Econometrics, **33(3)**, 341–365 (1986)
18. Kumar, V., Minz, S.: Feature selection: A literature review. In: SmartCR, **4(3)**, 211–229 (2014)
19. Wah, Y.B., Ibrahim, N., Hamid, H.A., Abdul-Rahman, S., Fong, S.: Feature selection methods: Case of filter and wrapper approaches for maximizing classification accuracy. In: Pertanika Journal of Science & Technology, **26(1)** (2018)
20. Li, J., Cheng, K., Wang, S., Morstatter, F., Trevino, R.P., Tang, J., Liu, H.: Feature selection: A data perspective. In: ACM Computing Surveys (CSUR), **50(6)**, 1–45 (2017)
21. Kira, K., Rendell, L.A.: A practical approach to feature selection. In: Machine Learning Proceedings 1992 (pp. 249–256). Elsevier (1992)
22. Dasgupta, N. Practical big data analytics: Hands-on techniques to implement enterprise analytics and machine learning using Hadoop, Spark, NoSQL and R. In: Packt Publishing Ltd (2018)
23. Hsieh, C.H., Barmish, B.R. On Kelly betting: Some limitations. In: 2015 53rd Annual Allerton Conference on Communication, Control, and Computing (Allerton) (pp. 165–172). IEEE. (2015)
24. Matej, U., Šourek, G., Hubáček, O., Železný, F.: Optimal sports betting strategies in practice: An experimental review. IMA Journal of Management Mathematics, **32(4)**, 465–489 (2021)
25. Jacot, B.P., Mochkovitch, P.V.: Kelly criterion and fractional Kelly strategy for non-mutually exclusive bets. Journal of Quantitative Analysis in Sports, **19(1)**, 37–42 (2023)

Bridging Route-Based Cycling Training with Digital Twins

Alen Rajšp and Iztok Fister Jr.

Abstract The use of digital technologies is revolutionizing sports training. Numerous approaches to sports training generation have been proposed, each solving a narrowly defined problem. The emerging issue in the smart sports training field is integrating these systems to create solutions to aid the athlete or trainer in the entire training cycle. Nowadays, solutions that can prepare training plans, record and evaluate the training, and forecast results exist, but they are commonly not integrated together. Since no standardization exists, each solution serves as an independent unit. This chapter aims to present the integration of a novel solution for generating highly adaptive cycling training routes with an artificial sports trainer (AST) that is based on the digital twin concept. The proposed solution allows the use of generated training routes as input in an AST monitor device, which can be used while cycling to record the training activity and the executed training. This is done by providing an export functionality to the cycling route generator and adapting the AST monitor to allow the import of training routes and provide navigation.

Keywords Digital twins · Artificial sports trainers · Smart sports training · Computational intelligence · IoT

1 Introduction

The sports training domain has been rapidly changing due to digital disruptions. The ubiquity of electronic devices and new data collected by them has allowed the use of Intelligent Data Analysis Methods in smart sports training [1]. According to predictions, by 2030, the majority of current Olympic records are expected

A. Rajšp (✉) · I. Fister Jr.
Faculty of Electrical Engineering and Computer Science, University of Maribor, Maribor, Slovenia
e-mail: alen.rajsp@um.si; iztok.fister1@um.si

M. J. Blondin et al. (eds.), *Artificial Intelligence, Optimization, and Data Sciences in Sports*, Springer Optimization and Its Applications 218,
https://doi.org/10.1007/978-3-031-76047-1_8

to be achieved with the help of information technologies in sports training [2]. Smart sports training approaches are being used in at least 31 different sports disciplines [1], showing digital technologies' impact on the field.

Sports training is a strictly organized routine between the trainer and the athlete that can be divided into four distinct phases: planning, realization, control, and evaluation. Sports training sessions are prepared in advance by a trainer and then executed by the athlete [3]. Solutions covering each of the named phases have been aiding trainers in smart sports training and for some sports, such as cycling, solutions in each phase of the sports training exist [1]. Connecting and integrating these solutions in large-scale systems can represent the next evolutionary step in sports training, where artificial/virtual sports trainers could potentially replace trainers.

Artificial/virtual sports trainers have been proposed and utilized to some degree in fields such as fitness [4], running [5], and cycling [6]. The main benefits of artificial sports trainers include lower costs than real trainers and their 24/7 availability. Artificial sports trainers could severely lower the costs of training sports and give each individual, even amateurs, the possibility of employing their own sports trainer.

Another concept in sports training has been the emergence of digital twins in the domain. These digital replications of physical entities enable data to be seamlessly transmitted between the physical and virtual worlds [7]. Another more straightforward explanation is that the digital twin is a contextualized software model of a real-world object [8]. In the sports training sense, they are virtual replicas of the athlete created from the previous history of athlete training data. So far [9], over 13 different studies utilizing the approach have been identified. In some cases [10, 11], this feedback, provided by the digital-twin model, may be available in real time, which can further help the athlete in their training outcomes.

The training history can be collected using IoT devices [12], wearable devices [9], and cameras [13]. A lot of proprietary devices, such as Fitbit, Garmin watches, and Apple Watch, have been created and used for general sports recording, with their yearly sales surpassing over 120 million units [14].

Cycling sports training is one such sport where the digital twin and artificial sports trainer concepts are highly applicable. This is possible because training is performed in real-world outdoor environments, meaning that geographical data can be collected, and the athlete needs to use a bicycle to perform training, where the bicycle can serve as a mount for various electronic devices needed for training that would otherwise be inconvenient to carry around for the athlete. This allows us to use cycling training as a case study for applying the digital twin and artificial sports trainer solution.

The artificial sports trainer monitor (AST monitor)[1] has been developed and presented in [6]. AST monitor represents a Raspberry Pi-based cycling computer. The device can aid the athlete by providing live feedback during the realization

[1] Public repository available at: https://github.com/firefly-cpp/AST-Monitor

phase of sports training. The usage of the in comparison to general-based fitness trackers is in its specialization in cycling training. The device and approach are presented in detail in the AST monitor section.

The contributions of this study are as follows:

- Adaptation and upgrade of a method for generating cycling sports training sessions
- Integration of the presented method with artificial sports trainer monitor
- Simple experimental case study and demonstration of the approach in practice

The remainder of this chapter is structured as follows. Section 2 presents the AST monitor, detailing its role and functionality in our integrated cycling training system. Section 3 delves into the methodology for generating cycling routes, highlighting its significance in tailoring the training experience. Section 4 describes our approach to integrating these generated cycling routes into the AST monitor, thereby enhancing its capabilities. Section 5 covers our solution's implementation and software architecture, discussing the technical aspects and design choices underpinning our system. Section 6 presents the results from a practical case study, demonstrating our integrated system in action and showcasing its real-world applicability. Finally, the chapter concludes in Sect. 7 with a summary of our findings and a discussion on the impact of our work in smart sports training.

2 AST Monitor

AST monitor is a bike computer prototype that allows running the artificial sports trainer for cycling training in practice [6]. The purpose of the bicycle-mounted device is to serve as a digital twin during the training exercise. This approach was first published in [15], where a general framework for the artificial sports trainer was proposed. The framework proposed an integration of computational intelligence algorithms in all phases of sports training to create a general solution. It should be noted that individual proprietary applications such as Strava [16], MapMyRide [17], and Bikemap [18] for monitoring and planning cycling training already exist. However, they do not offer any insight into their implementation, limiting their impact in the research field and demanding users pay for certain features. Open-source implementations, such as Golden Cheetah [19], that can be used in cycling training also exist but are limited to post-workout analysis. One of the main advantages of AST monitor over other solutions is its complete transparency in its implementation and its open-source nature. The current features include recording and displaying training parameters, i.e., distance, speed, ascent, and heart rate, and importing interval training plans.

2.1 Hardware

For this to be done, the hardware part of AST monitor was developed with the following goals in mind:

1. The device must be portable and lightweight enough to allow its use during sports training.
2. The device must be capable of being powered by a power bank, meaning it should not need too much power.
3. The components of the device should be readily available so that the device can be replicated.
4. The device should allow feedback during the training, meaning it needs a visual screen for it to convey this information.
5. The device must allow recording of the executed training done with the device.

The identified and used components that fulfilled the given parameters, thus allowing the use of developed AST monitor software, were [6]:

- **Raspberry Pi 4 model B micro-controller** [20]: the main computer that allows running the AST monitor software
- **Adafruit Ultimate GPS HAT for Raspberry Pi A+/B+/Pi 2/3/Pi 4** [21]: Global Positioning System (GPS) peripheral that allows real-time tracking and allows recording the geographical training data
- **USB ANT stick** [22]: A Universal Serial Bus (USB) peripheral that allows connection with compatible smartwatch devices to capture real-time athlete heart rate data
- **Five-inch HDMI LCD touch screen display**: for displaying the AST monitor software graphical user interface
- **Bicycle mount platform**: a simple platform that holds the hardware and has two straps that allow the device to be mounted on the bicycle

2.2 Software

The software executed from the Raspberry Pi CPU was developed in Python 3.9 [23]. The graphical user interface was developed in PyQt 6 [24], which is a Python library providing bindings for the Qt toolkit [25] open-source cross-platform software for creating graphical user interfaces. The application itself is composed of eight main parts. The whole application is wrapped inside and controlled by the *AST monitor* class as seen in Fig. 1.

The main class *AST monitor* controller interacts with all the dependent components as well as holds the state of the application. The data is generated by reading input from the sensors by the *GPS sensor* and *heart rate sensor* classes, which write their readings in real time inside of generated temporary files that are then retrieved and processed by the *basic data*. This is done to increase modularity

Fig. 1 Architectural design of AST monitor

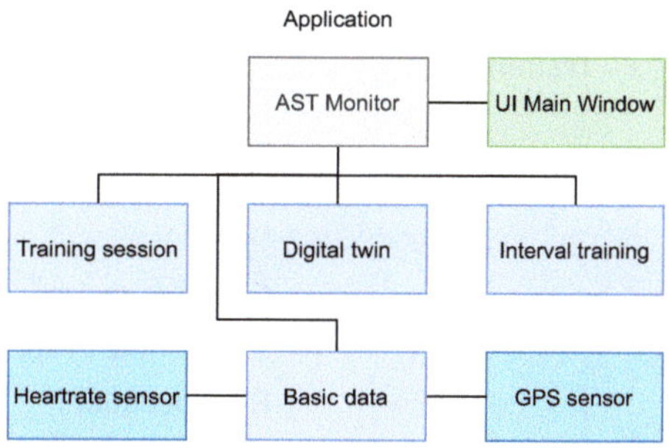

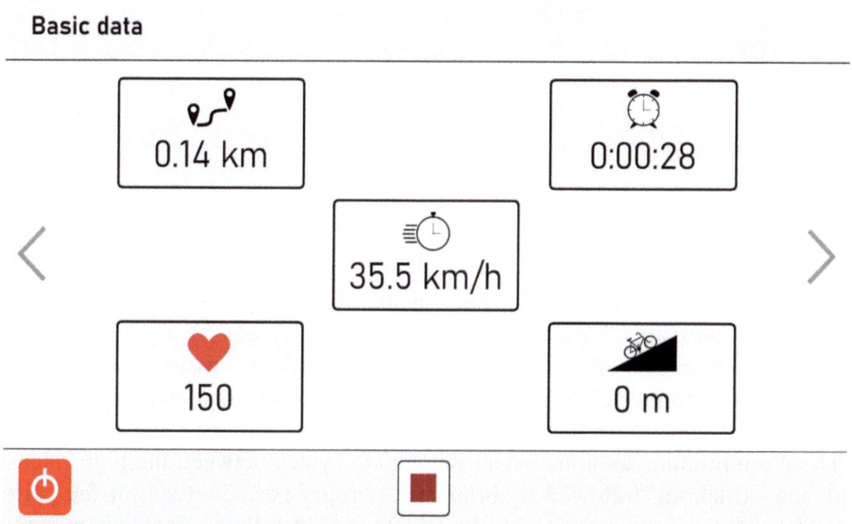

Fig. 2 AST monitor: basic data screen

and decrease coupling. The *heart rate sensor* module uses ANT communication channels for retrieving the data from the actual sensors. The *GPS sensor* uses a serial connection to communicate and retrieve data from the Adafruit GPS sensor. The retrieved training data is then processed inside *training session*, *interval training*, and *digital twin* classes whose output is then sent to the *user interface Main Window* class, which renders the visual components so that the user can receive visual cues regarding their training.

For the purpose of digital representation of the training and the digital twin, three interactive screens were designed to visually represent various metrics of a user's sports training. The basic data screen presentation Fig. 2 represents the user's current training session, where the cyclist can check their training duration, completed distance, ascent, and current heart rate. This allows the user to gain instant feedback regarding their progress during the training session. It should also be noted that the user can navigate between the screens by touching the left and right arrows that are

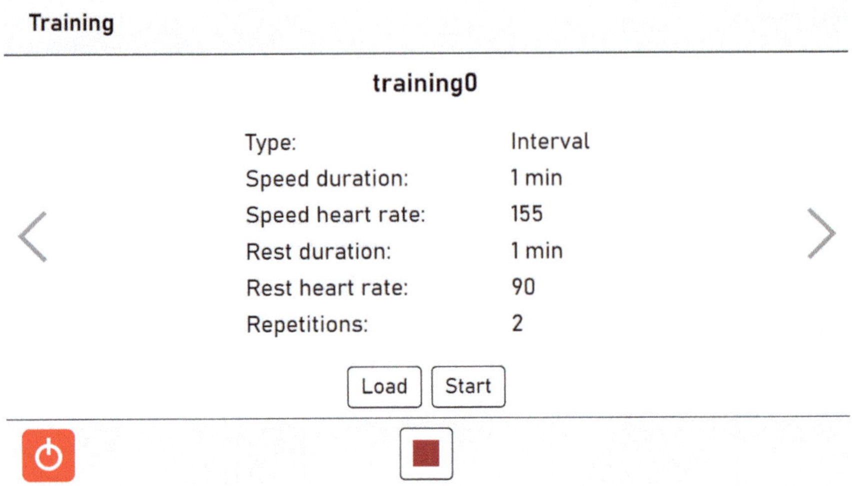

Fig. 3 AST monitor: training screen

present on any of the screens. The user can also start and pause the training session by using the "start/stop" button on the bottom of each screen.

The application also allows importing proposed training sessions where training goals can be determined (Fig. 3). Currently, this is used for interval training sessions in particular. The intervals have determined repetitions and rest durations.

These are training sessions where the athlete cycles between the high-intensity workouts, which are followed by brief rest or relief periods. Each of the periods has a determined heart rate, which the athlete should follow to achieve maximum training benefits. The primary goals of such sessions are increasing the strength of the heart muscle, improving the aerobic capacity, and allowing longer and more intense workouts [26].

Once the user starts the interval training, the user can view their interval training progress on the interval training screen as shown on Figs. 4 and 5. The screen displays the duration of the current phase, current heart rate, average heart rate, and the heart rate the user should try to achieve during the phase. In the middle, the difference between the planned and current heart rate is displayed, which provides a visual cue if the athlete should increase or decrease their effort to follow the proposed plan. The training plans can be imported inside the device by saving them on the Raspberry Pi micro SD card. The interval training plans are written in a simple JSON format.

The switches between the training phases are shown with push notifications, which briefly alert the user when the next phase has started so that the cyclist can adjust their behavior accordingly.

The presented solution has the capability of recording training devices as well as showing the progress of the determined goals in training. The AST monitor,

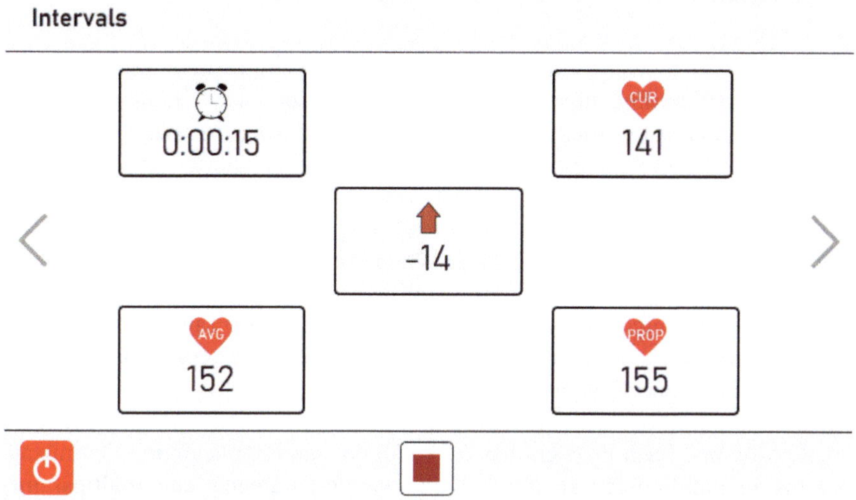

Fig. 4 AST monitor: interval screen during high-intensity phase

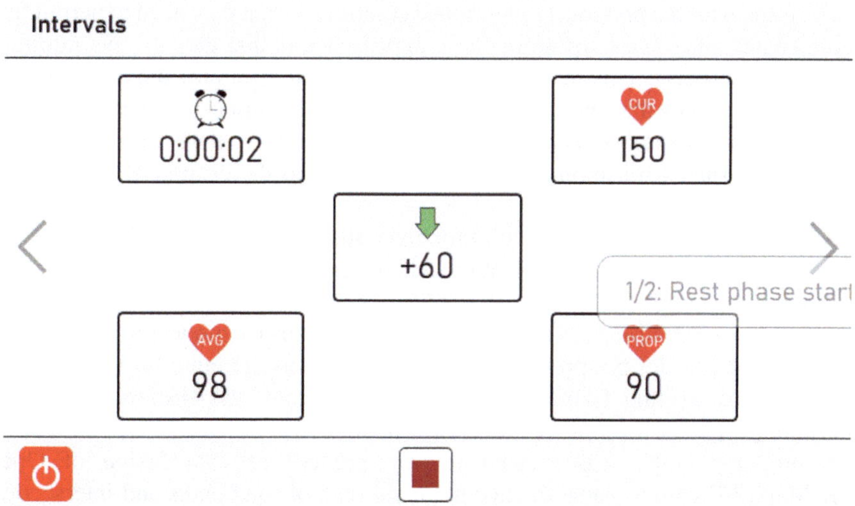

Fig. 5 AST monitor: interval screen during the rest phase

however, fell short in incorporating map integration. No functionalities and interfaces were previously implemented that would display the athlete's GPS position on a map, which meant that the data from the GPS sensor was used only for recording the training exercises but not for displaying the actual location. This also limited the usability of the device since there were no ways to display potential training routes.

3 Generating the Cycling Training Routes

Route finding is a widely known and researched optimization problem. Route finding and pathfinding algorithms have been widely used in the domains of robotics [27] and in logistics [28], as well as daily during the usage of GPS [29]. The goal of pathfinding is usually focused on finding the shortest or fastest path in a given search space. Pathfinding is most often performed on graphs G.

These are mathematical structures consisting of a set of vertices $V(G)$ and connections between them $E(G)$ [30]. When we add weights to the connections between points, indicating the cost of traveling between two connected vertices, we are talking about weighted graphs [31]. Travel in real space can thus be abstracted into pathfinding on a weighted graph, where intersections represent vertices and roads represent connections between them [32].

When talking about the problem of pathfinding, we are talking about several related problems. Each problem has its own goals and applicability. Optimization problems in pathfinding are divided into single-parameter and multiparameter optimized, with some of the most known representatives being the shortest path problem [33], vehicle routing problem [34], traveling salesman [35], and multi-objective shortest path [36].

The issue with the previously mentioned solutions is that they need to permit fine-tuning routes since the purpose of these approaches is that they always minimize their defined criteria. This means that paths generated minimize the costs in terms of the distance and time needed for their completion, minimize fuel costs, or try to find a compromise between them. However, generating sports training routes is mainly concerned with finding routes that conform to the criteria of distance and ascent.

A modified evolutionary algorithm for generating the cycling training routes [37] was developed together with an approach of generating a property graph from geographical data [38] to address this issue. The approach combines geographical data from OpenStreetMap [39] and EU-DEM digital surface model [40] and existing training records to form a property graph that serves as the input for the algorithm for generating cycling training routes. The algorithm was inspired by genetic algorithm [41] but with some distinct adjustments.

A property graph was created by the process of data fusion of OpenStreetMap [39] data to parse the geographical data of road paths and intersections between them as well as EU-DEM [40] to retrieve ascents of individual paths. The generated property graph contains the following elements with their attributes [38]:

- **Node** (*intersections*)
  - Latitude $\longrightarrow$ latitude coordinate of the intersection
  - Longitude $\longrightarrow$ longitude coordinate of the intersection
  - Node_id $\longrightarrow$ OpenStreetMap id of the node
  - Way_id $\longrightarrow$ OpenStreetMap way ids the node belongs to

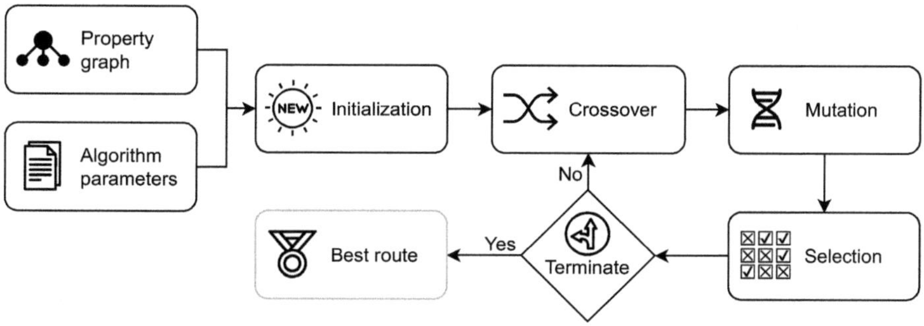

Fig. 6 Simplified execution flow of the modified evolutionary algorithm for generating cycling training routes

- **Relationship** (*paths*)
  - Ascent ⟶ ascent of the path, how much elevation did the athlete gain
  - Descent ⟶ descent of the path, how much elevation did the athlete lose
  - Distance ⟶ distance in meters
  - Type ⟶ OpenStreetMap path type
  - Relationship ⟶ node from-node to, which intersections does the path connect

In addition, each individual path has all its points saved in the form of a MongoDB collection to retrieve them later for visualizations.

This property graph serves as a search space, which can then be used for generating training routes by the algorithm [37]. The algorithm can search for routes with defined distances and total ascent within the given property graph.

Before initialization, the algorithm in Fig. 6 receives the start and end points and minimum and maximum distance and ascent the routes must conform to. In the first initialization step, the individual routes are generated from the previously presented property graph based on the start and end nodes from the route parameters.

This is followed by a crossover between routes, where special conditions are introduced. Most notably, the crossover between routes must result in a connected route, meaning only routes that share at least three points can be crossover, i.e., starting, ending, and one other point.

The constraint of route connectedness influenced the next mutation step, where in routes chosen for mutation, a random part of the route was deleted, and a new route was generated between the two nodes.

In the next step, routes were evaluated, and those with the worst grades were eliminated. This process continued until the termination criteria were reached: a route that conforms to the route parameters or the maximum number of iterations.

4 Proposed Method

The missing link between the three developed solutions was the integration of generated routes into the AST monitor so that generated routes could be used in practice, as shown in Fig. 7. The integration solution development followed the following steps presented in the following subsections: (1) identification of similar solutions and required functionalities, (2) user interface design, (3) exporting and importing routes, (4) selection of libraries and frameworks, and (5) software architecture and implementation.

4.1 Features and Requirement Identification

In the first step, mandatory and optional requirements and functionalities were identified. User interfaces of commonly used GPS routing mobile applications were inspected to do this. We investigated user interfaces of Google Maps [42], Strava [43], and Bikemap [44].

The universally found features of all the inspected applications were:

- **Map display** that displays a topographic view of the area
- **Marker** showing the user their current location
- **Route polyline** projected on the map showing the user where he must continue their route
- **Speed display** where the user can see their current speed
- **Distance**, either the remaining distance to the destination or distance traveled

Features identified but only present in some applications were the current heart rate display, completed ascent, turn-by-turn navigation explaining to the user exactly which turn to take, voice instructions, and duration of the exercise.

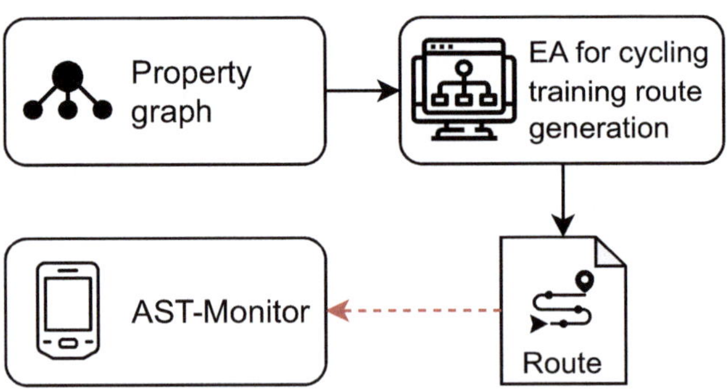

Fig. 7 Missing link between the generated routes and their use in the AST monitor

We have determined that universally found features were the mandatory requirements and determined which of the remaining features could be implemented inside the given framework. We have decided to incorporate a heart rate display to help the cyclist adjust their physical effort, completed ascent, and duration. Voice navigation was not pursued at this time because significant hardware and software changes would be required for the AST monitor device. Turn-by-turn navigation was also not pursued because the cycling path property graph only incorporates routes that can be used by the cyclist, which means that some intersections may be left out, which may turn out to be misleading to the cyclist.

4.2 User Interface Design

The user interface design phase was concerned with preparing a fundamentally straightforward sketch of the planned user interface for the map screen, as seen in Fig. 8. The planned screen needed to incorporate all of the planned features. The planned user interface was split into three parts, marked by shades of gray on the sketch. The map was selected as the central element, claiming a significant share of the available screen space. This was done to increase readability as well as compensate for the lack of turn-by-turn navigation.

We decided to display current physical and route parameters on the bottom side of the screen, where current speed, current heart rate, completed ascent, completed distance, and duration of the exercise were to be displayed. The right side of the screen was chosen to display the athlete's training progress. We have decided to add two elements, which display the remaining distance and ascent, as well as a bar chart that displays the route completion in regards to the distance completed versus the distance remaining. The bar chart was added because it has been found to be beneficial to user engagement and motivation [45].

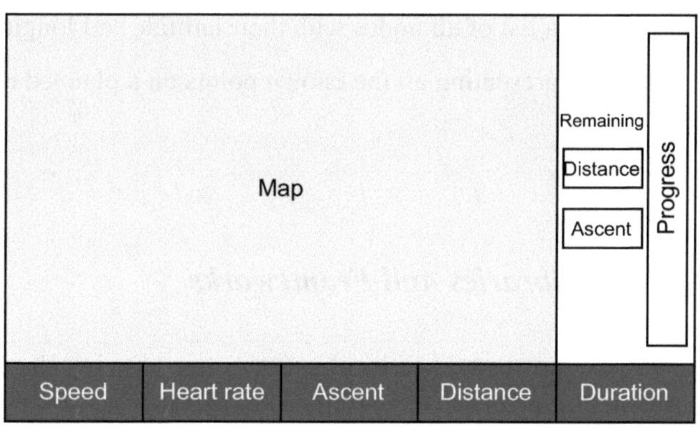

Fig. 8 Map user interface sketch

4.3 Exporting and Importing Routes

The algorithm-generated routes needed to be written in a format that could be later imported into the AST monitor. The first obstacle was choosing the suitable data representation/serialization format. We have identified three different approaches to solving this challenge, namely, (1) JSON (JavaScript Object Notation) [46], (2) XML (Extensible Markup Language) [47], and exporting/importing objects directly as binary pickle (native Python object serialization library) files since both the modified evolutionary algorithm for generating the cycling training routes and the AST monitor are implemented in the Python programming language.

We have eliminated the native Python object approach first because we wanted the files also to be human readable for the solution to be more easily debuggable as well as to maintain the modularity of the solution, meaning each part could be easily replaceable and should not lock in other parts of the system.

The remaining JSON and XML standards both offer readability. However, JSON does offer some advantages over XML in our case; the JSON files have been found to be smaller in size and interpreted faster for most cases over their XML counterparts [48]. XML offers advantages over advanced types, and validations are offered using schemas. This, however, was not needed in our case as the data was pretty straightforward.

The resulting JSON file had the following structure:

- route
  - nodes: [] ⟶ a list of node pairs representing relationships/paths between intersections
  - ascents: [] ⟶ a list of ascents on each part of the route node pairs
  - descents: [] ⟶ a list of descents on each part of the route node pairs
  - distances: [] ⟶ a list of distances on each part of the route node pairs
  - evaluation: { total_distance, total_ascent, total_descent } ⟶ displayed the total distance, ascent, and descent achieved by completing the route
- route_render: [] ⟶ a list of all nodes with their latitude and longitudes
  - id: ⟶ node id representing all the known points on a planned route
  - lat: ⟶ latitude
  - lon: ⟶ longitude

4.4 Selection of Libraries and Frameworks

Since the map display presented a critical point in our user interface, we needed a heavily adjustable and responsive solution. The dilemma was between choosing a PyQt6 plugin, such as Qt Location Open Street Map Plugin, and a solution not necessarily based on PyQt6. The search was expanded to JavaScript mapping libraries since no other solutions were discovered within the (Py)Qt ecosystem. This was done because PyQt6 offers a native way of embedding a web view, which can render a custom web page that can even be local as well as provide native functionality of calling the JavaScript functions natively from the PyQt application.

Another requirement was that the solution for providing maps should be free and open-source, provided under one of the compatible licenses with GNU Public License version 3 [49], in which the noncommercial version of PyQt6 is published. During the discovery phase, we identified JavaScript libraries for displaying maps. We have searched for libraries on the GitHub platform [50], which provides version control using git and also serves as a code repository for numerous projects. The search query, i.e., GitHub offers search by tags known as topics, was dome under the *maps* tag and filtered by language to JavaScript. The resulting query yielded over 1000 results. To limit the time spent searching to a manageable level, only projects containing over 1000 endorsements[2] were inspected. This yielded 12 results.

Out of the final identified results, we have decided to use *Leaflet.js* [51] because it has the most endorsements (over 38300), is small in size (42 kilobytes), and is well supported. The library itself supports the needed features for map display, such as different tile layers, markers, and polylines, as well as dynamic interactions with the rendered map.

The remainder of the selection of libraries was concerned with finding the right way to create the container for rendering the actual component together with the bottom and left displays that will show the training parameters as well as training progress. Since many objects are repeating with only minor differences in their visual look, we have decided that we need a solution that would allow the creation of reusable components that could then be used for individual displays, e.g., speed, heart rate, and distance. This is possible by numerous JavaScript frameworks such as Vue.js [52], React [53], and Angular [54]. We have chosen Vue.js (version 3) because we found a wrapper for the Leaflet.js library.

The last part was choosing the user interface front-end framework. In this case, the Bootstrap [55] framework was chosen, being one of the de facto standard frameworks for front-end design and also one of the most endorsed repositories on GitHub (over 162 thousand endorsements).

5 Implementation and Software Architecture

In the implementation stage, the development process and architecture of the solution were determined. We have decided first to create a standalone Vue.js application that can render the required components. To do this, we have created five Vue.js components: App, Map, Training metrics, Display card, and Route progress.

[2] GitHub tracks endorsements as *GitHub stars* which the users can award to projects they like.

The app component represents the entry point of the view; the component is the container for the map, training metrics, and route progress components. This component also serves as a communication point by exposing a function that can be called from the main PyQt app. The communication is made possible by creating and exposing a function that modifies this component state.

The map component implemented the map display using the *Vue-leaflet* wrapper library, which aids in using *Leaflet.js* in Vue.js-based applications. The data the component receives is the current latitude, longitude, and route, representing an ordered list of latitude-longitude pairs representing the planned route from which the polyline can be rendered.

The training metrics component represented the route and physical parameters. Because the component had five similar displays, another component, the Display card, was created, serving as a display for individual parameters.

The last component that was implemented was the route progress component, where a display for remaining distance and ascent was implemented, as well as the route progress bar, which shows the ratio of the completed route in regards to the total distance.

Vue.js [56], like many popular JavaScript frameworks, also uses the EcmaScript modules, which are a way of reusing JavaScript code introduced in EcmaScript specification 2015 [57]. Most modern browsers, e.g., Google Chrome, do not allow running these modules locally, which means that webpages must be served from a server and not locally if the module functionality is to be used [58].

This means that the Python application also had to create a simple server accessible only locally to serve the Vue.js application, for which we used the Python native library *socketserver*. The final architecture of the implemented solution is seen in Fig. 9 where a rough sketch of the system is seen.

The last part of the implementation process was building the application into a distribution version so that the user interface could be served quickly and without

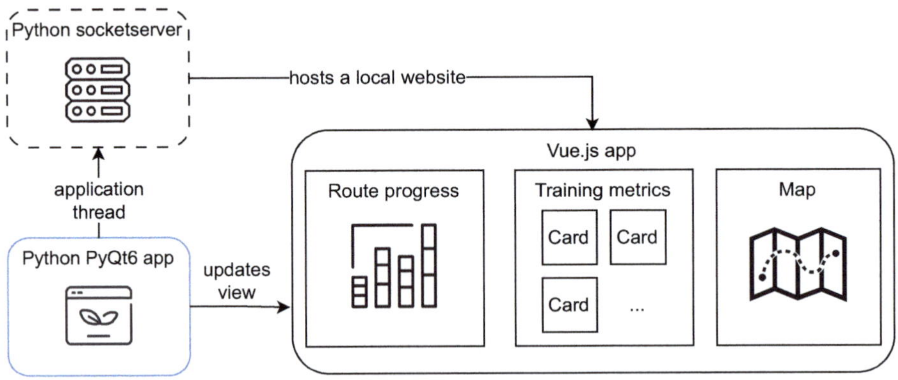

Fig. 9 Architecture of the map visualization solution in AST monitor

compiling at every start of the AST monitor, which would drastically increase the load times of the AST monitor since the device itself has limited processor and memory capabilities.

6 Results

The developed user interface solution was designed based on predefined criteria to enhance the user experience in digital twin model training. This enabled the use of the AST monitor device for navigation and training through unfamiliar terrains. The interface's design adheres to principles of clarity and functionality, ensuring ease of use across different user demographics. A significant feature of the interface is its capability to provide real-time feedback on various training metrics. As users engage with the route, the system dynamically displays metrics ranging from basic parameters like speed and distance to advanced metrics such as cadence and power output.

For empirical validation, the evolutionary algorithm for generating the cycling training routes was employed and used for generating a sample route between Ptuj and Murska Sobota. The chosen 44-km route, characterized by its 544 ascent meters, was designed to simulate varied terrains encountered in real-world cycling scenarios. Upon route determination, it was exported and integrated into the AST monitor for further analysis. A comprehensive simulation was conducted, wherein synthetic GPS data and heart rate metrics were generated to replicate cycling conditions. The screenshots from the recording are shown in Figs. 10, 11, 12, 13, and 14.

In the comprehensive visual representation of the training interface, the initial phase of the training is depicted in Fig. 10. This screen serves as an introduction to the user's journey, setting the stage for the subsequent stages of the training.

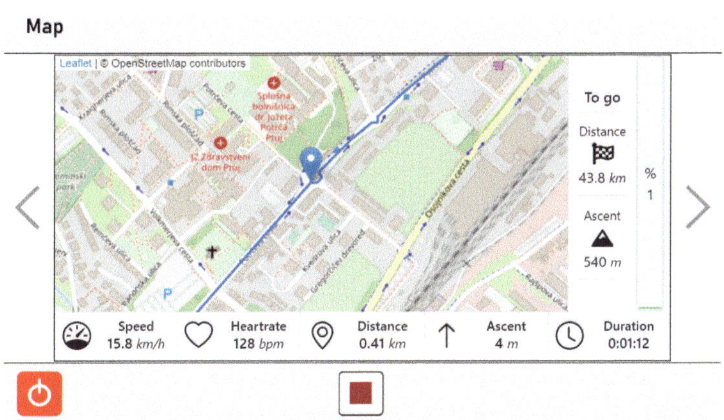

Fig. 10 Map display demo: start of the training session

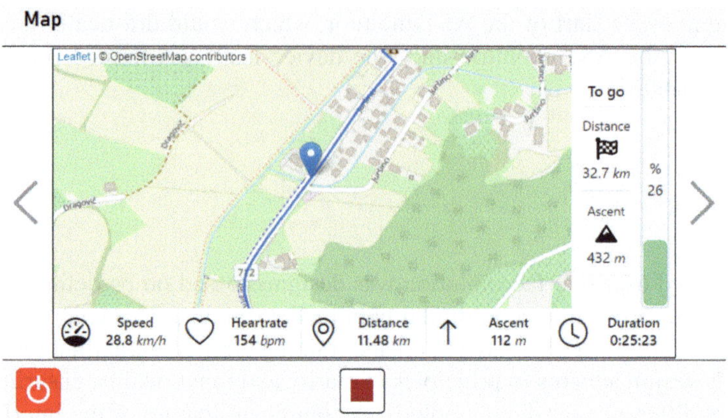

Fig. 11 Map display demo: training session 1

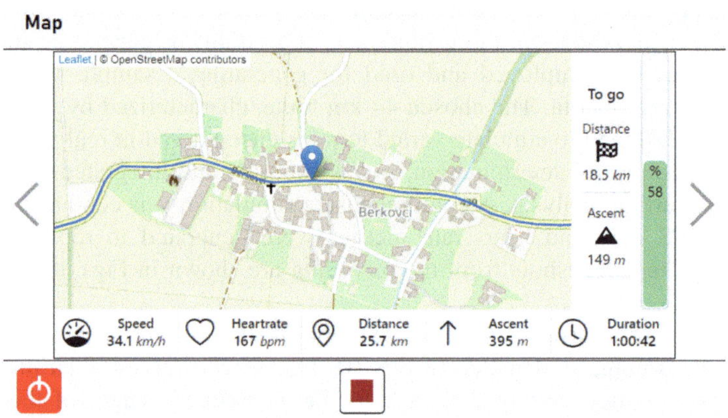

Fig. 12 Map display demo: training session 2

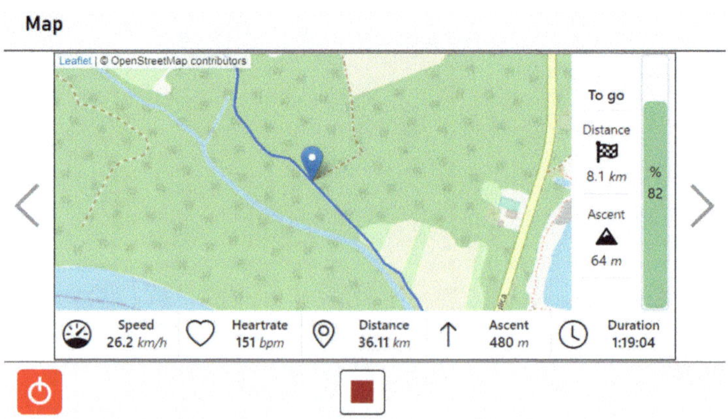

Fig. 13 Map display demo: training session 3

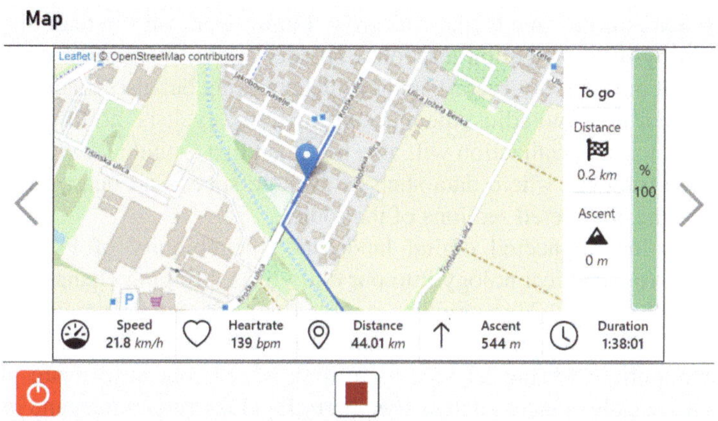

Fig. 14 Map display demo: end of the training session

As the training progresses, Figs. 11, 12, and 13 provide a detailed insight into the dynamic changes observed on the map display. One of the notable features captured in these figures is the progress bar, which incrementally fills up, reflecting the ratio of the completed distance to the total distance. Concurrently, these figures also highlight the decrement in the remaining distance and ascent, offering the user a real-time update on their performance and the challenges ahead. Concluding the visual journey, Fig. 14 captures the culmination of the training, signifying the achievement of the set objectives and the end of that particular training session. It can be seen that the solution displays the training metrics of speed, heart rate, distance, ascent, and duration, as well as the progress on the route in the form of a progress bar and distance and ascent parameters. The map display also shows the current position of the cyclist as well as the route followed.

In summary, the integration of a well-designed interface, advanced algorithms, and real-time data processing positions this system as a potential tool for advancing the field of cycling training. Future research will focus on refining the system based on empirical data and exploring additional functionalities to further its academic and practical contributions.

7 Conclusion

We have shown a use case of combining two well-designed solutions to create an integrated system that aids cyclists in training. Many intelligent solutions in sports training stay in the idea phase and never reach their full potential [1].

The provided case can not only serve as a use case for intelligent sports cycling training but also as a rough outline of how to plan, design, and implement intelligent sports training systems, as well as demonstrate the best practices in the process of creating such systems.

The system still has much space to grow. Future work on the map display will include turn-by-turn navigation and, on a broader scope, integration with systems that would not only import, record, and evaluate training but also generate pieces of training on their own and adjust them according to the cyclists' abilities. Regarding the AST monitor, a visualization will also be added that would allow visualizations of completed training, where data such as average speed and heart rate could be projected on the completed sections of the route.

In today's interconnected digital landscape, the ubiquity of IT devices has seamlessly introduced technology into our daily lives, bringing changes to how we interact with the world. This ubiquity, coupled with the advancements in modern technologies, has democratized innovation. Significant capital and resources are no longer required to bring an idea to fruition. Open-source software platforms, as well as accessible devices such as Raspberry Pi, allow researchers and innovators to swiftly transform their concepts into tangible products and systems. This accessibility means that niche solutions, such as sports training modules, can be developed, tested, and deployed without massive investments of time and money.

Furthermore, code repositories, digital marketplaces, and online communities ensure these solutions swiftly find their target audience. The confluence of IT device proliferation and cutting-edge technologies has catalyzed a dynamic environment where innovative ideas in smart sports training can be rapidly conceptualized, developed, and implemented, meeting the ever-evolving demands of sports science.

Acknowledgments This work was supported by the Slovenian Research Agency [Research Core Funding No. P2-0057].

References

1. A. Rajšp and I. Fister, "A Systematic Literature Review of Intelligent Data Analysis Methods for Smart Sport Training," *Applied Sciences*, vol. 10, p. 3013, Jan. 2020. Number: 9 Publisher: Multidisciplinary Digital Publishing Institute.
2. N. Frevel, D. Beiderbeck, and S. L. Schmidt, "The impact of technology on sports – A prospective study," *Technological Forecasting and Social Change*, vol. 182, p. 121838, Sept. 2022.
3. I. Fister, I. Fister, and D. Fister, "Fundamentals of Sports Training Theory," in *Computational Intelligence in Sports*, Adaptation, Learning, and Optimization, ch. Theory of, pp. 103–119, Cham: Springer International Publishing, 1 ed., 2019.
4. S. Lowe and G. ÓLaighin, "The age of the virtual trainer," *Procedia Engineering*, vol. 34, pp. 242–247, Jan. 2012.
5. F. Mulas, S. Carta, P. Pilloni, and M. Manca, "Everywhere run: a virtual personal trainer for supporting people in their running activity," in *Proceedings of the 8th International Conference on Advances in Computer Entertainment Technology*, ACE '11, (New York, NY, USA), pp. 1–2, Association for Computing Machinery, Nov. 2011.
6. I. Fister, A. Iglesias, A. Galvez, S. Deb, and D. Fister, "On Deploying the Artificial Sport Trainer into Practice," in 2021 8th International Conference on Soft Computing & Machine Intelligence (ISCMI), pp. 21–26, IEEE, nov 2021.

7. A. El Saddik, "Digital Twins: The Convergence of Multimedia Technologies," *IEEE MultiMedia*, vol. 25, pp. 87–92, apr 2018.
8. N. Crespi, A. T. Drobot, and R. Minerva, "What Is a Digital Twin and Why Is It Important?," in *The Digital Twin*, ch. The Digita, pp. 4–6, Cham: Springer International Publishing, 1 ed., 2023.
9. R. Gámez Díaz, Q. Yu, Y. Ding, F. Laamarti, and A. El Saddik, "Digital Twin Coaching for Physical Activities: A Survey," *Sensors*, vol. 20, p. 5936, oct 2020.
10. E. Wu and H. Koike, "Futurepose - mixed reality martial arts training using real-time 3d human pose forecasting with a rgb camera," in 2019 IEEE Winter Conference on Applications of Computer Vision (WACV), pp. 1384–1392, 2019.
11. P. L. Aouaidjia Kamel, Bowen Liu and B. Sheng, "An investigation of 3d human pose estimation for learning tai chi: A human factor perspective," *International Journal of Human–Computer Interaction*, vol. 35, no. 4-5, pp. 427–439, 2019.
12. L. Lukač, I. Fister, and I. Fister, "Digital Twin in Sport: From an Idea to Realization," *Applied Sciences*, vol. 12, p. 12741, dec 2022.
13. R. G. Diaz, F. Laamarti, and A. El Saddik, "DTCoach: Your Digital Twin Coach on the Edge During COVID-19 and Beyond," *IEEE Instrumentation & Measurement Magazine*, vol. 24, pp. 22–28, sep 2021.
14. C. G. Bender, J. C. Hoffstot, B. T. Combs, S. Hooshangi, and J. Cappos, "Measuring the fitness of fitness trackers," in 2017 IEEE Sensors Applications Symposium (SAS), pp. 1–6, 2017.
15. I. Fister, I. Fister, and D. Fister, "Design and Implementation of an Artificial Sports Trainer," in *Computational Intelligence in Sports*, pp. 121–135, Cham, Switzerland: Springer, Cham, 2019.
16. Strava, Inc., "Strava | Running, Cycling & Hiking App - Train, Track & Share." https://www.strava.com/. Accessed: 18-07-2023.
17. Under Armour, Inc., "Mapmyride." https://www.mapmyride.com/. Accessed: 17-07.2023.
18. Bikemap GmbH, "Bicycle tracks, route planner, bike computer app - bikemap." https://www.bikemap.net/. Accessed: 18-07-2023.
19. M. Dobiasch, "Training characteristics of athletes in golden cheetah open data," vol. 9, pp. 55–56, Dec. 2020.
20. R. P. T. Ltd., "Raspberry pi 4 computer model b product brief," tech. rep., Raspberry Pi Foundation, January 2021. Accessed: 11-08.2023.
21. Adafruit, "Adafruit ultimate gps hat for raspberry pi." https://learn.adafruit.com/adafruit-ultimate-gps-hat-for-raspberry-pi?view=all, 2023. Accessed: 12-08-2023.
22. Garmin, "Usb ant stick™," 2023. Accessed: 12-08-2023.
23. P. S. Foundation, *Python 3.9 Reference Manual*. Python Software Foundation, https://www.python.org/, 2020. Accessed: 12-08-2023.
24. R. C. Limited, "Pyqt6." https://www.riverbankcomputing.com/software/pyqt/, 2021. Accessed: 13-08-2023.
25. T. Q. Company, "Qt toolkit." https://www.qt.io/, 2023. Accessed: 14-08-2023.
26. P. Laursen and D. Jenkins, "The scientific basis for high-intensity interval training," *Sports Medicine*, vol. 32, no. 1, pp. 53–73, 2002.
27. A. Gasparetto, P. Boscariol, A. Lanzutti, and R. Vidoni, "Path Planning and Trajectory Planning Algorithms: A General Overview," in *Motion and Operation Planning of Robotic Systems* (G. Carbone and F. Gomez-Bravo, eds.), pp. 3–27, Cham, Switzerland: Springer, 1 ed., 2015.
28. X. Zhang, Z. Zhang, Y. Zhang, D. Wei, and Y. Deng, "Route selection for emergency logistics management: A bio-inspired algorithm," *Safety Science*, vol. 54, pp. 87–91, apr 2013.
29. V. Haria, Y. Shah, V. Gangwar, V. Chandwaney, T. Jain, Y. Dedhia, and A. Surendra Modi, "The Working of Google Maps, and the Commercial Usage of Navigation Systems," *International Journal of Innovative Research in Technology*, vol. 6, pp. 184–191, okt 2019.

30. R. Diestel, “Graphs and Their Representation,” in *Graph Theory* (S. Axler and K. Ribet, eds.), vol. 173 of *Graduate Texts in Mathematics*, pp. 1–12, Berlin, Heidelberg, Germany: Springer Berlin Heidelberg, 2017.
31. A. Bondy and U.S.R. Murty, *Weighted Graphs and Subgraphs*. Graduate Texts in Mathematics, Guildford, England: Springer, 2008 ed., dec 2007.
32. A. Rajšp, M. Heričko, and F. J. Iztok, “Preprocessing of roads in OpenStreetMap based geographic data on a property graph,” in *Proceedings of the Central European Conference on Information and Intelligent Systems*, pp. 193–199, Faculty of Organization and Informatics, University of Zagreb, 2021.
33. G. Yu and J. Yang, “On the Robust Shortest Path Problem,” *Computers & Operations Research*, vol. 25, pp. 457–468, June 1998.
34. J. E. Bell and P. R. McMullen, “Ant colony optimization techniques for the vehicle routing problem,” *Advanced Engineering Informatics*, vol. 18, pp. 41–48, Jan. 2004.
35. M. Jünger, G. Reinelt, and G. Rinaldi, “The traveling salesman problem,” in *Handbooks in Operations Research and Management Science*, vol. 7 of *Network Models*, pp. 225–330, Elsevier, Jan. 1995.
36. F. He, H. Qi, and Q. Fan, “An Evolutionary Algorithm for the Multi-objective shortest path problem,” in *Proceedings of the* 2007 International Conference on Intelligent Systems and Knowledge Engineering (ISKE 2007), pp. 1276–1280, Atlantis Press, Oct. 2007. ISSN: 1951-6851.
37. A. Rajšp and I. Fister, “A modified evolutionary algorithm for generating the cycling training routes,” *IEEE Access*, vol. 10, pp. 109743–109759, 2022.
38. A. Rajšp and I. F. Jr., “Neo4j graph dataset of cycling paths in slovenia,” *Data in Brief*, vol. 48, p. 109251, 2023.
39. Openstreetmap Foundation, “OpenStreetMap.” https://www.openstreetmap.org/, 2021. Accessed: 01-09-2023.
40. European Environment Agency, “EU-DEM 1.1.” https://www.eea.europa.eu/data-and-maps/data/copernicus-land-monitoring-service-eu-dem, dec 2017. Accessed: 12-08-2023.
41. D. Whitley, “A genetic algorithm tutorial,” *Statist. and Comput.*, vol. 4, no. 2, 1994. doi: 10.1007/BF00175354.
42. G. LLC, “Google maps.” https://play.google.com/store/apps/details?id=com.google.android.apps.maps, 2023. Accessed: 2023-07-11.
43. S. Inc., “Strava.” https://play.google.com/store/apps/details?id=com.strava, 2023. Accessed: 2023-07-12.
44. B. GmbH, “Bikemap.” https://play.google.com/store/apps/details?id=com.toursprung.bikemap, 2023. Accessed: 2023-07-12.
45. D. Liu, R. Santhanam, and J. Webster, “Toward meaningful engagement: A framework for design and research of gamified information systems,” *MIS Quarterly*, vol. 41, no. 4, pp. pp. 1011–1034, 2017.
46. D. Crockford, “The application/json media type for javascript object notation (json),” RFC 4627, Internet Engineering Task Force (IETF), 2006.
47. T. Bray, J. Paoli, C. M. Sperberg-McQueen, E. Maler, and F. Yergeau, “Extensible markup language (xml) 1.0 (fifth edition),” w3c recommendation, World Wide Web Consortium (W3C), 2008.
48. N. Nurseitov, M. Paulson, R. Reynolds, and C. Izurieta, “Comparison of json and xml data interchange formats: A case study,” in *ISCA International Conference on Computer Applications in Industry and Engineering*, 2009.
49. “Gnu general public license, version 3.” http://www.gnu.org/licenses/gpl.html, June 2007. Accessed: 11-07-2022.
50. github, “Github.” https://github.com/, 2020. Accessed: 11-08-2023.
51. V. Agafonkin, “Leaflet — an open-source JavaScript library for interactive maps.” https://leafletjs.com/, 2023. Accessed: 12-09-2023.
52. Evan You, “Vue.js - The Progressive JavaScript Framework | Vue.js.” https://vuejs.org/, 2023. Accessed: 15-09-2023.

53. Meta, "React." https://leafletjs.com/, 2023. Accessed: 15-09-2023.
54. Google, "Angular." https://angular.io/, 2023. Accessed: 15-09-2023.
55. Mark Otto, Jacob Thornton, Patrick H. Lauke, *et al.*, "Bootstrap." https://getbootstrap.com/, 2023. Accessed: 15-09-2023.
56. Evan You, "Quick Start | Vue.js." https://vuejs.org/guide/quick-start.html#using-vue-from-cdn. Accessed: 16-09-2023.
57. E. International, "Ecmascript 2015 language specification," Technical Report ECMA-262, 6th Edition, Ecma International, Rue du Rhone 114, CH-1204 Geneva, June 2015. Accessed: 16-09-2023.
58. Michal Zalewski, "Browser Security Handbook," tech. rep., Google Inc., 2008. Accessed: 16-09-2023.

Perspectives of Artificial Intelligence in Training and Exercise

Arnold Baca

Abstract As in many other fields of application, the availability of miniaturized sensors for measuring a wide variety of parameters, powerful computer technologies, and a wide range of artificial intelligence methods opens more and more support possibilities for training and exercise in sports. We present current developments and prospects for the use of respective systems and illustrate this with examples from our research. These include technologies and systems based on the automated recognition of movements, which can, for example, make recommendations in the event of incorrectly executed forms of movement, initiate immediate reactions in critical situations, or allow interaction in virtual environments. In addition, systems for providing feedback and recommending individual exercises based on personal performance levels, as well as data-based methods for training control or predicting performance gains through the type and amount of training that do not interfere with physical activity, are highlighted. The expected potential of such innovative approaches, limitations, and risks are critically reflected.

Keywords Feedback · Motion pattern · Training control · Virtual trainer

1 Introduction

Opportunities arising from the use of artificial intelligence methods in training and exercise are a focus of current sports- and fitness-related media coverage. Different systems for supporting and promoting physical activity, often developed by relevant startups such as Tempo, Tonal, Whoop, or HomeCourt, are presented and advertised. The range of respective paid and free apps, e.g., Aaptiv, Evolve AI, Fitbod, and FitnessAI, is continuously growing. For users, the best-known end devices are smartphones, smartwatches, and smart wristbands. However, smart mirrors, which

A. Baca (✉)
Centre for Sport Science and University Sports, University of Vienna, Vienna, Austria
e-mail: arnold.baca@univie.ac.at

M. J. Blondin et al. (eds.), *Artificial Intelligence, Optimization, and Data Sciences in Sports*, Springer Optimization and Its Applications 218,
https://doi.org/10.1007/978-3-031-76047-1_9

can display additional information such as real-time feedback alongside the mirror image, or smart home gyms, which contain intelligent training equipment for different goals and exercise and training options, are also becoming increasingly popular. The development of the latter has been driven by the limited usability of fitness centers during the pandemic.

The providers of smart mirrors for the fitness sector, e.g., lululemon Studio Mirror[1] or Forme Studio,[2] promise that exercisers can better integrate their workout into everyday life. Individual training programs can be suggested and executed interactively. An avatar or a real person can show suitable exercises on the screen according to certain criteria, such as the muscle groups to be trained, and pulse and calories burned can also be displayed. On some devices, an instructor can see the exerciser via an integrated camera and offer guidance. Real-time feedback may be provided on squats, presses, curls, and other movements. Personal profiles, such as Apple Watch, Strava, can be integrated to show additional personal performance development. Smart mirrors are often also a part of smart home gyms, e.g., Tempo Studio[3] or Tonal.[4] In addition, these include a range of other smart fitness equipment, such as a smart treadmill, or devices with adaptable weights. In particular, weight training workouts are supported. A special feature of Tonal, for example, is an adjustable digital weight system. The system determines performance during strength training and automatically increases the weight when a certain threshold is exceeded. During recovery exercises, the weight is automatically lowered and increased.

For the user, it is not always guaranteed that a benefit, and if so, what benefit, can be derived from such systems.

In addition to technological developments that will further increase the performance and range of applications of AI-based systems in training and competition, this chapter addresses development trends in the area of feedback systems and systems to support training design and control.

2 Technical and Methodological Developments

The performance of data-based approaches to support training and exercise not only depends on the AI technology applied but, moreover, is strongly influenced by the type and amount of available data.

A wide variety of sensor technologies are available for this purpose. Accelerometers determine the type and duration of physical activities; inertial sensors, GPS, and video sensors are used for tracking objects; strain gauges and piezoelectric

[1] https://prod-frontend.mirror.co/, accessed 22.1.2024.

[2] https://formelife.com, accessed 22.1.2024.

[3] https://tempo.fit/, accessed 22.1.2024.

[4] https://www.tonal.com/, accessed 22.1.2024.

and capacitive sensors are applied for measuring force; pulse oximeters or photoplethysmographs estimate oxygen saturation; and surface electromyographs determine muscular activity. These established technologies have recently been summarized in a book chapter of Baca [3]. In addition, novel sensors are addressed which can also be used for online training control or for assessing internal body processes and external influences during physical activities; for example, in smart textiles, the sensor technology is integrated directly into the textile fabric, allowing for the determination of respiratory rate by stretching, as well as humidity or temperature and other body parameters. There are several relevant products on the market, such as Hexoskin or Oxstren (cf. Scataglini et al. [21] for a list of smart clothing products in sports). Great potential is expected from implantable sensors that constantly measure certain parameters, e.g., vital signs, or biological biomarker sensors for determining parameters of biological processes. They could provide further insights into the training and recovery process by providing additional information about the physiological processes of athletes [22]. Athletes from endurance and team sports in particular could benefit from these technologies.

In recent years, the use of sleep sensors has also become increasingly widespread [10]. Different technologies and approaches are used. Polysomnography is used to continuously monitor very different body functions, while simpler approaches are based on motion sensors.

3 Feedback Systems

Kinematic parameter values are increasingly determined on the basis of video systems without the use of markers or sensors attached to the human body. AI-based methods using deep learning allow 2D and 3D whole-body pose reconstruction and estimation from video images. The open-source program OpenPose [6], for example, has achieved widespread use for this application. Figure 1 shows one frame of a motion sequence in climbing with the determined body key points.

Motion sequences recognized in this way are then compared or made comparable with predefined executions. This is done, for example, by superimposing the target movement and the executed movement on the screen or by providing verbal recommendations. One problem lies in the determination and specification of the individual target movements, particularly when the aim is to minimize joint loads.

One reason for this is the difficulty of estimating the forces and moments that occur and the resulting joint forces and loads from purely kinematic data. However, there are developments and approaches in which human movement dynamics are reconstructed from purely kinematic data using methods of deep learning and musculoskeletal simulation, and kinetic parameters are derived from this. Uhlrich et al. [26] present OpenCap, a respective system based on synchronized video recordings from smartphones.

Moreover, respective parameters can be assessed using instrumented exercise and sports equipment or sensors attached to the body, additionally.

Fig. 1 Body key points determined using OpenPose during a climbing movement

Already in 2012 and 2013, for example, approaches were presented by the author's research group [17, 18], in which movement patterns were derived from force and position data during exercise with a weight leg press machine equipped with a force sensor and a rotation encoder. In this way, significant force and displacement parameter values could be measured directly and then used to identify individual repetitions and extract other determinants such as duration, velocity, acceleration, or power.

Methods of AI (support vector machine and multilayer perceptron neural network, respectively) were used to distinguish between correct and two elementary faulty movement executions (faulty flexion and extension phase, respectively). Both approaches yielded similar results with high sensitivity and specificity. The future goal of such approaches must be to increase the number of different movement patterns to be distinguished to identify the types and causes of faulty movement execution and to provide the trainees with more specific feedback instructions. Different approaches for assessing the movement performed can be used to train such classifiers in the course of a supervised learning process. On the one hand, experts can evaluate the movement execution; on the other hand, the results of model-based biomechanical movement analyses, which are carried out during the training phase of the classifier for the executed movements, can be used. Like the method involving musculoskeletal simulation, the latter approach could allow measured values of input parameters to be related to values of joint loading parameters, thereby being able to detect injurious motion executions.

Sensor fusion, i.e., the combination of different measurement methods such as video-based technologies and force sensors, enables motion patterns to be identified even more reliably.

In any case, at the end of the process of identifying the movement pattern, there should be feedback information that contains concrete instructions for performing the movement again or further. This feedback can be provided by a personal trainer connected via the Internet or by a virtual personal trainer (cf. Baca et al. [4]). In order to simplify the creation of such feedback systems, a design paradigm and model for mobile feedback systems (direct mobile coaching) was developed with the lead participation of members of the author's working group [9].

Intelligent pose-estimation methods can also be applied for automated scoring of movement execution in technical-compositional sports disciplines. This allows athletes to receive feedback on the quality of movement execution during training without the need for the presence of real judges. Fujitsu in cooperation with the International Gymnastics Federation [14] has developed such a system for gymnastics. Based on skeleton recognition, AI identifies the performed elements using a database and checks quality criteria. An implementation of comparable systems to other sports involving rating of human judgers, such as water jumping or ski jumping, is conceivable. Studies [2, 16], however, suggest that the adoption of AI technologies for supporting decision-making or substituting human judgers is by no means straightforward.

In an ongoing project of the author's working group (Smart Sport Assistance[5]), assistance systems for physical activity and sports lessons for blind and visually impaired students are being developed. A system for swimming is designed to provide appropriate feedback to students before they reach the edge of the pool. One approach uses the potential of intelligent computer vision cameras to automatically track the swimmers. A prototype system has been trained by taking 3600 single images of swimming persons and shows a high recognition rate (0.95). Figure 2 shows how a swimming student is detected in the pool.

The early detection of overloads, possible injury risks and health problems, and the associated timely information to the athlete or adaptation of the further training program (cf. Fabbrizio et al. [12]) is another area of application of AI in training. In addition to the parameters characterizing the movement (kinematics, heart rate, etc.), the data set to be analyzed can also contain data on previous accidents and previous training programs. In a collaborative project that included the author's research group, for example, a system was developed that automatically provides athletes with live feedback during a marathon in the form of verbal instructions on how to change speed via smartphone and optimizes performance by avoiding over-

[5] https://smart.univie.ac.at/en/project/, accessed 22.1.2024.

Fig. 2 Automated detection of a swimmer approaching the pool edge

and underloads [25]. The heart rate and speed of the runners were used as input data for the integrated model.

4 Training Design and Control

Individual training programs can be compiled from extensive training databases. Commercial products include BodBot[6] and FitnessAI[7] among others. One objective of such developments is also to increase the motivation of the trainees.

Outdoor games were developed as part of EU-funded projects under the technical leadership of the author's working group. The pupils and students participating in these games were assessed according to their previously recorded physical abilities in order to achieve equal opportunities in the competition (Te(a)chIn, [27]), or the individual level of difficulty or effort of the movement tasks was adapted to these abilities (COMPASS[8]). Both approaches are intended to motivate young people who are less active in sports to engage in sports more often by improving the chances of winning in competition-like scenarios. In the Te(a)chIn approach, recorded movement activities in sports games are relativized to the predetermined physical level of the persons [20]; in COMPASS, which was developed for orienteering, the route to be completed can be adapted accordingly.

---

[6] https://www.bodbot.com/, accessed 22.1.2024.

[7] https://www.fitnessai.com/, accessed 22.1.2024.

[8] https://compass.orienteering.bg/, accessed 22.1.2024.

Methods from AI are expected to be applicable in training impact analysis in order to understand the relationship between cause and effect in training practice and to support the optimization of the individual training process [5].

We are partners in a respective project (AIRow[9]) that aims to measure training load in rowing, predict the effects of training stimuli, and optimize training management based on individual physiological responses of athletes using methods from AI. Information on training content and intensity, performance diagnostic parameters, and various biomarkers will thereby be processed using state-of-the-art machine learning techniques. In addition to rowing, the new approach is also aimed at application in other endurance sports. As a benchmark model, we adapt an antagonistic model (PerPot DoMo; [20]), which has demonstrated to be effective in process-accompanying training control in rowing.

With the help of AI, training in virtual environments can be combined with adaptive training by adjusting the simulation used to generate the VR to the athlete's abilities or physical condition [1, 28]. There are only a few studies on developments in this area yet. Gray [15] demonstrates the superiority of an adaptive baseball batting training in VR to other training forms including a nonadaptive training in VR. We can expect to see useful developments for training in the future.

5 Potentials and Risks

The expected continuous qualitative improvement of personalized training recommendations from AI-supported systems, based on fitness status, physical prerequisites, personal preferences, and existing equipment, suggests a high potential of corresponding assistance systems (cf. Bodemer [5]). This can lead to an ever better adaptation to the respective individual training needs. Increasingly, they represent powerful support tools for the trainer. Improved achievement of set goals as well as a reduction of accident risks can be expected.

Risks lie in the possible reduction in social contact, as physical activities can increasingly be carried out alone and at home. Another fear expressed is similar to that of earlier sports technology developments, namely, that the technology could replace real, human trainers. The latter is vehemently denied by developers and distributors of such systems; instead, the assistance and support character for trainers and supervisors is emphasized. In fact, an experienced coach has human skills such as motivation, emotional support, and tactical knowledge that are difficult to replicate with technology. In addition, a coach can respond individually to an athlete's needs and goals and support them in developing a holistic athletic performance.

Aspects of privacy and data security must also be considered. Ethical issues on how private data are used by foreign or unauthorized parties need to be

[9] https://airow.univie.ac.at/, accessed 22.1.2024.

thought on [24]. Data protection measures are currently being implemented by companies to varying degrees. Users should therefore familiarize themselves with the applicable data protection conditions. When collecting, transmitting, processing, and storing user-specific—especially health-related—data, appropriate protective measures must be taken in any case. Fabbrizio et al. [12] recommend the use of blockchain technology, in which data is stored in blocks that are interconnected by means of cryptographic processes in the form of blocks, to ensure security and privacy of user data.

It should also be noted that there are no validation studies for the majority of the available assistance systems and apps offered. But even if such studies are carried out, a new product version is often already on the market when they are published. This means that, on the one hand, the scientific community needs to conduct appropriate studies and, on the other, that people who recommend or use such systems in training and practice need to take a very critical approach.

Particularly in the case of AI-based processes, whose results are strongly based on the underlying data, great attention should be paid on data veracity [11] and quality. Appropriate accuracy of the sensors and their accurate use are essential for the quality of the recommendations and conclusions derived from the collected data. In the AIRow project mentioned previously, for example, performance tests are carried out on RP3 rowing machines. In order to obtain accurate performance data, the devices were equipped with additional sensors (force and position sensor), from which power is accurately calculated. Claudino et al. [7], as another example, emphasize the importance of data veracity on load monitoring of professional soccer players.

AI methods involve multidimensional statistical procedures and can be regarded as a black box approach, which can be problematic for use in sport, similar to medicine. Recommendations derived from the algorithm have a certain probability of error and can only be weakly justified, which also has ethical implications [8]. Sampedro [13] warns that most powerful AI is by now opaque to us, meaning a serious risk. Explainable AI (XAI) methods could provide a way out to a certain extent in the future. By means of different methods, algorithms provide not only results but also reasons how these results were obtained, such as which features extracted from the data are particularly relevant for the decision-making or result determination. An example of a relevant study on the classification of movements is the work of Slijepcevic et al. [23], which discusses the use of XAI methods in clinical gait analysis.

6 Conclusion

The potential of AI methods shows a broad spectrum of possibilities to support the training process and training planning but also to change it significantly. Feedback systems support the improvement of movement sequences, the assessment and monitoring of movement execution, and the prevention of injuries and can have a motivating effect. Completely new approaches to training planning and control are being developed and could lead to sustained changes. Significant risks and dangers can arise from the reduction of social contacts and the violation of ethical aspects, such as the misuse of data or the use of inferior data, and these should be taken into account in conceptual approaches.

References

1. Aguilar Reyes CI, Wozniak D, Ham A et al. (2023). Design and evaluation of an adaptive virtual reality training system. Virtual Real. 27:2509–2528. https://doi.org/10.1007/s10055-023-00827-7
2. Allen E, Fenton A, Parry K (2021). Computerised Gymnastics Judging Scoring System Implementation – An Exploration of Stakeholders' Perceptions. Sci. Gymnast 13(3):357–370. https://doi.org/10.52165/sgj.13.3.357-370
3. Baca A (2024). Outlook. In: Memmert D (ed) Computer Science in Sport: Modeling, Simulation, Data Analysis and Visualization of Sports-Related Data. Springer: In Press.
4. Baca A, Kornfeind P, Preuschl E, Bichler S, Tampier M, Novatchkov H (2010). A Server-Based Mobile Coaching System. Sensors 10(12):10640–10662. https://doi.org/10.3390/s101210640
5. Bodemer O (2023). Enhancing Individual Sports Training through Artificial Intelligence: A Comprehensive Review. TechRxiv. August 23, 2023. https://doi.org/10.36227/techrxiv.24005916.v1
6. Cao Z, Hidalgo G, Simon T, Wei S-E, Sheikh Y (2021). OpenPose: Realtime Multi-Person 2D Pose Estimation Using Part Affinity Fields. IEEE Trans. Pattern Anal. Mach. Intell. 43 (1):172-186. https://doi.org/10.1109/TPAMI.2019.2929257
7. Claudino JG, Cardoso Filho CA, Boullosa D, Lima-Alves A, Carrion GR, GianonI RLdS, Guimarães RdS, Ventura FM, Araujo ALC, Del Rosso S, et al. (2021) The Role of Veracity on the Load Monitoring of Professional Soccer Players: A Systematic Review in the Face of the Big Data Era. Appl. Sci. 11:6479. https://doi.org/10.3390/app11146479
8. Dindorf C, Bartaguiz E, Gassmann F, Fröhlich M (2023). Zentrale Herausforderungen, Limitationen und offene Forschungsfelder [Key challenges, limitations and open research areas]. In: Dindorf C, Bartaguiz E, Gassmann F, Fröhlich M (eds.)Künstliche Intelligenz in Sport und Sportwissenschaft [Artificial Intelligence in Sports and Sports Science], p 27–34. https://doi.org/10.1007/978-3-662-67419-2_4
9. Dobiasch M, Oppl S, Stöckl M, Baca A (2022). Direct Mobile Coaching as a Paradigm for the Creation of Mobile Feedback Systems. Appl. Sci. 12(11):5558. https://0-doi-org.brum.beds.ac.uk/10.3390/app12115558
10. Driller MW, Dunican IC, Omond SET, Boukhris O, Stevenson S, Lambing K, Bender AM (2023). Pyjamas, Polysomnography and Professional Athletes: The Role of Sleep Tracking Technology in Sport. Sports 11(14). https://doi.org/10.3390/sports11010014
11. Emani CK, Cullot N, Nicollle C (2015). Understandable Big Data: A survey. Comput Sci Rev 17:70–81. https://doi.org/10.1016/j.cosrev.2015.05.002
12. Fabbrizio A, Fucarino A, Cantoia M, De Giorgio A, Garrido ND, Iuliano E, Reis VM, Sausa M, Vilaça-Alves J, Zimatore G, Baldari C, Macaluso F (2023). Smart Devices for Health and Wellness Applied to Tele-Exercise: An Overview of New Trends and Technologies Such as IoT and AI. Healthc 11(12):1805. https://doi.org/10.3390/healthcare11121805
13. Sampedro A C (2021). The case of AI in sport: some ethical concerns at play. Diagoras: International Academic Journal on Olympic Studies, 5:18–29.
14. Fujitsu (2023). https://www.fujitsu.com/global/themes/data-driven/judging-support-system/ Retrieved 21.01.2024.

15. Gray R (2017). Transfer of training from virtual to real baseball batting. Front. Psychol 8:2183. https://doi.org/10.3389/fpsyg.2017.02183
16. Mazurova E, Standaert W, Penttinen E. et al. (2022). Paradoxical Tensions Related to AI-Powered Evaluation Systems in Competitive Sports. Inf Syst Front 24:897–922. https://doi.org/10.1007/s10796-021-10215-8
17. Novatchkov H, Baca A (2012). Machine learning methods for the automatic evaluation of exercises on sensor-equipped weight training machines. Procedia Engineering 34:562–567. https://doi.org/10.1016/j.proeng.2012.04.096
18. Novatchkov H, Baca A (2013). Artificial intelligence in sports on the example of weight training. J Sports Sci Med. 12(1):27-37. PMID: 24149722; PMCID: PMC3761781.
19. Perl J, Pfeiffer M (2011). PerPot DoMo: antagonistic meta-model processing two concurrent load flows. Int. J. Comput. Sci. Sport 10(2):85–92.
20. Preuschl E, Tampier M, Schermer T, Baca A (2015). Introduction to the Relative Activity Index: Towards a fair method to score school children's activity using smartphones. In Chung P, Soltoggio A, Dawson CW, Meng Q & M. Pain (eds.), Advances in Intelligent Systems and Computing 392, Proceedings of the 10th International Symposium on Computer Science in Sports (ISCSS). Cham: Springer, pp 153–156.
21. Scataglini S, Moorhead AP, Feletti F (2020). A systematic review of smart clothing in sports: Possible applications to extreme sports. Muscles, Ligaments Tendons J. 10(2):333–342. https://doi.org/10.32098/mltj.02.2020.19
22. Seshadri D., Li RT, Voos J. et al. (2019). Wearable sensors for monitoring the physiological and biochemical profile of the athlete. npj Digit. Med. 2:72. https://doi.org/10.1038/s41746-019-0150-9
23. Slijepcevic D, Horst F, Lapuschkin S, Horsak B, Raberger A-M, Kranzl A, Samek W, Breiteneder C, Schöllhorn W-I, Zeppelzauer M (2021). Explaining Machine Learning Models for Clinical Gait Analysis. ACM Trans. Comput. Healthcare 3, 2, Article 14. https://doi.org/10.1145/3474121
24. Sperlich B, Düking P, Leppich R and Holmberg H-C (2023) Strengths, weaknesses, opportunities, and threats associated with the application of artificial intelligence in connection with sport research, coaching, and optimization of athletic performance: a brief SWOT analysis. Front. Sports Act. Living 5:1258562. https://doi.org/10.3389/fspor.2023.1258562
25. Tampier M, Endler S, Novatchkov H, Baca A (2012). Development of an intelligent real-time feedback system. Int. J. Comput. Sci. Sport 11 (3):58–64.
26. Uhlrich SD et al. (2022). OpenCap: 3D human movement dynamics from smartphone videos, biorxiv. https://doi.org/10.1101/2022.07.07.499061
27. Yaneva A. et al. (2020). Guide for Physical Activity Through Innovative Technologies, University Publishing House "St. Kliment Ohridski" Sofia University.
28. Zhang Y, Tsai S-B (2021). Application of Adaptive Virtual Reality with AI-Enabled Techniques in Modern Sports Training, Mobile Information Systems, vol. 2021, Article ID 6067678. https://doi.org/10.1155/2021/6067678

A Fuzzy Model to Optimise the Football Rule Assuring Spectacle, Fair Play, Objectivity and Ethics

Jaime Gil-Lafuente and Domenico Marino

Abstract The history of football, as in most sports, has always been accompanied by controversy caused by refereeing decisions, based on a set of rules and their interpretation which, rightly or wrongly, has on countless occasions been decisive in the final results of matches in which justice was to be done.

Our proposal will provide, in an eminently theoretical way, those modifications to the football regulations that try to maximise the objectives that this sport requires, according to the experts interviewed, such as, among others, spectacle, fair play, objectivity and ethics.

Once all the proposals had been compiled from the experts interviewed, we asked them to assess the feasibility and coherence of each of the proposals using real or uncertain numbers. This information, duly aggregated by means of expertons, has allowed us to establish a ranking that determines those changes that are most recommendable at a general level. However, by means of the theory of forgotten effects, we will find to what degree each proposal directly and indirectly affects the objectives sought in football, so that we can really know which changes in the regulations would be a priority. Finally, and in order to avoid too drastic modifications in a short time, we have obtained by means of Pichat's affinity algorithm those 'packages' of necessary changes that make sense to carry them out together, thus being able to implement the improvements to the refereeing regulations gradually and adaptively.

Keywords Pichat's algorithm · Cause-effect · Criteria · Neglected effects · Spectacle · Expertise · Experton · Football · Objectives · Optimisation · Rules

---

J. Gil-Lafuente
Universitat de Barcelona, Barcelona, Spain
e-mail: j.gil@ub.edu

D. Marino (✉)
Un. Mediterranea di Reggio Calabria, Calabria, Italy
e-mail: dmarino@unirc.it

M. J. Blondin et al. (eds.), *Artificial Intelligence, Optimization, and Data Sciences in Sports*, Springer Optimization and Its Applications 218,
https://doi.org/10.1007/978-3-031-76047-1_10

1 Introduction

Over the past decades, football has been regarded in most Western countries as the most important sporting entertainment. In every match of the professional elite, there are still important controversies in specific plays in which a series of rules, previously and duly accepted by the regulatory bodies of the refereeing rules, affect and irritate fans, journalists, managers or coaches who consider unfair some interpretative applications of the rules or simply do not consider certain rules that regulate the game in this sport to be coherent [13–15].

Searching through the scientific literature, we have encountered articles that are concerned with identifying and locating the top referees [21], analysing their physical condition [2], assessing their perceived performance [18] or even evaluating their reputation [17]. However, it has proven challenging to find research articles directly related to the enhancement of refereeing regulations.

The consistency and objectivity of these decisions play a crucial role in the fairness and integrity of the game [3, 4, 6]. This literature review aims to provide an overview of the existing research on the problem of consistency and objectivity of refereeing decisions in football and to explore potential remedies. The literature on the topic is not very extensive. Studies on the consistency and objectivity of football referees' decisions suggest that there is room for improvement. Proposed remedies include more training, the use of video assistant referee (VAR) technology, clearer rules and mechanisms for reviewing and holding referees accountable. The home advantage in sports is well known, and crowd noise seems to play a crucial role. A quantitative study was conducted on football referees to assess how crowd noise and experience affect their decisions. The presence of crowd noise significantly influenced the referees, making them less confident and leading to fewer fouls against the home team. This suggests that crowd noise may influence referees to favour the home team. Nurcahya et al. [16] study the impact of video-based training on the performance of assistant football referees. The study involved 54 referees divided into experimental and control groups. The results indicate that video-based training improved the performance of referees by enabling them to analyse errors during matches. This underscores the importance of incorporating video technology into the training of referees to improve their performance. Spitz et al. [20] examined 2195 competitive football matches in 13 countries, and VAR conducted 9732 checks for potential match-changing incidents. The study found that after VAR intervention, the probability of making the correct decision increased significantly, with accuracy increasing from 92.1% to 98.3%. Samuel et al. [19] study the implementation of VAR in the Israeli Premier League, revealing the initial challenges and adjustments for referees. The introduction of video assistant referee (VAR) in football aimed to reduce referee errors and improve fairness. Although improvements were noted, referees faced scrutiny from various stakeholders, resulting in mixed perceptions of the change. Hüttermann et al. [8], studying the behaviour of assistant football referees, found that the ability to pay attention to multiple objects decreased as the visual angle between distant objects increased. Surprisingly, assistant referees made fewer errors when they were further away from the action, with a smaller visual angle. This pattern remained consistent even after accounting for individual differences in the spread of attention. This result suggests that fans sitting directly behind the assistant referee, further away from the players, may have an advantage in making correct calls due to the smaller visual angle formed by the players concerned.

In view of this latent need, we have considered developing a theoretical work, which we present below, in order to subsequently, in a later phase, develop it empirically in order to provide the international institutions that regulate European or world football with this scientifically based proposal.

Stages of the Algorithm to Be Followed

Considering that we are going to work with subjective expert opinions, since the convenient aggregation of subjective expert opinions is the closest to objectivity [9], we will go into the development of a macro-algorithm consisting of different models and algorithms already adapted by Kaufmann and Gil-Aluja to fuzzy logic [10].

The steps to be followed are as follows:

1. To obtain a base of ideas formed by nuances, modifications of different aspects of the football refereeing rules or even radical changes in them. This will be obtained both through simple interviews with people linked to or involved in the sport and through the STIM-5 creativity method [5].
2. Once we have obtained a battery of ideas with all the proposed changes, we will ask all the experts we have had to assess [9] the coherence of each of the proposals.
3. All these valuations will be duly aggregated by means of expertons [12], and, subsequently, the pertinent ordering of each and every one of them will be carried out, thus ascertaining those that are most accepted on the basis of their coherence.
4. Each and every one of our experts will be asked what the objectives and criteria of proper arbitration rules should be.
5. Once we have obtained all the criteria and their importance, we will contact our specialists again, so that they can indicate the degree of affectation between all the new proposals and all the objectives, but also those between the proposals and the proposals, and those between each of the criteria and all the others.
6. With the information obtained, we will develop the forgotten effects theory algorithm [10], thus obtaining the degree to which each of the rules actually affects the preferred objectives sought in football.
7. In order to provide coherent packages of standards and implement them gradually over time, we will describe each standard in fuzzy subsets and group them by different degrees of similarity using Pichat's algorithm.

We will now proceed to develop each of these steps.

2 Methodology

2.1 Algorithmic Approach

We employ a macro-algorithm combining various models adapted to fuzzy logic. The steps include:

1. *Idea generation*: Through interviews and the STIM-5 creativity method, we gather a base of proposals for modifying football rules.
2. *Expert assessment*: Experts evaluate the coherence of each proposal using fuzzy-Delphi methods, and the evaluations are aggregated through expertons.
3. *Impact analysis*: We determine the direct and indirect effects of proposals on football's objectives using the theory of forgotten effects.
4. *Proposal grouping*: Pichat's algorithm groups proposals into coherent packages for gradual implementation.

2.2 Practical Evaluation

To address the need for empirical validation, we propose, after this chapter, conducting simulations and assessments. This will include testing the proposed changes in controlled environments and gathering data to support the theoretical findings.

3 Obtaining Proposals

As we already know, there are a multitude of ways to get ideas, in this case, related to the legal regulation of the game. In the empirical work already well underway, we have sought out experts representing most aspects of the sport, interviewing retired referees, chairmen, active footballers, retired footballers, club presidents and top executives, coaches, former coaches, researchers, fans and leaders of football organisations. We wish to emphasise that the greater the heterogeneity in the profiles of those involved in this dynamic, the more productive and creative the ideas obtained will be [1].

In a first round of interviews, we asked them what rules they would change or add, and they came up with the following:

$$N = \{P, P_{12}, \ldots P\}_s$$

However, we believe that the STIM-5 method can add additional proposals of equal interest.

Its development can be obtained as follows [7]:

1. Either explicitly or implicitly, bisociations or associations of further elements are made which are to be assessed against the criterion of coherence. The degree of coherence of the ideas depends on the specification of the object (concrete or abstract) obtained. It is also possible to consider the time needed for this specification.
2. The possibility of the ideas arising from the associations of elements to become part of the idea bank is subject to the overcoming of thresholds established beforehand for the criteria introduced in the research. Only ideas that meet or exceed the threshold established for coherence will be valid. These thresholds act as a selection so that at the end of the process, only those ideas that have passed the sieve of the established criteria at the desired level are included in the bank. In the event that some ideas do not reach the required levels of coherence, they will be discarded from the ideas bank, although they will not be definitively eliminated in case it is interesting to take them up again at a later date and carry out a new analysis.
3. This criterion will be set between [0, 1] in such a way that the closer they are to 0, the lower the level required, and the closer they are to the maximum (valuation equal to 1), it will mean that the requirement of the criterion is high for the advanced idea.
4. Once the methodological basis for the development of the creativity circle has been established, it is necessary to choose the appropriate place for the development of the sessions. It is important that the creative process is carried out in a place that is detached from the usual work environment and, if possible, in a place that lends itself to a certain disconnection from the daily routine. In this sense, companies usually opt for spas, hotel complexes or places of relaxation that allow work sessions to be combined with some leisure activities. These stays do not usually last more than 2 or 3 days. The aim is to establish short but intense creativity sessions. The workplaces should be simply decorated so as not to distract the participants' attention and provide a quiet environment, away from the stress and worries of everyday life.
5. Work schedules can be set up in several shifts. It is usual to combine work sessions with recreational activities. Assuming that two working days are proposed, a conventional timetable would revolve around the following [7]:

   - 08h30–09h00: breakfast
   - 09h00–10h30: work
   - 10h30–11h00: coffee break
   - 11h00–12h30: work
   - 12h30–14h30: lunch
   - 14h30–16h00: work
   - 16h00–19h30: recreational activity
   - 19h30–21h00: dinner

6. It is important that outside the creativity sessions, there is no talk of work, as the participants should remain as far away as possible from the influences that the ideas of one could exert on the other. The important thing is that everyone

can express their thoughts freely and without restriction. Talking about work outside the sessions is therefore avoided. Another important aspect is that the participants enjoy a pleasant stay without boredom in order to have maximum motivation at the time of the sessions.

7. The timetable presented is, of course, variable depending on the season, location, customs, etc. and will be adapted according to the needs and results to be obtained.
8. Another fundamental aspect to take into account is the composition of the creativity group. It is usually made up of eight to ten people, three or four (sometimes only two) of whom are usually football professionals and the rest of whom are less involved but with a homogeneous education and cultural level. Their origin, age and profession may vary. The important thing is that the members of the group understand at all times the concepts and ideas to be developed. The importance of incorporating specialists is due to the need to evaluate the possibilities of developing the ideas that arise in the group. On the other hand, the role of non-professionals is based on contributing new concepts from perspectives that differ from the criteria usually followed by the company. In a way, it is a question of breaking the pre-established moulds and orienting arbitration rules differently.
9. In addition to the participants, the director and an assistant will be involved in the creativity circle. The leader is responsible for stimulating the creation of ideas, ensuring that the level of discussion is maintained, avoiding that some members of the group take the floor, while others act as mere spectators, varying the strategies and the pace at which the session is conducted, etc. In fact, the director is the one who knows the objectives and methods for the creation of ideas. On the other hand, the assistant is in charge of entering all the data, estimates, valuations, etc. into the computer and handling the programmes that will make it possible to obtain equiprobable random draws according to the morphology-forming set, calculation of maximum similarity subrelations in symmetrical and reflexive matrices, recording of the participants' votes, construction of the expertons, mathematical expectations and measured values, recording in memory of the results according to the director's wishes, etc.
10. The arrangement of the members of the creativity circle shall follow the following criteria:

    At the beginning of the sessions, two groups will be established among the members of the circle: those who belong to the active group (six or seven people) who must express their opinions out loud according to the director's instructions and those who are in the so-called purgatory. This group will be located in the same room but apart from the main table.

    The members of purgatory do not have the right to speak, although they can send written messages to the director. There is a rotation in the members of the two groups, taking into account that sometimes many ideas come better from the position of spectator than from the position of actor. Both specialist and non-specialist members can be relegated to purgatory.

In this context, the director must ensure that the factors that favour the mental entropy of the group, alternation and frequency adapted to the alternations, are fulfilled. Although the members of purgatory are not supposed to speak, they are allowed to express their opinions and votes by WhatsApp.

Once the dynamic has been completed, those proposals for regulations that are sufficiently coherent will be compiled and added to the bank of proposals obtained through interviews, thus obtaining the total number of proposals with which we are going to work:

$$P = \{P_1, P_2, \dots P\}_z$$

As an example, and for a clearer view of the process, we suggest, arbitrarily, these seven proposals, which could be as follows:

1. Prohibition of communicating with the referee unless he himself addresses the player to ask or comment on any aspect related to the game. Any other verbal contact will be punished with a red card. The number of suspension matches will be calculated based on the extent of aggressiveness in communication, if any. In short: maximum respect and education towards any of the match officials.
2. Expulsion in the first confrontation, even if minimal, between players (double penalty to the initiator, if it can be determined).
3. Any contact of the ball with the hand or arm is a direct foul.
4. The offside rule disappears.
5. 40 minutes of effective time (as in basketball) for each of the two halves that make up a match.
6. If during the game the ball stops moving for a second, it will be punished with an indirect free kick.
7. Unlimited substitutions during the match, as long as the player entering the field does so from the centre of the field on the side of his team's bench while the substituted player leaves from the same place (as in handball). These changes do not require stopping the game. Substituted players may re-enter if the coach deems it appropriate.

4 Assessment of the Liking or Coherence of Each Proposal

In a second round, all the experts are asked to make an assessment in the endecadary system [7] and, if they wish, by means of confidence intervals, the level of coherence for each of the proposals.

Example: $P_1 = [0.7, 0.9]$.

Given the impossibility of finding any literature review regarding how the composition of experts who will evaluate and analyse proposals and their consequences should be, we will consider that a significant percentage of them should have a direct relationship with professional football (referees, former referees, professional club and federation officials, etc.), while only a few should be indirectly linked to this sport.

On the other hand, there are different alternatives to conveniently aggregate the subjective opinions of all experts, such as fuzzy-Delphi [3, 5]. However, due to its easy adaptation to digitalisation, we believe that the evaluations given by each and every expert on coherence will be properly aggregated through expert systems [11].

For the example at hand, we present seven proposals, each evaluated by a panel of eight experts. Below are the tabulated results:

	Statistical		Frequencies normalised		Experton
Proposal 1: Prohibition of communicating with the referee					
0		0	0	0	1
0,1		0,1	0	0,1	1
0,2		0,2	0	0,2	1
0,3		0,3	0	0,3	1
0,4		0,4	0	0,4	1
0,5	1	0,5	0,125	0,5	1
0,6	1	0,6	0,125	0,6	0,875
0,7		0,7	0	0,7	0,75
0,8	1	0,8	0,125	0,8	0,75
0,9	2	0,9	0,25	0,9	0,625
1	3	1	0,375	1	0,375
Total	8		1	Mathematical expectation =	0,8375
Proposal 2: Expulsion for player confrontation					
0		0	0	0	1
0,1		0,1	0	0,1	1
0,2		0,2	0	0,2	1
0,3		0,3	0	0,3	1
0,4		0,4	0	0,4	1
0,5	1	0,5	0,125	0,5	1
0,6	2	0,6	0,25	0,6	0,875
0,7		0,7	0	0,7	0,625
0,8	2	0,8	0,25	0,8	0,625
0,9	2	0,9	0,25	0,9	0,375
1	1	1	0,125	1	0,125
Total	8		1	Mathematical expectation =	0,7625

(continued)

	Statistical		Frequencies normalised		Experton
Proposal 3: Hands, whenever there is contact					
0		0	0	0	1
0,1		0,1	0	0,1	1
0,2		0,2	0	0,2	1
0,3	1	0,3	0,125	0,3	1
0,4	1	0,4	0,125	0,4	0,875
0,5	1	0,5	0,125	0,5	0,75
0,6	1	0,6	0,125	0,6	0,625
0,7		0,7	0	0,7	0,5
0,8	2	0,8	0,25	0,8	0,5
0,9		0,9	0	0,9	0,25
1	2	1	0,25	1	0,25
Total	8		1	Mathematical expectation =	0,675
Proposal 4: Abolition of offside rule					
0	1	0	0,125	0	1
0,1	1	0,1	0,125	0,1	0,875
0,2	1	0,2	0,125	0,2	0,75
0,3	1	0,3	0,125	0,3	0,625
0,4		0,4	0	0,4	0,5
0,5		0,5	0	0,5	0,5
0,6		0,6	0	0,6	0,5
0,7	1	0,7	0,125	0,7	0,5
0,8	1	0,8	0,125	0,8	0,375
0,9	1	0,9	0,125	0,9	0,25
1	1	1	0,125	1	0,125
Total	8		1	Mathematical expectation =	0,5
Proposal 5: 40 minutes of effective time					
0		0	0	0	1
0,1		0,1	0	0,1	1
0,2		0,2	0	0,2	1
0,3		0,3	0	0,3	1
0,4		0,4	0	0,4	1
0,5		0,5	0	0,5	1
0,6		0,6	0	0,6	1
0,7	1	0,7	0,125	0,7	1
0,8	1	0,8	0,125	0,8	0,875
0,9	5	0,9	0,625	0,9	0,75
1	1	1	0,125	1	0,125
Total	8		1	Mathematical expectation =	0,875

(continued)

	Statistical		Frequencies normalised		Experton
Proposal 6: Ball must always be in motion					
0		0	0	0	1
0,1		0,1	0	0,1	1
0,2		0,2	0	0,2	1
0,3		0,3	0	0,3	1
0,4		0,4	0	0,4	1
0,5		0,5	0	0,5	1
0,6		0,6	0	0,6	1
0,7		0,7	0	0,7	1
0,8	5	0,8	0,625	0,8	1
0,9	1	0,9	0,125	0,9	0,375
1	2	1	0,25	1	0,25
Total	8		1	Mathematical expectation =	0,8625
Proposal 7: Unlimited substitutions					
	Statistical		Frequencies normalised		Experton
0		0	0	0	1
0,1		0,1	0	0,1	1
0,2		0,2	0	0,2	1
0,3		0,3	0	0,3	1
0,4		0,4	0	0,4	1
0,5	1	0,5	0,125	0,5	1
0,6		0,6	0	0,6	0,875
0,7		0,7	0	0,7	0,875
0,8	1	0,8	0,125	0,8	0,875
0,9	3	0,9	0,375	0,9	0,75
1	3	1	0,375	1	0,375
Total	8		1	Mathematical expectation =	0,875

Of course, we can also weigh the strength of each expert, should this be the case.

Ranking from the 'Best' to the Least Accepted Proposal

In order to establish a ranking among all the expertons that represent the coherence considered by all the experts for each proposal, we will proceed to find the experton supremum, which is the one that optimises the totality of the expertons under comparison.

By comparing each of the expertons with the supremum, we will know that the one whose distance from the supremum is smaller will be the highest experton and, consequently, the most coherent proposal for each and every one of the specialists who have analysed it.

		P1			**P2**			**P3**			**P4**			**P5**			**P6**			**P7**			**Supremum**
	0	1		0	1		0	1		0	1		0	1		0	1		0	1		0	1
	0,1	1		0,1	1		0,1	1		0,1	0,875		0,1	1		0,1	1		0,1	1		0,1	1
	0,2	1		0,2	1		0,2	1		0,2	0,75		0,2	1		0,2	1		0,2	1		0,2	1
	0,3	1		0,3	1		0,3	1		0,3	0,625		0,3	1		0,3	1		0,3	1		0,3	1
	0,4	1		0,4	1		0,4	0,875		0,4	0,5		0,4	1		0,4	1		0,4	1		0,4	1
[S] =	0,5	1	∨	0,5	1	∨	0,5	0,75	∨	0,5	0,5	∨	0,5	1	∨	0,5	1	∨	0,5	1	=	0,5	1
	0,6	0,875		0,6	0,875		0,6	0,625		0,6	0,5		0,6	1		0,6	1		0,6	0,875		0,6	1
	0,7	0,75		0,7	0,625		0,7	0,5		0,7	0,5		0,7	1		0,7	1		0,7	0,875		0,7	1
	0,8	0,75		0,8	0,625		0,8	0,5		0,8	0,375		0,8	0,875		0,8	1		0,8	0,875		0,8	1
	0,9	0,625		0,9	0,375		0,9	0,25		0,9	0,25		0,9	0,75		0,9	0,375		0,9	0,75		0,9	0,75
	1	0,375		1	0,125		1	0,25		1	0,125		1	0,125		1	0,25		1	0,375		1	0,375

		P1			P2			P3			P4			P5			P6			P7			Supremum
	0	1		0	1		0	1		0	1		0	1		0	1		0	1		0	1
	0,1	1		0,1	1		0,1	1		0,1	0,875		0,1	1		0,1	1		0,1	1		0,1	1
	0,2	1		0,2	1		0,2	1		0,2	0,75		0,2	1		0,2	1		0,2	1		0,2	1
	0,3	1		0,3	1		0,3	1		0,3	0,625		0,3	1		0,3	1		0,3	1		0,3	1
	0,4	1		0,4	1		0,4	0,875		0,4	0,5		0,4	1		0,4	1		0,4	1		0,4	1
[S] =	0,5	1	∨	0,5	1	∨	0,5	0,75	∨	0,5	0,5	∨	0,5	1	∨	0,5	1	∨	0,5	1	=	0,5	1
	0,6	0,875		0,6	0,875		0,6	0,625		0,6	0,5		0,6	1		0,6	1		0,6	0,875		0,6	1
	0,7	0,75		0,7	0,625		0,7	0,5		0,7	0,5		0,7	1		0,7	1		0,7	0,875		0,7	1
	0,8	0,75		0,8	0,625		0,8	0,5		0,8	0,375		0,8	0,875		0,8	1		0,8	0,875		0,8	1
	0,9	0,625		0,9	0,375		0,9	0,25		0,9	0,25		0,9	0,75		0,9	0,375		0,9	0,75		0,9	0,75
	1	0,375		1	0,125		1	0,25		1	0,125		1	0,125		1	0,25		1	0,375		1	0,375

Therefore, we obtain the following order of priority:

1. 0.375: 40 minutes of effective playing time.
2. 0.375: Unlimited substitutions.
3. 0.5: The ball must always be in motion.
4. 0.75: Prohibition of communicating with the referee.
5. 1.5: Expulsion for player confrontations.
6. 2.375: Handball whenever there is contact.
7. 4.125: Elimination of the offside rule.

5 Criteria for Acceptance of Proposals

For the third time, we will ask the experts what criteria they are looking for in this sport to ensure that the regulations governing matches are as close to optimal as possible.

Normally, the answers tend to be: favouring the spectacle, being objective, being fair, favouring fair play, etc., although we also believe it is interesting to add some other more forgotten ones, such as ethics and setting an example for children or increasing the responsibility of football players and limiting their egomania.

Once we have collected all the objectives sought: $O = \{O_1, O_2, \ldots O_m\}$, we will ask our experts to make assessments, also on the endecadary scale and, if they wish, using confidence intervals, of the impact of each proposal P towards each objective O.

Thus, we obtain the matrix of direct effects between each proposed regulatory change and each objective to be achieved.

		O_1	O_2	...	O_m
	P_1	$\mu\ P_1O_1$	$\mu\ P_1O_2$	...	$\mu\ P_1O_m$
$[M] =$	P_2	$\mu\ P_2O_1$	$\mu\ P_2O_2$	...	$\mu\ P_2O_m$
	...	...	...	...	...
	P_z	$\mu\ P_zO_1$	$\mu\ P_zO_2$	...	$\mu\ P_zO_m$

Continuing with our example, if the chosen criteria were:

- Justice
- Objectivity
- Entertainment
- Fair play
- Ethical example
- Athlete's self-responsibility

we would establish the matrix of impacts, duly assessed by the experts.

However, in order to know the indirect (or second-generation) effects, we must also obtain, according to the theory of forgotten effects, the effects of the proposals on the proposals and of the objectives on the objectives.

Again asking our experts, we will have:

		P_1	P_2	...	P_z
	P_1	$\mu\ P_1P_1$	$\mu\ P_1P_2$	...	$\mu\ P_1P_m$
$[P] =$	P_2	$\mu\ P_2P_1$	$\mu\ P_2P_2$	...	$\mu\ P_2P_m$
	...	...	...	...	...
	P_z	$\mu\ P_zP_1$	$\mu\ P_zP_2$	...	$\mu\ P_zP_z$

For our example:

	Justice	Objectivity	Entertainment	Fair play	Ethical example	Athlete's self-responsibility
40 minutes of effective time	0,9	1	0	0	0	0
Unlimited substitutions	0,5	1	1	0,2	0,2	0,2
Ball must always be in motion	0,6	0,7	1	0,4	0,3	0,3
Prohibition of communicating with the referee	0,2	0,7	0	1	1	1
Player expulsion for confrontation	0,7	0,8	0,2	1	1	1
Hands, whenever there is contact	0	1	0,6	0,5	0,4	0,3
Offside rule disappears	0,5	1	0,5	0,6	0,6	0,5

The matrix of proposal effects with proposals:

		O_1	O_2	...	O_m
	O_1	$\mu\ O_1O_1$	$\mu\ O_1O_2$	...	$\mu\ O_1O_m$
$[O] =$	O_2	$\mu\ O_2O_1$	$\mu\ O_2O_2$	...	$\mu\ O_2O_m$
	...	...	...	...	...
	O_m	$\mu\ O_mO_1$	$\mu\ O_mO_2$	...	$\mu\ O_mO_m$

In our case:

	40 minutes of effective time	Unlimited substitutions	Ball must always be in motion	Prohibition of communicating with the referee	Player expulsion for confrontation	Hands, whenever there is contact	Offside rule disappears
40 minutes of effective time	1	1	0,8	0,1	0,6	0	0
Unlimited substitutions	1	1	0	0	0,2	0	0
Ball must always be in motion	1	0	1	0	0	0	0
Prohibition of communicating with the referee	0	0	0	1	0,5	0,5	0,1
Player expulsion for confrontation	0	0,4	0,1	0,8	1	0,7	0,1
Hands, whenever there is contact	0	0	0,7	0	0,5	1	0
Offside rule disappears	0	0,7	0,8	0,8	0,1	0	1

The matrix of target affectations with targets:

	Justice	Objectivity	Entertainment	Fair play	Ethical example	Athlete's self-responsibility
Justice	1	0,5	0,9	1	0,9	0,9
Objectivity	1	1	0,8	1	1	1
Entertainment	0,5	0,7	1	0,8	0,6	0
Fair play	1	1	0,6	1	1	1
Ethical example	1	1	0,7	1	1	1
Athlete's self-responsibility	1	1	0,1	1	1	1

By performing the maxmin convolution, between $[P] \circ [M] \circ [O]$ we will obtain the second-generation matrix $[M]^*$, which will inform us of the real effects of all the proposals on the objectives we are looking for, so that those with the highest value will be the ones that will have the most importance in the desired football.

		P_1	P_2	...	P_z			O_1	O_2	...	O_m			O_1	O_2	...	O_m
	P_1	$\mu_{P_1P_1}$	$\mu_{P_1P_2}$	...	$\mu_{P_1P_m}$		P_1	$\mu_{P_1O_1}$	$\mu_{P_1O_2}$	...	$\mu_{P_1O_m}$		O_1	$\mu_{O_1O_1}$	$\mu_{O_1O_2}$	...	$\mu_{O_1O_m}$
$[M]^* =$	P_2	$\mu_{P_2P_1}$	$\mu_{P_2P_2}$	...	$\mu_{P_2P_m}$	$\circ$	P_2	$\mu_{P_2O_1}$	$\mu_{P_2O_2}$	...	$\mu_{P_2O_m}$	$\circ$	O_2	$\mu_{O_2O_1}$	$\mu_{O_2O_2}$	...	$\mu_{O_2O_m}$
	...	...	...	...	...		...	...	...	...	...		...	...	...	...	...
	P_z	$\mu_{P_zP_1}$	$\mu_{P_zP_2}$	...	$\mu_{P_zP_z}$		P_z	$\mu_{P_zO_1}$	$\mu_{P_zO_2}$	...	$\mu_{P_zO_m}$		O_m	$\mu_{O_mO_1}$	$\mu_{O_mO_2}$	...	$\mu_{O_mO_m}$

For our example, the first- and second-generation impacts are:

	Justice	Objectivity	Entertainment	Fair play	Ethical example	Athlete's self-responsibility
40 minutes of effective time	1	1	1	1	1	1
Unlimited substitutions	1	1	1	1	1	1
Ball must always be in motion	1	1	1	1	1	1
Prohibition of communicating with the referee	1	1	0,7	1	1	1
Player expulsion for confrontation	1	1	0,8	0,7	1	1
Hands, whenever there is contact	1	1	0,8	1	1	1
Offside rule disappears	0,8	0,8	0,8	0,8	0,8	0,8

We would like to point out that with a simple subtraction of $[M]^* - [M]$, we will know the effects that we forgot in the first of the matrices.

The Grouping of Proposals by Similarity

In this last phase, in order to find the maximum similarity subrelationships, we propose to describe each and every proposal by means of fuzzy subsets, based on coherence, but also on the objectives it fulfils.

The first proposal would therefore be described as follows:

	Justice	Objectivity	Entertainment	Fair play	Ethical example	Athlete's self-responsibility
$P_1 =$	0,9	1	0	0	0	0

We would then find the distances between each and every one of the proposals. These can be the Hamming distance, the Euclidean distance, etc., depending on the nature of the problem to be solved. In this case, we will opt for the relative Hamming distance.

The dissimilarity matrix $[D]$:

		*P*1	*P*2	*P*3	*P*4	*P*5	*P*6	*P*7
	*P*1	0	0,23333333	0,43333333	0,66666667	0,6	0,45	0,43333333
	*P*2	0,23333333	0	0,45	0,66666667	0,6	0,25	0,26666667
	*P*3	0,43333333	0,45	0	0,51666667	0,5	0,25	0,26666667
[*D*]=	*P*4	0,66666667	0,66666667	0,51666667	0	0,7	0,48333333	0,4
	*P*5	0,6	0,6	0,5	0,7	0	0,78333333	0,33333333
	*P*6	0,45	0,25	0,25	0,48333333	0,78333333	0	0,55
	*P*7	0,43333333	0,26666667	0,26666667	0,4	0,33333333	0,55	0

We can transform it into the similarity matrix $1 - [D]$:

		$P1$	$P2$	$P3$	$P4$	$P5$	$P6$	$P7$
	$P1$	1	0,76666667	0,56666667	0,33333333	0,4	0,55	0,56666667
	$P2$	0,76666667	1	0,55	0,33333333	0,4	0,75	0,73333333
	$P3$	0,56666667	0,55	1	0,48333333	0,5	0,75	0,73333333
[S] =	$P4$	0,33333333	0,33333333	0,48333333	1	0,3	0,51666667	0,6
	$P5$	0,4	0,4	0,5	0,3	1	0,21666667	0,66666667
	$P6$	0,55	0,75	0,75	0,51666667	0,21666667	1	0,45
	$P7$	0,56666667	0,73333333	0,73333333	0,6	0,66666667	0,45	1

In order to obtain the maximum similarity sub-correlations, we will work with Pichat's algorithm. It is now a matter of making the important decision of choosing the level of homogeneity that is considered sufficient for the clustering of proposals. Suppose we choose a minimum similarity $\alpha \geq 0.7$. From this fuzzy matrix, we could obtain a binary matrix like this:

		$P1$	$P2$	$P3$	$P4$	$P5$	$P6$	$P7$
	$P1$	1	1	0	0	0	0	0
	$P2$	1	1	0	0	0	1	1
	$P3$	0	0	1	0	0	1	1
[S_]=	$P4$	0	0	0	1	0	0	0
	$P5$	0	1	1	0	1	0	0
	$P6$	0	1	1	0	0	1	0
	$P7$	0	1	1	0	0	0	1

This calculation procedure follows the following path:

1. For each of the rows of the Boolean matrix, all the 1's above the main diagonal are considered.
2. In the cells where there is no 1, we look at which column P_i corresponds to, and once this has been determined, we assign a Boolean variable x_i or simply an i.
3. Once the Boolean variables x_i or the indices i have been assigned, the Boolean variable (or index) corresponding to each row is added (+) to the Boolean product of all the Boolean variables (or indices) that have been found in this row.
4. From all the representative binomials of each of the rows, the Boolean product is calculated, taking into account that:

$$x + x = x$$
$$x - x = x$$
$$xy + x = x$$

5. When in a row there is no empty cell (theoretically with a 0) to the right of the main diagonal, this row is assigned the number 1.
6. When the corresponding products are carried out, a polynomial S will be obtained. It must be taken into account that when numerical indices are used, these symbols are not numbers but Boolean variables.
7. Each of the monomials that form the polynomial S is transformed into its complementary, i.e. those variables (or indices) that are not in the monomial are considered. A new polynomial S' is obtained.
8. The monomials that constitute complements provide the products that form the maximum similarity subrelations.

In this way we will have 'packages' of regulations that can be implemented gradually over time in consecutive or alternating years.

For the example we are considering:

$$S = (1 + 34, 567)\ (2 + 345)\ (3 + 45)\ (4 + 567)\ (5 + 67)\ (6 + 7)$$
$$S = 12, 456 + 13, 456 + 12, 457 + 13, 457 + 123, 567 + 34, 567 + 123, 467$$
$$\mathbf{S' = P3\ P7 + P2\ P6 + P3\ P6 + P2\ P6 + P4 + P1\ P2 + P5}$$

These would be the possible groupings that could be made from the proposed regulations, considering that some of them have sufficient similarities (always ≥ 0.7) with other proposals. We will thus have full freedom to establish the groupings we want based on these options.

6 Conclusions

In recent decades, football has stood as one of the most significant sporting entertainments across Western countries. Yet, controversies persist in elite professional matches, with fans, journalists, administrators and coaches often finding fault with interpretations of the rules or perceived inconsistencies in how the game is regulated. This theoretical work serves as a foundation for a subsequent empirical phase, offering a scientifically based proposal to international football governing bodies. The literature reviewed here reveals that consistency and objectivity in refereeing decisions are areas that warrant improvement. Proposed remedies include enhanced training, the implementation of video assistant referee (VAR) technology, clearer rule formulations and mechanisms for reviewing and holding referees accountable. Studies suggest that crowd noise can influence referee decisions, potentially favouring the home team. VAR has shown promise in increasing the accuracy of decisions.

To address these challenges and make informed changes to football regulations, a comprehensive algorithmic approach involving experts and fuzzy logic is proposed. This approach will gather, evaluate and prioritise ideas for rule modifications and assess their impact on predefined objectives and criteria. A fuzzy decisional model will be used to determine how each rule affects the desired objectives. This comprehensive approach seeks to create coherent sets of rules that align with the overarching goals of football. By implementing changes based on scientific analysis and expert consensus, the aim is to improve the fairness and integrity of the sport while addressing the concerns and controversies that have persisted for years.

The future development of the research will consist of gathering the opinions of the practitioners and going into detail about the impact of individual regulatory changes needed to ensure greater efficiency and fairness for the game of football.

References

1. Ananth, P., & Harvey, S. (2023). Ideas in the Space Between: Stockpiling and Processes for Managing Ideas in Developing a Creative Portfolio. Administrative Science Quarterly, 68(2), 465–507.
2. Castillo-Rodríguez, A., Alejo-Moya, E. J., Figueiredo, A., Onetti-Onetti, W., & González-Fernández, F. T. (2023). Influence of physical fitness on decision-making of soccer referees throughout the match. Heliyon, 9(9).
3. Galarza, M. J. (2023). Impacto social derivado de la Economía Circular en Gipuzkoa: propuesta de categorías e indicadores empleando el método Fuzzy Delphi. [Título del artículo]. Nombre de la revista científica de computación, Volumen (Número), páginas.
4. Gil Aluja, J. (2000). Génesis de una teoría de la incertidumbre. Encuentros multidisciplinares.
5. Gil Lafuente, J. (1997). Marketing para el nuevo milenio. Nuevas técnicas para la gestión comercial en la incertidumbre. Ediciones Pirámide. Madrid.
6. Gil-Aluja, J. (Ed.). (2013). Handbook of management under uncertainty, 55. Springer Science & Business Media.
7. Gil-Lafuente, A. M., & Gil-Lafuente, J. (2007). Modelos y algoritmos para el tratamiento de la creatividad en la gestión empresarial. Milladoiro: Vigo, Spain.
8. Hüttermann, S., Noël, B., & Memmert, D. (2017). Evaluating erroneous offside calls in soccer. PLoS One, 12(3), e0174358.
9. Kaufmann, A., & Aluja, J. G. (1983). Técnicas especiales para la gestión de expertos. Milladoiro.
10. Kaufmann, A., Aluja, J. G., & Pirla, J. M. F. (1986). Introducción de la teoría de los subconjuntos borrosos a la gestión de las empresas. Milladoiro.
11. Kaufmann, A., & Gil-Aluja, J. (1988). Modelos para la investigación de efectos olvidados. Milladoiro.
12. Kaufmann, A., Gil Aluja, J., & Gil Lafuente, A. M. (1994). La creatividad en la gestión de las empresas. Ed. Pirámide, Madrid.
13. Marino, D. (2010). Governance and Sporting Success of Top 20 Football Clubs After Economic Crisis. In: Butenko, S., Gil-Lafuente, J., Pardalos, P. (eds) Optimal Strategies in Sports Economics and Management. Springer, Berlin, Heidelberg. https://doi.org/10.1007/978-3-642-13205-6_5
14. Marino D., Stilo P., 2024, An evolutionary game to describe doping dynamics, Heliyon, Vol. 10, Issue 10, https://doi.org/10.1016/j.heliyon.2024.e31089.

15. Marino, D., Stilo, P. (2017). Sport and Doping: A Theoretical Approach. In: Peris-Ortiz, M., Álvarez-García, J., Del Río-Rama, M. (eds) Sports Management as an Emerging Economic Activity. Springer, Cham. https://doi.org/10.1007/978-3-319-63907-9_7
16. Nurcahya, Y., Awwaludin, P., Rosdiana, F., & Stiadi, D. (2021). Improving the performance of football assistant referees through video-based training. Journal of Physics: Conference Series, 1806, 012212.
17. Pajuelo, Á. G. (2023). Algunos apuntes y debates sobre la reputación del colectivo arbitral en el fútbol español. Revista española de derecho deportivo, (52), 91–113.
18. Robazza, C., Sivilli, P., Bortoli, L., & Ruiz, M. C. (2023). Perfection behind the whistle: Perfectionism and perceived performance in soccer referees. Heliyon, 9(12).
19. Samuel, R. D., et al. (2020). Implementation of the Video Assistant Referee (VAR) as a Career Change-Event: The Israeli Premier League Case Study. Frontiers in Psychology, 11. https://doi.org/10.3389/fpsyg.2020.564855
20. Spitz, J., Wagemans, J., Memmert, D., Williams, A., & Helsen W. (2020). Video assistant referees (VAR): The impact of technology on decision making in association football referees. Journal of Sports Sciences, 39, 1–7.
21. Webb, T., O'Gorman, J., & Markham, L. (2023). Striving for excellence: talent identification and development in English football refereeing. European Sport Management Quarterly, 23(2), 351–369.

Physical Efficiency in Soccer: Relevance, Correlations, and Impacts Using AI Methods

Daniel Capanema, Adriano Alves, João Claudino, and Adriano Pereira

Abstract Physical efficiency plays a pivotal role in the success of both individual soccer players and teams. Soccer is a multidimensional sport that demands a harmonious blend of technical, tactical, and physical skills. Essentially, physical efficiency can be defined as a player's capacity to perform technical and tactical maneuvers at a consistently high level throughout a match while expending the least physiological effort. In essence, physical efficiency in soccer can be viewed as the ability to achieve superior physical performance while minimizing the physiological strain over the course of an entire game. This implies that players must engage in rigorous preparation to ensure their bodies can effectively recuperate post-match. Within this chapter, we aim to provide a more comprehensive elucidation of the concept of physical efficiency, drawing upon findings from pertinent literature. Additionally, we will delve into an evaluation of physical efficiency within the context of a professional soccer team. Leveraging artificial intelligence techniques, we will explore the relationships between different variables, examining how they may correlate with injuries and other game-related factors. This analysis will shed light on the intricate web of connections within soccer's physical efficiency landscape.

D. Capanema (✉)
Federal Center for Technological Education of Minas Gerais, Computing Department, Belo Horizonte, Brazil

A. Alves
Fortaleza Esporte Clube, Fortaleza, Brazil

J. Claudino
Group of Research, Innovation and Technology Applied to Sport (GSporTech), Department of Physical Education, Center for Health Sciences, Federal University of Piauí, Piauí, Brazil
e-mail: claudinojg@ufpi.edu.br

A. Pereira
Federal University of Minas Gerais (UFMG), Computer Science Department (DCC), Belo Horizonte, Brazil
e-mail: adrianoc@dcc.ufmg.br

M. J. Blondin et al. (eds.), *Artificial Intelligence, Optimization, and Data Sciences in Sports*, Springer Optimization and Its Applications 218,
https://doi.org/10.1007/978-3-031-76047-1_11

Keywords Physical efficiency · Soccer performance · Artificial intelligence · Training load management · Injury correlation

1 Physical Efficiency in Soccer

Soccer is a sport in which performance depends on physical, technical, tactical, and psychological factors [1]. It is characterized by intermittent and high-intensity efforts [2]. High-intensity actions consist of sprints, accelerations, and decelerations, accounting for up to 25% of the distance covered in a match [3]. Such physical demands can lead to skeletal muscle damage, resulting in a decline in neuromuscular performance and the release of cellular proteins into the bloodstream. Consequently, more than three days after a match are needed for the musculoskeletal system, muscle soreness, and performance recovery [4, 5, 6]. Nonetheless, players compete in a congested schedule of games [7, 8], with insufficient time for full recovery between matches, potentially leading to both acute and chronic fatigue [4, 9, 10]. In this context, data collection and analysis for load monitoring purposes have become a common practice to assess fatigue, subsequent adaptations, performance, and injury risk [11, 12].

Training load management can reduce the risk of negative training adaptation, such as excessive fatigue and injury [13, 14]. In elite team sports like soccer, coaches and performance departments have used external and internal loads to develop various strategies for training load management [15, 16]. Monitoring load in professional soccer players is the subject of study in various research papers [17]. According to [18], an increase in mechanical action execution capacity, alongside a reduction in physiological stress, may indicate a better training state. Consequently, the use of the global positioning system (GPS) and heart rate (HR) monitor has become common in team sports [19, 20, 21, 16, 22], facilitating the assessment of internal and external loads during competition.

With the aim of evaluating athletes during their own activities, whether competitive or not, and obtaining more information about their physical condition periodically, the use of the external-to-internal load ratio, known as the Physical Efficiency Index (PEI), has recently been introduced as a tool for assessing physical condition, training readiness, and fatigue in sports like soccer [23, 16, 24]. Furthermore, the PEI can be utilized for specific assessment during competition and in response to a congested schedule.

In recent years, some shuttle run performance tests have been suggested to evaluate aerobic power in soccer, including the Yo-Yo Endurance Test Level 2 (YYETL2) [25] and the Yo-Yo Intermittent Recovery Test Level 1 (YYIRTL1) [26]. These tests are considered suitable for assessing soccer players, but they can only predict physical performance in the game. In contrast, the PEI allows for the evaluation of physical performance during actual competition. Unlike individual sports like cycling, where external load is closely related to training status and measurable through direct physical metrics (e.g., distance, speed), in team sports such as soccer, both external and internal loads are subject to the dynamic and unpredictable nature of the game. Contextual factors like the opponent's tactics and game intensity can significantly affect these loads, making the assessment of physical condition more complex [27, 28, 29]. This interference can have a greater impact on internal and external loads than physical condition itself [30]. Using the PEI can reduce the impact of contextual factors on physical condition assessment during training [31] or competition [24].

The PEI comprises external load metrics (dividend), including total distance, the number of accelerations, and high-intensity distance (HID), in its calculation [24, 32]. For the internal load (divisor), variables have been limited to subjective session perception (sPSE) and variables derived from HR [33, 24, 35], such as the training impulse (TRIMP) and maximum heart rate (HRmax). Some studies have identified a decrease in the PEI during official soccer matches [24], a significant correlation between the PEI and anaerobic threshold [34, 35], an association between individual fatigue perception and the PEI [33], and a negative impact of previous training load on the PEI [32]. These characteristics, as presented in the literature, indicate that the PEI could be a promising tool for assessing the physical condition of soccer players during the competitive season.

The assessment of the training status and subsequent performance level of soccer players is crucial for planning and training load managements [1]. However, conducting these evaluations is challenging due to the congested competitive schedule [7]. Additionally, considering that athletes may not be fully recovered between games [36, 4, 37], conducting a formal physical test at this time, especially one that requires maximum effort, might be inadequate.

An insufficient solution proposed in the literature for assessing the physical performance of soccer players is the evaluation of external load through metrics such as the quantity of actions, total distance covered, distance covered at high intensity, and sprints during the actual game [9]. However, this type of assessment may provide limited insights into the players' training status due to contextual factors that are not related to fatigue and can affect this evaluation, such as tactical instructions and the level of the opponent, among others [38, 28]. Therefore, the use of a parameter that considers the response of a specific internal load in the presence of an external load could be employed as a constant assessment of the training status of active soccer players, minimizing the impact of contextual factors [39, 40]. Furthermore, it would provide higher-quality information than merely counting the number of actions performed during a match [34].

Initially, load assessments in soccer matches were presented without considering internal load [38] or external load [4], and when both were presented, no PEI was calculated [41, 42]. With technological advancements and increasing access to global positioning systems (GPS), camera systems, and heart rate monitors, more studies involving these technologies have emerged in soccer, futsal, and beach soccer [43, 44]. However, these studies have yet to calculate a PEI, which is the ratio of internal and external loads for these activities.

The literature hints at the potential sensitivity of the PEI to fluctuations in physical condition, and it suggests a potential correlation with the level of physical fitness. However, despite these indications, scientific evidence concerning the applicability of the PEI in the context of high-level soccer training load management remains incomplete. To advance our understanding in this area, it is imperative to chart a path for future research.

One promising direction for future inquiry is the inclusion of more studies involving elite soccer players. Such investigations can help us delve deeper into the relationships between physical efficiency and performance in specific field tests, providing a more nuanced understanding. Moreover, there is a need for research with extended data collection periods, allowing for the exploration of the PEI's dynamics in the context of a competitive season, particularly its correlation with various recovery variables. These avenues for further study promise to enrich our knowledge of the PEI's role in optimizing high-level soccer training and performance management. In the next section we will discuss about physical efficiency analysis.

2 Physical Efficiency Analysis

The field of sports science is currently witnessing a surge in interest and a burgeoning trend in the application of data science and artificial intelligence, as underscored by Claudino et al. [45]. Their recent study aimed to analyze the influence of physical efficiency on various significant variables within the sports domain. The research utilized data gathered from athletes competing in the Brazilian first division soccer league, meticulously collected over an eight-month period. This dataset includes comprehensive information from games, training sessions, and laboratory data, meticulously compiled by the club's physiologist, ensuring a holistic view of the athletes' performance and physical condition.

This extensive dataset consisted of 8,300 rows and spanned 63 diverse variables, covering observations from 46 athletes during the collection period, making it a valuable resource for sports science research. However, it is crucial to note that physical efficiency data was not uniformly collected across all activities within the dataset. In their meticulous data processing phase, the researchers worked diligently to eliminate data anomalies, such as records of players with limited activities and fields with missing or incomplete data. This rigorous data preparation process was pivotal in ensuring the reliability and quality of information used in their study. Together, these paragraphs highlight the vital intersection of research in sports science and the ongoing exploration of the PEI's significance in high-level soccer.

In an initial analysis, the K-means method was used to classify players into groups based on characteristics deemed important by the club's physiologist, including physical efficiency and other relevant variables collected. The elbow and silhouette methods in Python were employed to determine the number of groups to be considered in the algorithm. It was observed that the ideal number of groups for classifying the players ranged between 3 and 5, with the decision to use 5 groups for a more detailed analysis.

Table 1 Athletes by K-mean groups

Group	Num athletes	Num injured
0	14	0
1	9	4
2	4	0
3	9	3
4	2	0

In applying the K-means method, players with fewer than 10 observations during the season were removed, and the following parameters collected at each athlete's training were considered: physical efficiency, total distance covered, TRIMP, high-speed distance, deceleration, acceleration, total training minutes, and the athlete's perception before and after activities. For each of these parameters, the average value and standard deviation per player were calculated. After running the K-means algorithm in the Python tool, the players were classified into groups identified from 0 to 4. To better understand the characteristics of the players in each group, the number of injured players listed in each group was calculated, and the result is presented in Table 1. As can be observed, injured players were allocated only in two groups, and no injury occurrence was noted in the group with the largest number of players.

The group to which each player belonged was included in the database so that the characteristics of the players in each group could be analyzed based on the information collected from all training sessions and games of the season. By conducting a comparative analysis between the groups with injured players, 1 and 3, and the other groups without injured players, it is observed that the attributes related to the player's perception before and after activities have higher values in groups where there were injuries. This raises an important question that deserves further analysis: is there a relationship between a player feeling better and getting injured? Other attributes also differ between the groups with and without injuries, such as distances and time. Groups with players who tend to get injured apparently cover greater distances, whether at high intensity or not, in shorter periods than in other groups, which can lead to hypotheses that this greater effort may tend to result in injuries. No differences were observed in the other parameters, including physical efficiency.

Given that no significant correlations between injuries and physical efficiency were observed, further analyses were conducted to delve deeper into how training influences an athlete's efficiency. We focused on specific correlations that could elucidate the relationship between their training regimen and performance outcomes, aiming to uncover underlying patterns that might explain the variations in physical efficiency across different training and gameplay scenarios. Therefore, we constructed a correlation graph that illustrates the connection between the athlete's training load and their physical efficiency. This was chosen because these variables are central to evaluating the effectiveness of training programs in enhancing game-day performance. Figure 1 showcases this relationship, highlighting how training

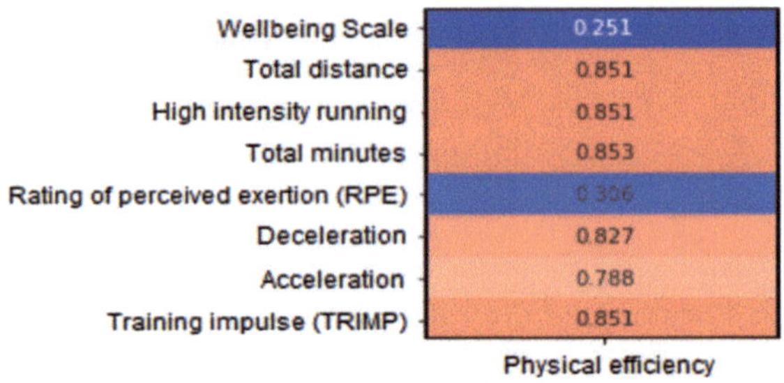

Fig. 1 Correlation between physical efficiency and averages of other variables from recent workouts. It is possible to observe that the correlation of the variables concerning the athlete's perception of their physical state, whether before or after activities, Wellbeing Scale and RPE, shows a low correlation with efficiency. The remaining variables exhibited a strong correlation with the physical efficiency variable

intensity and volume directly correlate with the athlete's ability to maintain high levels of performance throughout competition.

In light of the high correlation between physical efficiency and other variables, we analyzed the efficiency averages for each athlete, identifying the two athletes with the highest (Best1 and Best2) and the two with the lowest (Worst1 and Worst2) efficiency averages. These athletes were selected to compare how different variables relate to their physical efficiency. In Figs. 2, 3, and 4, the data for the athletes with the best averages are presented in the first row, while those with the worst averages are in the second row. Physical efficiency is indicated with yellow points in each graph. Blue points represent the variable being analyzed in each figure, with the color shifting to red for training sessions to reflect the significant variability between game and training performances. All variables were normalized to a scale from 0 to 1 to facilitate comparison across different scales. It was observed that variables such as acceleration, deceleration, and activity time did not show significant differences between the groups with the best and worst averages, underscoring the importance of focusing on the efficiency metrics for a nuanced analysis.

Figure 2 presents the total distance covered by players from both groups, with the distance covered in games being greater than that in non-game activities for both groups. One difference between the groups is that players with better averages seem to cover longer distances, ranging from 0.7 to 1.0 on game days and up to 0.6 in non-game activities, while the group with players with lower averages covers distances ranging from 0.6 to 0.8 in games and with variations up to 0.5 when not in games.

Here we can observe that players with better averages are the ones who cover the longest distances in games, but this distance varies between games, while players with lower averages tend to cover the same distance in games with little variation. Among athletes with better averages, there is a large dispersion of data, with games and training sessions having similar distances, whereas players with lower averages clearly cover longer distances in games. It is necessary to analyze whether this difference between training and games influences physical efficiency, assisting coaches in adjusting the distances covered in training sessions.

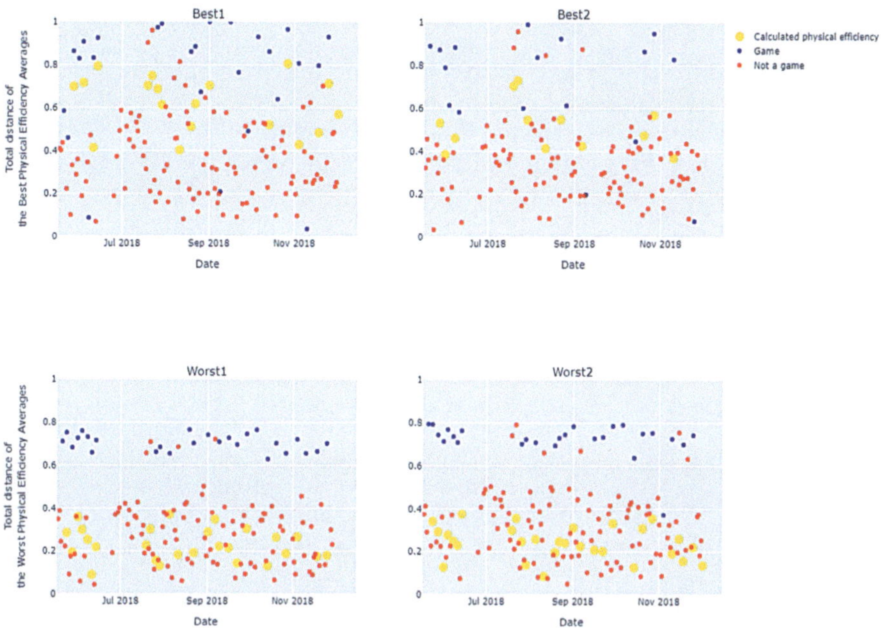

Fig. 2 Best and worst efficiency averages by distance traveled

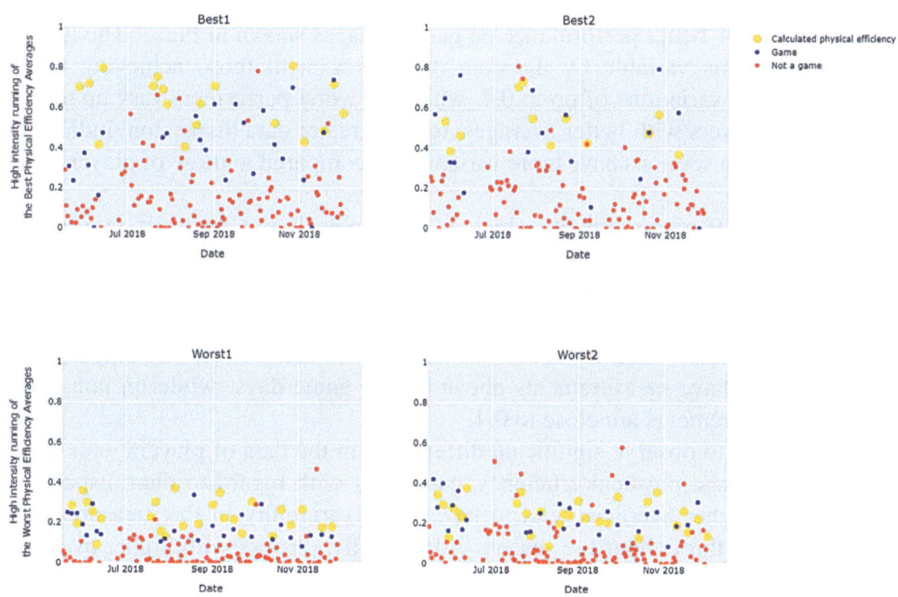

Fig. 3 Best and worst efficiency averages by high-intensity distances

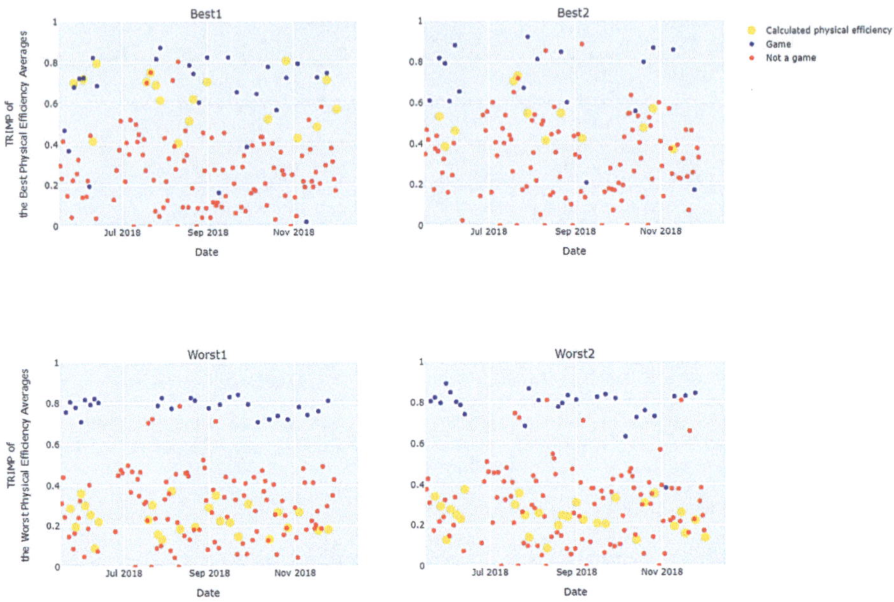

Fig. 4 Best and worst efficiency averages by TRIMP

Just like in the total distance covered, when analyzing high-intensity distances, both groups show better performance on game days, as shown in Fig. 3. The groups also differ in this variable by the data dispersion, with those achieving better averages having variations of up to 0.7, while the worst performers vary up to 0.4. Once again, players with better averages exhibit greater data dispersion, indicating that their training sessions have more varied loads compared to those of players with lower averages.

In the TRIMP measurement, which is a variable based on heart rate and is presented in Fig. 4, both groups show good data dispersion, but the variation in measurements in gaming and non-gaming activities is higher in the group with players having lower average physical efficiency. In other words, players with better averages have similar TRIMP values in both types of activities, while players with lower averages have measurements above 0.8 on game days, while on non-game days, the measurements are close to 0.4.

It is possible to observe significant differences in the data of players with lower and higher average physical efficiency, especially with regard to data dispersion. Players with higher averages tend to have more variability in the measurements of the variables throughout the season. Additionally, players with higher averages exhibit less variability between game activities and training. In contrast, the group of players with lower averages seemingly have higher measurements during game activities.

3 Final Remarks

In conclusion, the significance of physical efficiency in soccer cannot be overstated. It serves as a linchpin for both individual players and entire teams, bridging the gap between technical finesse and tactical acumen. Defined as the capacity to execute technical and tactical maneuvers at a consistently high level while conserving physiological effort, physical efficiency embodies the essence of soccer performance.

This paramount attribute demands unwavering dedication and rigorous preparation, especially considering the exhaustive demands of a full match. A player's ability to deliver superior physical performance while minimizing physiological strain is a testament to their commitment and conditioning.

Our exploration in this chapter sought to provide a comprehensive understanding of physical efficiency, drawing from relevant literature to illuminate its intricacies. Furthermore, we delved into the practical evaluation of physical efficiency within the context of professional soccer teams. Leveraging cutting-edge artificial intelligence techniques, we unveiled intricate relationships between various variables, shedding light on how they intertwine with injuries and influence the broader spectrum of game-related factors.

Ultimately, this analysis serves to enhance our knowledge of the complex web of connections that underlie physical efficiency in soccer. As the sport evolves and the pursuit of excellence continues, understanding these dynamics becomes increasingly vital for athletes, coaches, and sports scientists seeking to unlock the full potential of this beautiful game.

References

1. Stolen, T.; Chamari, K.; Castagna, C.; Wisloff, U. Physiology of soccer: an update. Sports Med, v. 35, n. 6, p. 501–536, 2005.
2. Bangsbo, J. The physiology of soccer–with special reference to intense intermittent exercise. Acta Physiol Scand Suppl, v. 619, n., p. 1–155, 1994.
3. Chmura, P.; Konefa 1, M.; Andrzejewski, M.; Kosowski, J.; Rokita, A.; Chmura, J. Physical activity profile of 2014 FIFA World Cup players, with regard to different ranges of air temperature and relative humidity. Int J Biometeorol, v. 61, n. 4, p. 677–684, 2017.
4. Ispirlidis, I.; Fatouros, I. G.; Jamurtas, A. Z.; Nikolaïdis, M. G.; Michailidis, I.; Douroudos, I.; Margonis, K.; Chatzinikolaou, A.; Kalistratos, E.; Katrabasas, I.; Alexiou, V.; Taxildaris, K. Time-course of changes in inflammatory and performance responses following a soccer game. Clin J Sport Med, v. 18, n. 5, p. 423–431, 2008.
5. Alves, A. L.; Garcia, E. S.; Morandi, R. F.; Claudino, J. G.; Pimenta, E. M.; Soares, D. D. Individual analysis of creatine kinase concentration in Brazilian elite soccer players. Revista Brasileira De Medicina Do Esporte, v. 21, n. 2, p. 112–116, 2015.
6. Silva, J. R.; Rumpf, M. C.; Hertzog, M.; Castagna, C.; Farooq, A.; Girard, O.; Hader, K. Acute and Residual Soccer Match-Related Fatigue: A Systematic Review and Meta-analysis. Sports Med, v. 48, n. 3, p. 539–583, 2018.

7. Carling, C.; McCall, A.; Le Gall, F.; Dupont, G. What is the extent of exposure to periods of match congestion in professional soccer players? J Sports Sci, v. 33, n. 20, p. 2116–2124, 2015.
8. Pino-Ortega, J.; Rojas-Valverde, D.; Gomez-Carmona, C. D.; Bastida-Castillo, A.; Hernandez-Belmonte, A.; Garcia-Rubio, J.; Nakamura, F. Y.; Ibanez, S. J. Impact of Contextual Factors on External Load During a Congested-Fixture Tournament in Elite U'18 Basketball Players. Front Psychol, v. 10, n., p. 1100, 2019.
9. Mohr, M.; Draganidis, D.; Chatzinikolaou, A.; Barbero-Alvarez, J. C.; Castagna, C.; Douroudos, I.; Avloniti, A.; Margeli, A.; Papassotiriou, I.; Flouris, A. D.; Jamurtas, A. Z.; Krustrup, P.; Fatouros, I. G. Muscle damage, inflammatory, immune and performance responses to three football games in 1 week in competitive male players. Eur J Appl Physiol, v. 116, n. 1, p. 179–193, 2016.
10. Owen, A. L.; Djaoui, L.; Dellal, A.; Ates, O.; Mendes, B. Biochemical Response Comparisons of a Competitive Microcycle Vs. Congested Fixture Periods in Elite Level European Champions League Soccer Players. J Complement Med Alt Healthcare, v. 10, n. 1, p. 9, 2019.
11. Claudino, J.G., Capanema, D.d., de Souza, T.V. et al. Current Approaches to the Use of Artificial Intelligence for Injury Risk Assessment and Performance Prediction in Team Sports: a Systematic Review. Sports Med—Open 5, 28, 2019.
12. Thornton, H. R.; Delaney, J. A.; Duthie, G. M.; Dascombe, B. J. Developing Athlete Monitoring Systems in Team Sports: Data Analysis and Visualization. Int J Sports Physiol Perform, v. 14, n. 6, p. 698–705, 2019.
13. Bourdon, P. C.; Cardinale, M.; Murray, A.; Gastin, P.; Kellmann, M.; Varley, M. C.; Gabbett, T. J.; Coutts, A. J.; Burgess, D. J.; Gregson, W.; Cable, N. T. Monitoring Athlete Training Loads: Consensus Statement. Int J Sports Physiol Perform, v. 12, n. Suppl 2, p. S2161-s2170, 2017.
14. Burgess, D. J. The Research Doesn't Always Apply: Practical Solutions to Evidence-Based Training-Load Monitoring in Elite Team Sports. Int J Sports Physiol Perform, v. 12, n. Suppl 2, p. S2136-s2141, 2017.
15. Kempton, T.; Sirotic, A. C.; Coutts, A. J. An integrated analysis of match-related fatigue in professional rugby league. J Sports Sci, v. 33, n. 1, p. 39–47, 2015.
16. Suarez-Arrones, L.; Torreno, N.; Requena, B.; Saez de Villarreal, E.; Casamichana, D.; Barbero-Alvarez, J. C.; Munguia-Izquierdo, D. Matchplay activity profile in professional soccer players during official games and the relationship between external and internal load. J Sports Med Phys Fitness, v. 55, n. 12, p. 1417–1422, 2015.
17. Claudino, J.G.; Cardoso Filho, C.A.; Boullosa, D.; Lima-Alves, A.; Carrion, G.R.; GianonI, R.L.d.S.; Guimarães, R.d.S.; Ventura, F.M.; Araujo, A.L.C.; Del Rosso, S.; et al. The Role of Veracity on the Load Monitoring of Professional Soccer Players: A Systematic Review in the Face of the Big Data Era. Appl. Sci. 2021.
18. Jaspers, A.; Brink, M. S.; Probst, S. G.; Frencken, W. G.; Helsen, W. F. Relationships Between Training Load Indicators and Training Outcomes in Professional Soccer. Sports Med, v. 47, n. 3, p. 533–544, 2017.
19. Antonacci, L.; Mortimer, L. F.; Rodrigues, V. M.; Coelho, D. B.; Soares, D. D.; Silami-Garcia, E. Competition, estimated, and test maximum heart rate. J Sports Med Phys Fitness, v. 47, n. 4, p. 418–421, 2007.
20. Dupont, G.; Nedelec, M.; McCall, A.; McCormack, D.; Berthoin, S.; Wisløff, U. Effect of 2 soccer matches in a week on physical performance and injury rate. Am J Sports Med, v. 38, n. 9, p. 1752–1758, 2010.
21. Scribbans, T. D.; Berg, K.; Narazaki, K.; Janssen, I.; Gurd, B. J. Heart rate during basketball game play and volleyball drills accurately predicts oxygen uptake and energy expenditure. J Sports Med Phys Fitness, v. 55, n. 9, p. 905–913, 2015.
22. Wilke, C. F.; Ramos, G. P.; Pacheco, D. A.; Santos, W. H.; Diniz, M. S.; Gonçalves, G. G.; Marins, J. C.; Wanner, S. P.; Silami-Garcia, E. Metabolic Demand and Internal Training Load in Technical-Tactical Training Sessions of Professional Futsal Players. J Strength Cond Res, v. 30, n. 8, p. 2330–2340, 2016.

23. Suarez-Arrones, L.; Núñez, J.; Munguía-Izquierdo, D.; Portillo, J.; Mendez-Villanueva, A. Impact of several matches in a day on physical performance in rugby sevens referees. Int J Sports Physiol Perform, v. 8, n. 5, p. 496–501, 2013.
24. Torreno, N.; Munguia-Izquierdo, D.; Coutts, A.; De Villarreal, E. S.; Asian-Clemente, J.; Suarez-Arrones, L. Relationship Between External and Internal Loads of Professional Soccer Players During Full Matches in Official Games Using Global Positioning Systems and Heart-Rate Technology. Int J Sports Physiol Perform, v. 11, n. 7, p. 940–946, 2016.
25. Castagna, C.; Impellizzeri, F. M.; Chamari, K.; Carlomagno, D.; Rampinini, E. Aerobic fitness and yo-yo continuous and intermittent tests performances in soccer players: a correlation study. J Strength Cond Res, v. 20, n. 2, p. 320–325, 2006.
26. Bangsbo, J.; Iaia, F. M.; Krustrup, P. The Yo-Yo intermittent recovery test : a useful tool for evaluation of physical performance in intermittent sports. Sports Med, v. 38, n. 1, p. 37–51, 2008.
27. Boullosa, D.; Casado, A.; Claudino, J. G.; Jiménez-Reyes, P.; Ravé, G.; Castaño-Zambudio, A.; Lima-Alves, A.; De Oliveira, S. A., Jr.; Dupont, G. Granacher, U.; Zouhal, H. Do you Play or Do you Train? Insights From Individual Sports for Training Load and Injury Risk Management in Team Sports Based on Individualization. Front Physiol, v. 11, n., p. 995, 2020.
28. Rampinini, E.; Coutts, A. J.; Castagna, C.; Sassi, R.; Impellizzeri, F. M. Variation in top level soccer match performance. Int J Sports Med, v. 28, n. 12, p. 1018–1024, 2007.
29. Brito, J.; Hertzog, M.; Nassi, G. P. Do Match-Related Contextual Variables Influence Training Load in Highly Trained Soccer Players? J Strength Cond Res, v. 30, n. 2, p. 393–399, 2016.
30. Lacome, M.; Simpson, B.; Broad, N.; Buchheit, M. Monitoring Players' Readiness Using Predicted Heart Rate Responses to Football Drills. Int J Sports Physiol Perform, v., n., p. 1–25, 2018.
31. Delaney, J. A.; Duthie, G. M.; Thornton, H. R.; Pyne, D. B. Quantifying the relationship between internal and external work in team sports: development of a novel training efficiency index. Sci Med Football, v. 2, n. 2, p. 149–156, 2018.
32. Grünbichler, J.; Federolf, P. Workload efficiency as a new tool to describe external and internal competitive match load of a professional soccer team: A descriptive study on the relationship between pre-game training loads and relative match load. Eur J Sport Sci, v. 20, n. 8, p. 1034–1041, 2020.
33. Gallo, T. F.; Cormack, S. J.; Gabbett, T. J.; Lorenzen, C. H. Pre-training perceived wellness impacts training output in Australian football players. J Sports Sci, v. 34, n. 15, p. 1445–1451, 2016.
34. AKUBAT, I.; BARRETT, S.; ABT, G. Integrating the internal and external training loads in soccer. Int J Sports Physiol Perform, v. 9, n. 3, p. 457–462, 2014.
35. Akubat, I.; Barrett, S. The Validity of External:Internal Training Load Ratios in Rested and Fatigued Soccer Players. Sports (Basel), v. 6, n. 2, p., 2018.
36. Andersson, H.; Raastad, T.; Nilsson, J.; Paulsen, G.; Garthe, I.; Kadi, F. Neuromuscular fatigue and recovery in elite female soccer: effects of active recovery. Med Sci Sports Exerc, v. 40, n. 2, p. 372–380, 2008.
37. Fatouros, I. G.; Chatzinikolaou, A.; Duroudos, I.; Nikolaïdis, M. G.; Kyparos, A.; Margonis, K.; Michailidis, Y.; Vantarakis, A.; Taxildaris, K.; Katrabasas, I.; Mandalidis, D.; Kouretas, D.; Jamurtas, A. Z. Time course of changes in oxidative stress and antioxidant status responses following a soccer game. J Strength Cond Res, v. 24, n. 12, p. 3278–3286, 2010.
38. BARROS, R. M.; MISUTA, M. S.; MENEZES, R. P.; FIGUEROA, P. J.; MOURA, F. A.; CUNHA, S. A.; ANIDO, R.; LEITE, N. J. Analysis of the distances covered by first division brazilian soccer players obtained with an automatic tracking method. J Sports Sci Med, v. 6, n. 2, p. 233–242, 2007.
39. Bradley, P. S.; Noakes, T. D. Match running performance fluctuations in elite soccer: indicative of fatigue, pacing or situational influences? J Sports Sci, v. 31, n. 15, p. 1627–1638, 2013.
40. Kalapotharakos, V. I. Contextual Factors and Aerobic Fitness Influence Match Running Performance in Elite Soccer. Science Performance and Science Reports, v. 113, n., p. 3, 2020.

41. Helgerud, J.; Engen, L. C.; Wisloff, U.; Hoff, J. Aerobic endurance training improves soccer performance. Med Sci Sports Exerc, v. 33, n. 11, p. 1925–1931, 2001.
42. Krustrup, P.; Mohr, M.; Ellingsgaard, H.; Bangsbo, J. Physical demands during an elite female soccer game: importance of training status. Med Sci Sports Exerc, v. 37, n. 7, p. 1242–1248, 2005.
43. Barbero-Alvarez, J. C.; Soto, V. M.; Barbero-Alvarez, V.; Grandavera, J. Match analysis and heart rate of futsal players during competition. J Sports Sci, v. 26, n. 1, p. 63–73, 2008.
44. Castellano, J.; Casamichana, D. Heart Rate and Motion Analysis by GPS in Beach Soccer. J Sports Sci Med, v. 9, n. 1, p. 98–103, 2010.
45. Claudino, J.G., Capanema, D.d.O., Santiago, P.R.P. (2022). AIM in Sports Medicine. In: Lidströmer, N., Ashrafian, H. (eds) Artificial Intelligence in Medicine. Springer, Cham.

A PageRank-Based Method for College Football Recruiting Rankings

Sergiy Butenko, Andrew Johnson, Erick Moreno-Centeno, and Justin Yates

Abstract This chapter applies an analytical method, PageRank, to rank the recruiting classes for teams in the Football Bowl Subdivision (FBS) based upon two pieces of data: (1) the list of teams reported to have extended formal offers to each player that signed a commitment letter and (2) the actual signing team for each player. Despite the limited inputs, PageRank develops similar rankings as those developed using expert opinions and published by major media outlets without the need for costly experts or national tours. The performance of the method is demonstrated using the National Signing Day data for three consecutive years.

Keywords College football · Ranking · PageRank method

1 Introduction

In the United States, the first Wednesday of February is college football's National Signing Day, the first day high school seniors are allowed to sign a binding letter declaring their choice of college or university where they will participate in football activities. These binding letters are used by major sports news outlets to develop college football recruitment (CFR) quality rankings for the recruiting classes of Football Bowl Subdivision (FBS) teams (or some subset such as the top 25). CFR rankings are highly publicized and are important for several reasons. In particular, they are viewed as a performance measure of head coach/staff recruiting abilities. In addition, these rankings influence the opinions of other coaches and the media and

S. Butenko (✉) · A. Johnson · E. Moreno-Centeno
Wm. Michael Barnes '64 Department of Industrial and Systems Engineering, Texas A&M University, College Station, TX, USA
e-mail: butenko@tamu.edu; ajohnson@tamu.edu; emc@tamu.edu

J. Yates
American Airlines, Fort Worth, TX, USA
e-mail: justin.yates@aa.com

M. J. Blondin et al. (eds.), *Artificial Intelligence, Optimization, and Data Sciences in Sports*, Springer Optimization and Its Applications 218,
https://doi.org/10.1007/978-3-031-76047-1_12

thus indirectly influence the national rankings of college football teams, national publicity, potential broadcasting decisions, and alumni donations [15, 14, 27, 3]. There have also been attempts of using the recruitment rankings as a predictor of success on the field [20, 10, 21, 17]. Therefore, objective methods to evaluate the quality of schools' annual recruiting classes are needed.

With thousands of high school football players joining more than 130 FBS teams each year, it is unrealistic to expect that any one analyst/expert can gather information on all players and all teams. As a result, the individual player rankings are developed as an aggregation of different people's opinions, all of whom have varying levels of expertise [13, 2, 1]. Assuming that the subjective opinions of experts may provide additional insight for the rankings, aggregation methods which impose extensive assumptions to rectify discrepancies in the experts' opinions may overwhelm the value of the experts insight. Thus, aggregation methods with minimal axiomatic foundations are preferred.

Ranking problems arise in a variety of contexts. In sports, rankings are commonly needed to evaluate teams that are not able to directly compete either because the number of teams is too large or because of geography or time restrictions [25]. In other contexts such as industry, corporate managers often would like to rank establishments or employees to inform resource allocation decisions [19]. A topic familiar to much of the readership, in academics journals are often ranked to help determine the quality of faculty publications [26]. The method we describe in this chapter is applied to the ranking of college football recruiting classes, but it could easily be extended to ranking universities based on admissions or ranking investments based on product offerings and investor decisions.

However, ranking and aggregate evaluation are challenging research topics for a variety of reasons. First, there is no directly observable true ranking. Thus, it is impossible to empirically test the quality of methods. For a discussion of the challenges see [16]. The "best" method utilizes a reasonable set of assumptions which, being consistent with Occam's razor, should be a minimal set [22]. Thus, the only information used in the developed approach is the data on scholarship offers/acceptances available in the public domain [1]. Second, ranking can be hard in general because there may be multiple dimensions to good performance. In this setting we are strictly focused on quality of the recruiting classes as "voted" by the parties involved (teams and players) on National Signing Day, assuming that (1) recruiters tend to extend offers to better players, and (2) by selecting a scholarship offer from team T_1 over team T_2, a player improves the quality of T_1's recruiting class relative to that of T_2, regardless of the motives leading to this choice. This well-defined scope avoids many issues of multi-criteria aggregation that would arise in other contexts.

This chapter proposes a method for developing recruiting class rankings using data available on team offers and actual commitments. The major benefit of the proposed method is that it relies only on the reported university and player decisions themselves without considering expert opinion and the subjectivity such opinions introduce.[1] The information used comes directly from the participants themselves who clearly have a high incentive to identify the "best" among available alternatives (teams or players). To obtain the rankings, the choices made by universities and players are analyzed using a known method, PageRank, which was originally developed for ranking websites in a search engine. For comparison, the rankings developed by several media outlets, including ESPN, Rivals, and Scout (now a part of 247Sports), are presented. The benefits and limitations of the proposed method are discussed and relative advantages identified.

The remainder of this chapter is organized as follows. Section 2 describes the method proposed for ranking recruiting classes. Section 3 discusses the results obtained from the 2011, 2012, and 2013 National Signing Day data and the comparisons to the rankings published by media outlets and recruiting agencies. Section 4 concludes the chapter. Finally, Appendix 5 contains tables with complete rankings obtained using the proposed method.

2 The Method

Originally developed by Brin and Page [8] as the method to rank webpages in Google's search engine, the PageRank method has gained considerable popularity largely due to its effectiveness and simplicity. In particular, PageRank and similar eigenvalue-based methods have been used for sports rankings based on results of competitions [18, 6, 24, 28, 11]. PageRank assigns a numerical score to each page based on the scores of webpages with a hyperlink to the page being scored. We replace the webpages with schools and players and interpret the hyperlinks as scholarship offers and team signings. As [9] suggest, the PageRank method can be interpreted as a democracy where links are votes in favor of the best website; each website is allowed to vote for a finite number of webpages by dividing the voting website's score evenly over the set of webpages for which it is voting. A similar interpretation can be developed for the rankings of recruiting classes and recruited players. A university can vote for a player to be a top recruit by offering a scholarship. As a result, the university's (player's) score can be viewed as a weight assigned to the university's (player's) vote. By casting its vote for several players, a

[1] Note that another dimension of the quality of the recruiting class is how well the class fills the needs of the team. For example, if the team needs to improve at the cornerback position, does the recruiting class provide talent at that position? The proposed method does not capture this dimension of quality directly, only to the extent that a top cornerback might be more likely to join a team in need of a cornerback because they will likely receive playing time sooner. Expert opinion might be able to include the issue of team needs. Thus, the method is offered as a supplement to the rankings developed by experts. However, given that the incubation period for high school recruits to become contributing players to a college football team is often several years and given that high school recruits often change positions at the college level, this type of positional need is difficult to evaluate, even by experts.

team uniformly distributes the weight of its vote among these players. As a result, the weight of a player's vote is given by the sum of the partial weights obtained from each university that votes for him. The university's weight is given by the weighted sum of the players' votes, i.e., the players signing letters of intent. Note that the underlying assumption of the method is that a team offering a scholarship to a player would benefit from signing this player. Thus, by choosing team T_1 over team T_2, the player strengthens the recruiting class of T_1 relative to that of T_2.

Mathematically, the problem of determining scores that will satisfy these conditions can be cast as the problem of computing the principal eigenvector of a certain matrix as follows. Assume that there are N teams $T_1, \ldots, T_N$ and M players, $P_1, \ldots, P_M$. Let O_p be the set of teams offering a scholarship to player p ($p = 1, \ldots, M$), and let A_t be the set of players choosing to play for a team t ($t = 1, \ldots, N$). Obviously, $p \in A_t$ implies that $t \in O_p$. Let n_t be the total number of scholarship offers that a team t extends, i.e., n_t is the number of elements in the set $\{p : t \in O_p\}$. Denoting by w_t, the score of team T_t, $t = 1, \ldots, N$, and by w_p the score of player P_p, $p = 1, \ldots, M$, we obtain the following relationship:

$$w_p = \sum_{t \in O_p} \frac{w_t}{n_t}, \quad p = 1, \ldots, M; \tag{1}$$

$$w_t = \sum_{p \in A_t} w_p, \quad t = 1, \ldots, N. \tag{2}$$

Given these definitions, we can summarize the scores in a column vector w of length $N+M$, in which the first N components correspond to the teams and the remaining M components correspond to the players. Next, we introduce a matrix A of size $(N + M) \times (N + M)$, in which the element a_{ij} in the ith row and jth column is given by

$$a_{ij} = \begin{cases} 1, & \text{if } 1 \leq i \leq N, N+1 \leq j \leq N+M, \text{ and } (j-N) \in A_i; \\ 1/n_j, & \text{if } N+1 \leq i \leq N+M, 1 \leq j \leq N, \text{ and } j \in O_i; \\ 0, & \text{otherwise.} \end{cases} \tag{3}$$

Then the relations (1) and (2) can be written in matrix form as follows:

$$w = Aw. \tag{4}$$

Thus, the problem of finding universities' scores reduces to computing the eigenvector of matrix A that corresponds to the eigenvalue equal to 1. Note that the matrix A is column-stochastic (i.e., all of its elements are nonnegative and each column sums up to 1), and it is known that every column-stochastic matrix has 1 as one of its eigenvalues. After calculating the scores of the players and the teams, we construct the rankings for the teams by ordering the scores from highest to lowest and assigning the corresponding ranking with the highest score receiving a rank of

1. The reader is referred to [9] for a detailed discussion of the linear algebra behind the PageRank method and for a discussion of related literature.

To illustrate the method, consider the following hypothetical situation involving 3 teams (T_1, T_2, and T_3) and 6 players ($P_1, \ldots, P_6$). Assume that team T_1 offers scholarships to all six players: team T_2, to players P_3, P_4, and P_5, and team T_3, to player P_3 only. Suppose that P_1, P_2, and P_6 sign for T_1; P_4 and P_5 sign for T_2; and P_3 signs for T_3. Then, even though team T_1 signs most players (3) and team T_3 signs only 1 player, according to PageRank, T_3 is ranked the highest and T_1 is ranked the lowest of the three teams. This is due to the fact that the only player T_3 signs had offers from all three teams, and T_2 beats out T_1 in recruiting both players it signs.

One of the main advantages of using the PageRank method for analyzing college football recruiting data is that it can produce very sensible class rankings based on minimal input. However, should additional useful information regarding each team–player pair become available, the method can be further enhanced by including this information. This could be done by, e.g., allowing one to grade the teams' level of interest in each player according to a predetermined scale. Denoting by g_{pt} the grade given by team t to player p, we can modify the definition of a_{ij}, $i, j = 1, \ldots, N+M$ in Eq. (3) as follows:

$$a_{ij} = \begin{cases} 1, & \text{if } 1 \le i \le N, N+1 \le j \le N+M, \text{ and } (j-N) \in A_i; \\ \frac{g_{(i-N)j}}{\sum_{p=1}^{M} g_{(i-N)p}}, & \text{if } N+1 \le i \le N+M, 1 \le j \le N; \\ 0, & \text{otherwise.} \end{cases} \tag{5}$$

Then the resulting matrix A can be analyzed as in the PageRank method to obtain team and player rankings. For illustration, assume that each team in the above example expresses its level of interest in each player using a 0–5 scale, with 0 representing complete lack of interest and 5 the highest level of interest, as given in the following Table 1:

With this additional information, we can observe that the players signed by team T_1 were of some interest to the other teams, even though no offers have been extended. As a result, the enhanced PageRank method ranks the teams in the following order: 1, T_1; 2, T_2; and 3, T_3. In addition, since a complete evaluation of players by each team is given, the method can be used to obtain ranking of players, which is given by the following: 1, P_3; 2, P_2; 3, P_6; 4, P_1; and 5–6, P_4 and P_5.

Table 1 Data for PageRank example

	P_1	P_2	P_3	P_4	P_5	P_6
T_1	5	5	3	2	2	4
T_2	1	2	5	4	4	2
T_3	2	3	5	2	2	3

3 Results

3.1 Recruitment Data

The data used in the PageRank CFR rankings was obtained from the Scout website [1] in the beginning of February 2011, 2012, and 2013, immediately after the National Signing Day.[2] In those years, the Scout site provided data on students' commits to a university through letter of intent signings. Scout.com also reported detailed information on each prospect, including lists of schools competitively pursuing a recruit either by declaring interest or extending an offer. Note that the offer data obtained from Scout.com was self-reported by the high school players being recruited and is subject to the problems of recruitment ambiguity; see, for example, [23, 4, 5]. Further, the completeness of recruiting information for all FBS universities is difficult to ensure because offer lists are not provided by the universities and smaller schools tend to be underreported by the high school recruits. Nevertheless, during the considered years Scout.com provided the most comprehensive openly available set of recruiting data. Table 2 summarizes the basic characteristics of the National Signing Day data supporting the analysis in this chapter.

3.2 CFR Rankings and Discussion

Tables 3, 4, and 5 show the university recruitment rankings constructed using the PageRank as well as the rankings from ESPN [13], Rivals [2], and Scout [1]. The column *Median* contains the results from aggregating the ESPN, Rivals, and Scout rankings. As the name suggests, this aggregate ranking was obtained as follows: the ranking of each university is the median of its ESPN, Rivals, and Scout ranking. ESPN and Rivals did not disclose their ranking methodologies, whereas Fox Sports explained its ranking methodology as:

Table 2 The number of FBS teams, recruited players, scholarship offers, and average number of offers per team and per player for 2011–2013

Year	Teams	Players	Offers	Offers/team	Offers/player
2011	120	2,737	10,592	88.27	3.87
2012	124	2,787	12,879	103.86	4.62
2013	126	2,886	14,066	111.63	4.87

[2] In 2013 Scout was sold by Fox Sports to North American Membership Group, rebranded as Scout Media, and then acquired by CBS in 2017. Hence, 2013 was the last year Scout football recruiting data was available.

Table 3 Comparison of the 2011 rankings obtained using the PageRank (PR) method with the rankings published by ESPN, Rivals, and Scout

Team	ESPN	Rivals	Scout	PR	Median
Florida State	1	2	1	3	1
Alabama	2	1	7	7	2
Auburn	3	7	2	5	3
USC	4	4	5	10	4
Texas	5	3	4	2	4
Georgia	6	5	6	1	6
Ohio State	7	11	3	4	7
Clemson	8	8	13	12	8
Notre Dame	9	10	8	6	9
LSU	10	6	10	17	10
Oklahoma	11	14	15	23	14
Florida	12	12	18	14	12
Tennessee	13	13	12	9	13
Oregon	14	9	9	11	9
South Carolina	15	18	11	8	15
North Carolina	16	16	16	13	16
Nebraska	17	15	22	16	17
California	18	17	14	15	17
Arkansas	19	24	17	18	19
Virginia	20	25	25	21	25
Stanford	21	22	21	24	21
Louisville	22	29	33	38	29
Oklahoma State	23	28	20	28	23
Washington	24	23	23	25	23
Mississippi	25	19	24	19	24

"Team Rankings are a math formula based on a player's rating and his rankings. 5-Star is a rating, No. 1 quarterback is his ranking.

5 Star = 200 points
4 Star = 120 points
3 Star = 40 points
2 Star = 20 points

The No. 1 player at a position is worth 100 points, counting down to the last ranked player at his position to 0."

Tables 3, 4, and 5 list the schools in order of their ESPN ranking (we use the ESPN ranking to determine table layout as ESPN only provided its top 25 ranked universities in 2011 and 2012 and top 75 ranked classes in 2013). Note that in 2011 and 2012 the PageRank method's top-25 list differs from ESPN's by only two teams, with both differences occurring outside of the top 20. Namely, in 2011 #22 Louisville and #23 Oklahoma State in ESPN's ranking are replaced with #20 Miami (FL) and #22 Iowa in PageRank method's top 25, both of which are unranked

Table 4 Comparison of the 2012 rankings obtained using the PageRank (PR) method with the rankings published by ESPN, Rivals, and Scout

Team	ESPN	Rivals	Scout	PR	Median
Alabama	1	1	2	2	1
Florida State	2	6	10	8	6
Texas	3	2	1	1	2
Florida	4	3	5	3	4
Georgia	5	11	13	10	11
Ohio State	6	4	3	4	4
Michigan	7	7	4	9	7
Miami	8	9	8	11	8
Clemson	9	14	16	18	14
Notre Dame	10	22	18	13	18
Oklahoma	11	10	9	12	10
Stanford	12	5	6	5	6
USC	13	8	19	14	13
LSU	14	17	7	16	14
Texas A&M	15	15	20	19	15
South Carolina	16	19	12	6	16
Auburn	17	12	14	7	14
Oregon	18	16	15	22	16
UCLA	19	13	11	17	13
Texas Tech	20	26	24	24	24
Tennessee	21	18	26	20	21
California	22	23	33	29	23
Washington	23	20	22	34	22
Rutgers	24	24	35	21	24
Virginia Tech	25	21	27	15	25

in ESPN's top 25 (it should be noted that Louisville does not appear in the top 25 in any of the other methods reported in Table 3). Interestingly, in 2012 #22 and #23 in ESPN's ranking are, once again, not included in PageRank's top 25. This time, ESPN's #22 California and #23 Washington are replaced with PageRanks's #23 Mississippi State and #25 Baylor. In 2013, ESPN's PageRank's top 75 differ by 5 teams (ESPN's #67 Colorado, #68 Hawaii, #70 Fresno State, #71 SMU, and #72 Connecticut versus PageRanks's #51 Louisiana, #64 Southern Miss, #67 Syracuse, #68 Western Kentucky, and #69 Tulane).

The PageRank method can be extended to rank the players recruited (e.g., by weighing the importance score allocations based on their timing); however several challenges arise. In particular, players who make their intentions clear early to attend a particular school and only receive an offer from one school are limited in how highly they can be ranked regardless of their ability. Players who consider multiple schools and receive offers from more than one school tend to rank higher. One may argue that considering the date of commitment could help to address this issue, assuming players who commit earlier are better players and adjusting the grades g_{pt}

Table 5 Comparison of the 2013 rankings obtained using the PageRank (PR) method with the rankings published by ESPN, Rivals, and Scout for top 75 recruiting classes according to ESPN

Team	ESPN	Rivals	Scout	PR	Median
Alabama	1	1	3	10	1
Florida	2	4	7	2	4
Ohio State	3	2	1	7	2
Notre Dame	4	3	4	1	4
Mississippi	5	7	10	12	7
Michigan	6	5	2	8	5
LSU	7	6	8	6	7
Texas A&M	8	11	6	5	8
Florida State	9	10	16	18	10
Georgia	10	12	9	3	10
Auburn	11	8	14	16	11
UCLA	12	8	5	4	8
Clemson	13	14	12	11	13
USC	14	13	18	15	14
Texas	15	23	22	9	22
Oklahoma	16	15	15	23	15
South Carolina	17	16	23	17	17
Washington	18	18	13	19	18
Virginia Tech	19	22	33	30	22
North Carolina	20	40	29	26	29
Miami (FL)	21	45	34	20	34
Vanderbilt	22	19	19	13	19
Nebraska	23	17	11	22	17
Penn State	24	41	44	28	41
Mississippi State	25	25	21	36	25
Oregon	26	21	17	14	21
Oklahoma State	27	34	24	25	27
Baylor	28	29	25	34	28
Tennessee	29	20	35	21	29
California	30	30	28	37	30
Arkansas	31	26	32	29	31
West Virginia	32	24	27	33	27
Wisconsin	33	56	48	42	48
Virginia	34	27	39	27	34
Michigan State	35	38	43	59	38
Kentucky	36	28	38	32	36
Rutgers	37	44	31	49	37
Missouri	38	39	36	38	38
Arizona	39	36	26	61	36
Stanford	40	63	57	35	57
Pittsburgh	41	33	20	45	33
Arizona State	42	32	30	31	32

(continued)

Table 5 (continued)

Team	ESPN	Rivals	Scout	PR	Median
Louisville	43	57	55	43	55
Indiana	44	42	49	50	44
Maryland	45	31	47	39	45
Oregon State	46	37	40	44	40
Illinois	47	47	41	57	47
Utah	48	42	53	54	48
TCU	49	35	46	24	46
Purdue	50	55	56	58	55
Northwestern	51	52	45	66	51
USF	52	49	62	40	52
Boise State	53	62	42	60	53
Kansas State	54	64	68	75	64
Kansas	55	45	54	47	54
Iowa State	56	58	64	73	58
Texas Tech	57	51	58	41	57
Washington State	58	52	37	55	52
Iowa	59	52	51	63	52
Wake Forest	60	58	60	65	60
Marshall	61	58	71	56	61
Houston	62	50	59	46	59
Cincinnati	63	71	50	53	63
Georgia Tech	64	84	72	48	72
Duke	65	67	70	62	67
NC State	66	47	52	52	52
Colorado	67	67	67	98	67
Hawaii	68	92	82	80	82
San Diego State	69	77	66	70	69
Fresno State	70	64	77	91	70
SMU	71	67	75	83	71
Connecticut	72	64	63	76	64
BYU	73	71	61	74	71
UCF	74	73	89	72	74
Minnesota	75	61	73	71	73

in (5) accordingly; however, such an assumption could be questionable. Moreover, a recent empirical study conducted by Bricker and Hanson [7] concludes that

> "In one season and at a typical average early signing date, early-committing athletes played in 0.21 fewer games per season, or about four percent of the average number of games played."

While quantifying the value of a verbal commitment is an interesting open question, it is clear that regardless of how reasonable the corresponding method is, it will inevitably introduce additional subjectivity into the analysis of recruiting classes.

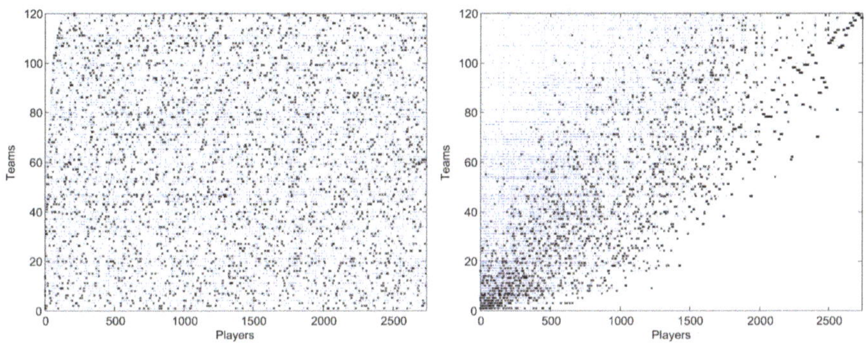

Fig. 1 A graphical representation of the scholarship offer data for year 2011. A dot with coordinates (x, y) represents an offer from university y to player x. An accepted offer is described by an x-mark. In the left picture, the players are ordered according to their file IDs in the Scout database (can be considered a random order), and the teams are arranged in the order of their first appearance in a player's scholarship offer list. In the right picture, both the players and the teams are ordered according to their PageRank rankings

Recognizing the aforementioned weaknesses of PageRank as a method for ranking individual players, we can still argue that such a ranking is useful in providing a "big picture" description of the college football recruiting data. This point is illustrated in Fig. 1, which visualizes the scholarship offer/acceptance data for 2011. One can clearly see from the figure that higher-ranked players tend to receive/accept offers from higher-ranked teams.

While the PageRank method is used to derive independent rankings, the median ranking aggregates all three of the major recruitment rankings in its assessment. Thus, the median ranking can be interpreted as a *middle ground* ranking, aggregating the ESPN, Rivals, and Scout rankings. To illustrate the similarity among the top-25 (top-75 for 2013) rankings reported in Tables 3, 4, 5, 6, 7, and 8 report the pairwise values of Kendall's τ-rank correlation coefficient, the pairwise values of the sum of absolute differences, and the pairwise values of the sum of squared differences, respectively. These metrics of rank similarity are reported by taking the average over the years 2011, 2012, and 2013. For example, Kendall's τ-coefficient reported between ESPN and Rivals (0.73) was obtained by averaging Kendall's τ-coefficients between ESPN and Rivals in 2011, 2012, and 2013 (which were 0.77, 0.66, and 0.77, respectively). As expected, under all similarity metrics, the median ranking is highly correlated with (alternatively very close to) the three expert rankings.[3]

[3] We note that the use of offer data to provide player valuation contains some inherent bias based on player/school location. Players from high schools located in areas with a higher concentration of FBS conference football teams will likely attract more offers than a similarly skilled player in a remote region [12].

Table 6 Average Kendall's τ-rank correlation coefficient for the 2011, 2012, and 2013 rankings reported in Tables 3, 4, and 5, respectively

	Rivals	Scout	PR	Median
ESPN	0.73	0.71	0.67	0.83
Rivals		0.71	0.66	0.87
Scout			0.68	0.82
PR				0.70

Table 7 Average sum (rounded to nearest integer) of absolute differences for the 2011, 2012, and 2013 rankings reported in Tables 3, 4, and 5, respectively

	Rivals	Scout	PR	Median
ESPN	9925	10119	11618	5986
Rivals		10086	13013	5759
Scout			13844	6175
PR				11999

Table 8 Average square-rooted sum (rounded to nearest integer) of squared differences for the 2011, 2012, and 2013 rankings reported in Tables 3, 4, and 5, respectively

	Rivals	Scout	PR	Median
ESPN	289	291	341	186
Rivals		290	376	171
Scout			386	192
PR				344

Table 9 Average Pearson product-moment correlation coefficient between the statistics reported in Tables 10, 11, and 12. Entries that are statistically significant are marked by asterisk significant (i.e., the corresponding p-value is less than 0.05)

	OM	PS	PS/OM	ARO	ARS
PRrank	−0.469*	−0.097	0.415*	0.948*	0.983*
OM		0.226*	−0.767*	−0.501*	−0.390*
PS			0.296*	0.095	0.027
PS/OM				0.563*	0.395*
ARO					0.955*

To further analyze the PageRank ranking, we analyze its relation to the following relevant team performance statistics: number of offers each team made (OM), the number of players each team signed (PS), the signing success ratio(PS/OM) of each team, and the average rank of the players given an offer and signed (ARO and ARS) by each team. In particular, Table 9 gives the Pearson product-moment correlation coefficients between the PageRank ranking and these statistics (averaged over the three years considered, 2011, 2012, and 2013). Tables 10, 11, and 12 in Appendix 5 report, for each team, its PageRank ranking and its other performance statistics previously mentioned (OM, PS, PS/OM, ARO and ARS). As Table 9, shows, not surprisingly, the recruiting class ranking is highly correlated with the average rankings of the players offered a scholarship/signed by a team. Notice the PageRank and number of offers made is negatively correlated and significant. An assumption of the PageRank method is when a player accepts an offer from school

A when having an offer from school B, the player prefers school A over school B, and thus school A is ranked higher. However, if the freshman class size of any team is limited, then making excess offers assures a team will not be able to recruit all players to which offers are made, thus the negative correlation between number of offers and ranking. It should be noted that the average figures reported in this table were nearly identical to their counterparts for each individual year. Moreover, most of the correlation figures were significant.

4 Conclusion

This chapter is the first attempt to use analytical tools, such as the PageRank method, for ranking college football recruiting classes. Unlike the existing approaches, which use some simple formulae to aggregate subjective evaluations of prospects by unidentified scouts, this method is based solely on publicly available information on offers made/accepted. The PageRank method for ranking college football recruiting classes is highly correlated with expert opinion. The PageRank method proposed in this chapter is easily implementable in standard software packages supporting matrix operations, such as MATLAB®, and thus are an inexpensive alternative to gathering expert opinions.

While our primary objective in this chapter was to rank recruiting classes, the PageRank method provides rankings of players as a byproduct. The player rankings could be improved by taking into account early commitments, regional preferences, and team need. Future research could improve player rankings and compare them to NFL draft outcomes to determine the quality of the rankings.

5 Detailed Results

Table 10 NCAA FBS teams listed according to their 2011 PageRank rankings. The remaining columns show the number of offers made by the team (OM), the number of players signed (PS), the average ranking of players offered a scholarship by the team (ARO), and the average ranking of players signed (ARS)

Rank	Team	OM	PS	PS/OM	ARO	ARS
1	Georgia	82	26	0.32	124.0	171.3
2	Texas	31	22	0.71	73.0	91.7
3	Florida State	101	28	0.28	181.8	258.7
4	Ohio State	50	23	0.46	124.9	175.7
5	Auburn	102	24	0.24	202.4	246.3
6	Notre Dame	90	23	0.26	166.0	233.9

(continued)

Table 10 (continued)

Rank	Team	OM	PS	PS/OM	ARO	ARS
7	Alabama	123	23	0.19	166.9	267.0
8	South Carolina	124	31	0.25	325.5	455.0
9	Tennessee	184	27	0.15	262.7	424.8
10	USC	127	30	0.24	268.2	486.3
11	Oregon	95	24	0.25	322.3	411.0
12	Clemson	96	29	0.30	276.8	529.7
13	North Carolina	115	25	0.22	301.6	454.5
14	Florida	111	18	0.16	152.9	286.7
15	California	82	22	0.27	291.1	408.7
16	Nebraska	110	20	0.18	360.8	598.2
17	LSU	117	21	0.18	291.5	613.0
18	Arkansas	156	30	0.19	505.2	923.4
19	Mississippi	120	26	0.22	443.0	629.2
20	Miami (FL)	111	15	0.14	274.4	500.5
21	Virginia	81	26	0.32	497.2	694.7
22	Iowa	83	24	0.29	484.0	670.1
23	Oklahoma	66	17	0.26	266.5	517.7
24	Stanford	124	19	0.15	451.9	688.5
25	Washington	96	23	0.24	572.8	804.4
26	Rutgers	96	24	0.25	498.5	821.3
27	Virginia Tech	49	21	0.43	393.7	599.1
28	Oklahoma State	77	25	0.32	485.7	806.2
29	Texas Tech	109	29	0.27	512.3	843.0
30	TCU	52	26	0.50	558.8	752.3
31	Michigan	131	19	0.15	365.6	684.2
32	Wisconsin	99	21	0.21	516.7	840.7
33	Georgia Tech	80	22	0.28	462.9	868.9
34	Texas A&M	79	23	0.29	510.5	943.3
35	South Florida	115	21	0.18	566.1	906.2
36	Penn State	54	16	0.30	422.3	767.1
37	UCF	96	29	0.30	709.4	1061.1
38	Louisville	152	20	0.13	590.0	887.8
39	Kentucky	107	22	0.21	681.5	943.2
40	Vanderbilt	105	22	0.21	639.5	964.3
41	Boston College	86	23	0.27	585.1	1032.1
42	SMU	79	28	0.35	803.3	1126.2
43	Michigan State	106	21	0.20	501.8	962.8
44	Mississippi State	96	22	0.23	644.9	1041.0
45	UCLA	109	16	0.15	414.9	914.9
46	Illinois	170	28	0.16	752.9	1341.7

Table 10 (continued)

Rank	Team	OM	PS	PS/OM	ARO	ARS
47	Cincinnati	128	25	0.20	693.5	1143.8
48	Utah	83	19	0.23	826.3	1116.9
49	Colorado	97	18	0.19	676.1	1117.6
50	Oregon State	90	25	0.28	753.0	1252.2
51	Marshall	109	30	0.28	1063.2	1356.0
52	Missouri	58	17	0.29	627.3	1106.6
53	Arizona	133	18	0.14	611.4	1186.5
54	West Virginia	178	22	0.12	585.8	1193.4
55	Baylor	97	18	0.19	700.8	1308.9
56	Northwestern	58	17	0.29	674.1	1008.1
57	Kansas	133	26	0.20	807.0	1386.4
58	Pittsburgh	89	20	0.22	687.9	1210.9
59	Indiana	83	20	0.24	876.9	1381.1
60	Syracuse	104	26	0.25	808.7	1427.5
61	Maryland	154	21	0.14	641.9	1509.5
62	Duke	150	20	0.13	641.4	1410.7
63	Purdue	83	16	0.19	681.7	1328.1
64	Kansas State	83	25	0.30	899.6	1542.3
65	Louisiana Tech	69	24	0.35	1130.6	1427.6
66	Boise State	55	18	0.33	965.7	1368.4
67	East Carolina	91	25	0.27	949.4	1621.3
68	Wake Forest	87	15	0.17	659.5	1360.1
69	North Carolina State	107	20	0.19	668.3	1550.4
70	Memphis	173	29	0.17	1122.9	1859.5
71	Southern Miss	102	25	0.25	994.1	1627.3
72	Toledo	91	19	0.21	1165.9	1585.5
73	Troy	58	26	0.45	1260.6	1656.3
74	Louisiana	58	28	0.48	1413.8	1684.5
75	Houston	73	22	0.30	940.8	1615.5
76	Arizona State	93	15	0.16	672.7	1428.3
77	Rice	43	19	0.44	1113.9	1448.4
78	Navy	69	54	0.78	1830.5	1986.2
79	Tulane	65	22	0.34	1270.7	1743.1
80	Western Kentucky	76	26	0.34	1271.2	1766.0
81	Minnesota	160	24	0.15	929.2	1948.3
82	Middle Tennessee	84	19	0.23	1148.8	1792.1
83	New Mexico	78	19	0.24	1170.9	1871.4
84	Western Michigan	67	18	0.27	1361.3	1793.3
85	San Jose State	61	23	0.38	1401.1	1779.5
86	UTEP	45	22	0.49	1422.6	1697.5

(continued)

Table 10 (continued)

Rank	Team	OM	PS	PS/OM	ARO	ARS
87	San Diego State	72	23	0.32	1246.8	1937.6
88	Colorado State	110	26	0.24	1264.3	2159.0
89	Nevada	57	25	0.44	1429.1	1857.1
90	Florida International	106	24	0.23	1122.3	2126.7
91	Connecticut	59	16	0.27	1001.2	1761.3
92	Washington State	103	28	0.27	1192.9	2001.4
93	Ohio	75	23	0.31	1316.2	1901.4
94	Tulsa	99	24	0.24	1200.3	2109.9
95	Fresno State	40	18	0.45	1196.6	1597.5
96	UAB	62	27	0.44	1393.8	1974.1
97	Arkansas State	59	24	0.41	1573.6	2044.9
98	Hawaii	56	25	0.45	1366.3	1968.8
99	Army	62	42	0.68	1905.3	2197.0
100	Bowling Green	82	22	0.27	1342.9	2033.9
101	Kent State	43	16	0.37	1420.5	1932.6
102	Temple	40	21	0.53	1571.4	2040.5
103	UNLV	46	20	0.43	1536.6	2111.8
104	Northern Illinois	45	21	0.47	1668.4	2067.7
105	BYU	29	18	0.62	1501.8	1911.6
106	Air Force	90	29	0.32	1673.2	2292.7
107	Iowa State	88	21	0.24	1164.5	2215.6
108	Idaho	57	23	0.40	1719.1	2322.5
109	North Texas	59	21	0.36	1647.4	2177.8
110	Utah State	60	22	0.37	1719.1	2258.7
111	Buffalo	84	20	0.24	1469.7	2467.6
112	ULM	49	20	0.41	1721.0	2319.9
113	Central Michigan	71	25	0.35	1556.9	2324.0
114	Akron	73	24	0.33	1669.6	2417.4
115	Wyoming	81	23	0.28	1542.4	2439.8
116	Eastern Michigan	41	18	0.44	1851.7	2511.2
117	Miami (OH)	60	19	0.32	1675.6	2539.5
118	Ball State	83	19	0.23	1533.0	2564.1
119	New Mexico State	66	19	0.29	1662.5	2570.6
120	Florida Atlantic	46	22	0.48	1774.1	2651.0

Table 11 NCAA FBS teams listed according to their 2012 PageRank rankings. The remaining columns show the number of offers made by the team (OM), the number of players signed (PS), the average ranking of players offered a scholarship by the team (ARO), and the average ranking of players signed (ARS)

Rank	Team	OM	PS	PS/OM	ARO	ARS
1	Texas	48	28	0.58	104.3	145.7
2	Alabama	149	27	0.18	200.8	381.4
3	Florida	147	23	0.16	196.1	281.3
4	Ohio State	85	26	0.31	225.9	362.1
5	Stanford	109	22	0.20	287.3	342.7
6	SouthCarolina	123	25	0.20	279.7	383.0
7	Auburn	160	20	0.13	199.4	283.9
8	Florida State	118	19	0.16	204.7	315.7
9	Michigan	130	25	0.19	257.5	426.0
10	Georgia	102	19	0.19	217.2	343.3
11	Miami (FL)	152	33	0.22	311.1	659.9
12	Oklahoma	86	25	0.29	232.9	445.4
13	Notre Dame	131	16	0.12	189.2	312.6
14	USC	97	15	0.15	174.2	350.7
15	Virginia Tech	89	28	0.31	448.1	696.5
16	LSU	101	23	0.23	246.5	589.6
17	UCLA	148	28	0.19	501.3	643.4
18	Clemson	109	21	0.19	294.6	620.6
19	Texas A&M	89	21	0.24	367.2	627.5
20	Tennessee	199	21	0.11	319.8	650.9
21	Rutgers	104	19	0.18	472.4	606.2
22	Oregon	105	21	0.20	339.5	670.5
23	Mississippi State	142	28	0.20	543.7	828.7
24	Texas Tech	104	27	0.26	488.8	884.5
25	Baylor	96	24	0.25	421.6	746.3
26	Vanderbilt	210	22	0.10	511.9	645.4
27	Nebraska	133	17	0.13	330.8	543.2
28	Mississippi	150	17	0.11	507.1	682.2
29	California	91	17	0.19	361.6	608.3
30	Oklahoma State	83	24	0.29	499.6	813.2
31	South Florida	129	19	0.15	467.7	778.4
32	Missouri	94	18	0.19	477.3	775.4
33	Virginia	120	26	0.22	511.7	1055.2
34	Washington	123	25	0.20	547.0	872.9
35	WestVirginia	191	26	0.14	589.3	976.0
36	TCU	71	24	0.34	484.2	834.1
37	Utah	122	28	0.23	668.3	1008.9
38	Arkansas	171	24	0.14	434.4	919.4
39	North Carolina	145	23	0.16	418.3	935.6

(continued)

Table 11 (continued)

Rank	Team	OM	PS	PS/OM	ARO	ARS
40	Iowa	117	26	0.22	582.1	1072.7
41	Colorado	169	29	0.17	679.2	1070.9
42	Georgia Tech	94	17	0.18	437.3	843.6
43	Pittsburgh	148	18	0.12	626.8	911.7
44	Louisville	164	22	0.13	594.9	1003.7
45	Michigan State	139	18	0.13	445.5	844.4
46	Penn State	85	19	0.22	432.8	927.3
47	Cincinnati	199	30	0.15	744.6	1303.8
48	Wisconsin	65	12	0.18	440.4	724.8
49	Maryland	108	23	0.21	590.9	1198.9
50	North CarolinaState	131	22	0.17	582.1	1212.8
51	Arizona State	164	23	0.14	659.2	1179.6
52	Arizona	188	24	0.13	621.7	1120.9
53	Purdue	150	25	0.17	626.8	1318.6
54	Oregon State	115	24	0.21	771.9	1328.8
55	Southern Miss	123	26	0.21	808.1	1339.3
56	Kentucky	133	29	0.22	753.8	1328.7
57	Northwestern	90	21	0.23	645.5	1205.5
58	Illinois	226	18	0.08	621.3	1300.1
59	Boise State	92	23	0.25	747.0	1244.4
60	BYU	34	16	0.47	809.1	1037.7
61	Minnesota	135	27	0.20	809.8	1571.9
62	Houston	76	24	0.32	939.8	1365.3
63	Washington State	123	26	0.21	867.5	1362.0
64	Wake Forest	129	19	0.15	658.1	1256.8
65	Miami (OH)	68	25	0.37	1127.7	1374.0
66	Duke	112	20	0.18	636.7	1339.5
67	Boston College	127	17	0.13	658.3	1472.5
68	Syracuse	105	22	0.21	706.5	1371.0
69	Florida International	117	31	0.26	1047.2	1606.4
70	Louisiana Tech	77	15	0.19	952.0	1215.1
71	Western Kentucky	69	15	0.22	951.8	1305.1
72	Memphis	165	26	0.16	1013.0	1719.0
73	Florida Atlantic	77	28	0.36	1202.5	1649.1
74	Tulane	68	16	0.24	953.7	1305.2
75	Indiana	151	25	0.17	820.3	1533.9
76	San Diego State	72	21	0.29	954.5	1389.5
77	Marshall	96	22	0.23	1164.8	1737.6
78	Troy	56	34	0.61	1395.6	1662.6
79	Iowa State	112	20	0.18	899.4	1664.5
80	Arkansas State	127	28	0.22	1202.0	1742.5
81	SMU	95	22	0.23	895.5	1598.5

Table 11 (continued)

Rank	Team	OM	PS	PS/OM	ARO	ARS
82	South Alabama	70	23	0.33	1204.5	1758.0
83	East Carolina	72	22	0.31	1057.9	1848.4
84	Temple	104	35	0.34	1380.3	2122.5
85	Louisiana	71	27	0.38	1260.4	1834.6
86	New Mexico	72	21	0.29	1231.4	1819.5
87	UNLV	48	19	0.40	1322.5	1773.3
88	Middle Tennessee	68	19	0.28	1315.5	1897.9
89	Toledo	165	27	0.16	1314.2	2188.0
90	UTEP	52	23	0.44	1473.0	1953.0
91	Kansas	101	17	0.17	926.5	1933.9
92	Kansas State	100	27	0.27	1128.9	2211.9
93	Ohio	85	20	0.24	1290.4	2027.5
94	Hawaii	50	16	0.32	1309.6	1888.7
95	Texas State	39	13	0.33	1326.5	1739.8
96	UAB	68	28	0.41	1527.0	2229.6
97	Rice	61	24	0.39	1467.4	2062.0
98	Connecticut	95	25	0.26	1078.4	2123.4
99	FresnoState	44	15	0.34	1322.5	1920.1
100	Navy	51	22	0.43	1638.0	2089.3
101	Western Michigan	141	29	0.21	1563.2	2450.5
102	Bowling Green	117	21	0.18	1346.2	2165.0
103	Massachusetts	44	26	0.59	1664.0	2142.4
104	Air Force	104	28	0.27	1660.8	2259.0
105	Colorado State	64	20	0.31	1374.9	2093.3
106	New Mexico State	67	24	0.36	1562.1	2234.6
107	San Jose State	52	20	0.38	1451.3	2116.7
108	UCF	98	15	0.15	854.9	2298.1
109	ULM	62	22	0.35	1628.4	2209.9
110	Eastern Michigan	75	24	0.32	1673.8	2338.6
111	Northern Illinois	83	31	0.37	1605.4	2363.0
112	Texas-San Antonio	38	21	0.55	1642.0	2215.7
113	Nevada	61	13	0.21	1195.8	2017.9
114	Utah State	50	9	0.18	1263.2	2099.1
115	Ball State	104	24	0.23	1578.5	2353.5
116	Wyoming	62	17	0.27	1547.4	2326.0
117	Central Michigan	68	23	0.34	1718.6	2382.9
118	Idaho	61	23	0.38	1634.1	2378.8
119	Tulsa	53	20	0.38	1502.2	2427.8
120	North Texas	81	24	0.30	1551.0	2549.7
121	Kent State	76	22	0.29	1625.4	2507.6
122	Buffalo	100	20	0.20	1445.0	2509.8
123	Akron	106	24	0.23	1592.1	2624.6
124	Army	49	22	0.45	1842.2	2624.4

Table 12 NCAA FBS teams listed according to their 2013 PageRank rankings. The remaining columns show the number of offers made by the team (OM), the number of players signed (PS), the average ranking of players offered a scholarship by the team (ARO), and the average ranking of players signed (ARS)

Rank	Team	OM	PS	PS/OM	ARO	ARS
1	Notre Dame	144	24	0.17	171.9	231.5
2	Florida	146	28	0.19	206.5	304.6
3	Georgia	114	32	0.28	241.6	388.3
4	UCLA	124	26	0.21	288.5	311.5
5	Texas A&M	122	32	0.26	281.8	433.2
6	LSU	93	27	0.29	244.4	351.3
7	Ohio State	127	25	0.20	200.9	314.3
8	Michigan	119	27	0.23	232.6	463.7
9	Texas	42	15	0.36	154.1	193.1
10	Alabama	129	25	0.19	242.3	431.2
11	Clemson	144	23	0.16	331.5	532.0
12	Mississippi	254	26	0.10	396.7	551.8
13	Vanderbilt	183	27	0.15	371.1	580.5
14	Oregon	85	19	0.22	287.4	452.4
15	USC	82	12	0.15	129.3	139.0
16	Auburn	137	25	0.18	301.7	511.3
17	South Carolina	132	23	0.17	323.2	517.3
18	Florida State	142	22	0.15	271.2	476.2
19	Washington	124	23	0.19	368.3	582.1
20	Miami (FL)	120	17	0.14	293.0	544.7
21	Tennessee	237	22	0.09	335.5	676.6
22	Nebraska	201	26	0.13	370.5	672.3
23	Oklahoma	148	24	0.16	264.7	700.4
24	TCU	82	22	0.27	393.3	590.3
25	Oklahoma State	93	23	0.25	412.7	733.7
26	North Carolina	119	17	0.14	346.3	595.6
27	Virginia	108	22	0.20	445.1	758.3
28	Penn State	101	17	0.17	381.0	722.4
29	Arkansas	179	23	0.13	436.2	856.1
30	Virginia Tech	59	18	0.31	370.4	648.9
31	Arizona State	196	25	0.13	512.6	940.7
32	Kentucky	147	22	0.15	561.8	846.1
33	West Virginia	275	27	0.10	517.1	950.7
34	Baylor	106	25	0.24	502.8	912.9
35	Stanford	57	12	0.21	234.5	469.5
36	Mississippi State	198	23	0.12	552.1	950.9
37	California	134	26	0.19	494.8	1001.4
38	Missouri	115	20	0.17	505.9	878.8
39	Maryland	114	22	0.19	532.1	1000.9
40	South Florida	172	23	0.13	621.8	966.7

Table 12 (continued)

Rank	Team	OM	PS	PS/OM	ARO	ARS
41	Texas Tech	159	24	0.15	584.3	1078.1
42	Wisconsin	97	20	0.21	498.0	926.8
43	Louisville	127	16	0.13	551.0	920.5
44	Oregon State	129	25	0.19	658.2	1110.2
45	Pittsburgh	116	27	0.23	648.4	1163.6
46	Houston	131	26	0.20	776.3	1206.1
47	Kansas	119	25	0.21	726.8	1270.2
48	Georgia Tech	85	14	0.16	435.4	888.0
49	Rutgers	134	23	0.17	621.3	1217.1
50	Indiana	121	22	0.18	708.8	1197.3
51	Louisiana	70	23	0.33	929.8	1195.7
52	North Carolina State	146	23	0.16	694.2	1248.6
53	Cincinnati	182	20	0.11	789.0	1211.1
54	Utah	161	28	0.17	742.1	1368.3
55	Washington State	162	27	0.17	736.5	1325.9
56	Marshall	87	25	0.29	1008.8	1255.6
57	Illinois	186	25	0.13	712.1	1469.8
58	Purdue	233	24	0.10	642.0	1475.3
59	Michigan State	111	18	0.16	496.5	1191.2
60	Boise State	96	26	0.27	706.9	1239.3
61	Arizona	171	25	0.15	627.0	1476.5
62	Duke	122	20	0.16	616.4	1211.0
63	Iowa	100	22	0.22	749.0	1317.6
64	Southern Miss	101	28	0.28	986.3	1532.3
65	Wake Forest	140	25	0.18	752.9	1479.5
66	Northwestern	90	19	0.21	676.4	1309.9
67	Syracuse	133	19	0.14	798.8	1386.8
68	Western Kentucky	133	33	0.25	1149.0	1640.9
69	Tulane	57	23	0.40	1048.1	1377.7
70	San Diego State	136	23	0.17	957.9	1691.4
71	Minnesota	119	23	0.19	818.6	1488.1
72	UCF	101	18	0.18	747.6	1266.2
73	Iowa State	143	25	0.17	939.9	1635.5
74	BYU	54	30	0.56	1196.8	1552.0
75	Kansas State	92	18	0.20	822.1	1479.3
76	Connecticut	104	25	0.24	911.2	1625.1
77	Navy	94	38	0.40	1490.5	1825.3
78	East Carolina	81	24	0.30	976.9	1680.9
79	Temple	106	21	0.20	1153.0	1850.2
80	Hawaii	111	32	0.29	1187.0	1976.2
81	Toledo	139	22	0.16	1260.9	1858.9
82	Texas State	55	21	0.38	1408.3	1695.0
83	SMU	87	24	0.28	979.4	1757.3

(continued)

Table 12 (continued)

Rank	Team	OM	PS	PS/OM	ARO	ARS
84	UAB	66	28	0.42	1384.1	1851.0
85	Memphis	139	34	0.24	1263.8	2170.9
86	South Alabama	73	22	0.30	1253.7	1895.4
87	Tulsa	71	25	0.35	1179.7	1892.0
88	Bowling Green	145	23	0.16	1350.7	2014.2
89	Middle Tennessee	71	19	0.27	1264.8	1849.6
90	Florida International	118	21	0.18	1204.3	2104.8
91	Fresno State	84	30	0.36	1347.2	2073.7
92	Troy	57	20	0.35	1414.9	2097.7
93	Colorado State	107	24	0.22	1325.9	2120.8
94	Northern Illinois	77	24	0.31	1463.6	2178.9
95	Louisiana Tech	63	21	0.33	1354.2	2066.8
96	Nevada	118	20	0.17	1280.3	2147.2
97	Kent State	67	20	0.30	1419.1	2040.0
98	Colorado	160	19	0.12	835.4	2002.3
99	New Mexico	99	27	0.27	1560.7	2273.3
100	Rice	51	19	0.37	1449.5	2041.9
101	Massachusetts	79	19	0.24	1438.2	2136.2
102	San Jose State	56	16	0.29	1466.9	2166.6
103	Boston College	160	15	0.09	822.5	2277.6
104	UTEP	35	15	0.43	1477.2	2019.7
105	Air Force	119	46	0.39	1896.6	2471.5
106	Miami (OH)	71	25	0.35	1617.2	2227.9
107	Ohio	89	22	0.25	1518.9	2292.9
108	Akron	85	17	0.20	1503.7	2274.0
109	Florida Atlantic	79	26	0.33	1523.6	2394.7
110	Texas San Antonio	59	25	0.42	1677.9	2362.1
111	Georgia State	55	25	0.45	1850.6	2391.6
112	ULM	39	17	0.44	1657.6	2130.5
113	North Texas	68	16	0.24	1318.6	2226.8
114	New Mexico State	53	12	0.23	1489.2	2296.9
115	UNLV	47	16	0.34	1518.7	2262.2
116	Old Dominion	59	20	0.34	1606.6	2393.3
117	Arkansas State	93	18	0.19	1393.7	2524.5
118	Eastern Michigan	72	21	0.29	1788.7	2536.6
119	Central Michigan	63	20	0.32	1784.2	2483.5
120	Western Michigan	102	18	0.18	1569.0	2433.7
121	Ball State	96	21	0.22	1622.2	2542.2
122	Utah State	107	24	0.22	1384.5	2644.3
123	Idaho	61	32	0.52	2127.2	2700.5
124	Army	69	30	0.43	2021.0	2753.9
125	Buffalo	81	18	0.22	1725.3	2772.8
126	Wyoming	78	20	0.26	1768.9	2810.7

References

1. 247sports (formerly Scout). http://scout.com/, redirects to https://247sports.com,. Accessed November 2023.
2. Rivals. http://www.rivals.com/. Accessed November 2023.
3. S. A. Bergman and T. D. Logan. Revenue per quality of college football recruit. *Journal of Sports Economics*, 2020. https://doi.org/10.1177/1527002520921223.
4. K. G. Bigsby, J. W. Ohlmann, and K. Zhao. Online and off the field: Predicting school choice in college football recruiting from social media data. *Decision Analysis*, 14(4), 2017.
5. K. G. Bigsby, J. W. Ohlmann, and K. Zhao. The turf is always greener: Predicting decommitments in college football recruiting using Twitter data. *Decision Support Systems*, 116:1–12, 2019.
6. V. Boginski, S. Butenko, and P. M. Pardalos. Matrix-based methods for college football rankings. In S. Butenko, J. Gil-Lafuente, and P. M. Pardalos, editors, *Economics, Management and Optimization in Sports*, pages 1–13. Springer, 2004.
7. J. Bricker and A. Hanson. The impact of early commitment on games played: Evidence from college football recruiting. *Southern Economic Journal*, 79:971–983, 2013.
8. S. Brin and L. Page. The anatomy of a large-scale hypertextual Web search engine. *Computer Networks*, 30:107–117, 1998.
9. K. Bryan and T. Leise. The $25,000,000,000 eigenvector: The linear algebra behind Google. *SIAM Review*, 48:569–581, 2006.
10. C. A. Caro. College football success: The relationship between recruiting and winning. *International Journal of Sports Science & Coaching*, 7(1):139–152, 2012.
11. A. Daud, F. Muhammad, H. Dawood, and H. Dawood. Ranking cricket teams. *Information Processing & Management*, 51(2):62–73, 2015.
12. J. M. Dumond, A. K. Lynch, and J. Platania. An economic model of the college football recruiting process. *Journal of Sports Economics*, 9:67–87, 2008.
13. ESPN. College football recruiting. https://www.espn.com/college-sports/football/recruiting/. Accessed November 2023.
14. A. Farmer and P. Pecorino. Is the coach paid too much? coaching salaries and the NCAA cartel. *Journal of Economics & Management Strategy*, 19:841–862, 2010.
15. B. Feldman. Recruiting's importance keeps growing. ESPN, https://www.espn.com/college-sports/recruiting/football/columns/story?columnist=feldman_bruce&id=4453458, 2009. Accessed November 2023.
16. T. J. Herda, E. D. Ryan, J. M. DeFreitas, P. B. Costa, A. A. Walter, K. M. Hoge, J. P. Weir, and J. T. Cramer. Can recruiting rankings predict the success of NCAA Division I football teams? an examination of the relationships among Rivals and Scouts recruiting rankings and Jeff Sagarin end-of-season ratings in collegiate football. *Journal of Quantitative Analysis in Sports*, 5:1–11, 2009.
17. K. T. Hornbeak. The Efficacy of College Football Recruiting Services 247sports.com and Rivals.com. Doctor of Education Dissertation, United States Sports Academy, 2023.
18. J. P. Keener. The Perron–Frobenius theorem and the ranking of football teams. *SIAM Review*, 35(1):80–93, 1993.
19. P. Korhonen and M. Syrjänen. Resource allocation based on efficiency analysis. *Management Science*, 50:1134–1144, 2004.
20. G. Langelett. The relationship between recruiting and team performance in division 1a college football. *Journal of Sports Economics*, 4:240–245, 2003.
21. J. A. Mankin, J. Rivas, and J. J. Jewell. The effectiveness of college football recruiting ratings in predicting team success: A longitudinal study. *Research in Business and Economics Journal*, 14:1–19, 2020.
22. K. Menger. A counterpart of Ockham's Razor in pure and applied mathematics: Ontological uses. *Synthese*, 12:415, 1960.

23. M. Navarro. Are college football players being promised big NFL bucks and being short-changed in the classroom and on the field? *Hastings Communications and Entertainment Law Journal*, 42(1):1–16, 2020.
24. F. Radicchi. Who is the best player ever? A complex network analysis of the history of professional tennis. *PLoS One*, 6(2):e17249, 2011.
25. G. Ryan and K. Jerome. Assessing methods for college football rankings. *Journal of Quantitative Analysis in Sports*, 5:1–21, 2009.
26. A. Serenko and M. Dohan. Comparing the expert survey and citation impact journal ranking methods: Example from the field of Artificial Intelligence. *Journal of Informetrics*, 5:629–648, 2011.
27. A. Traugutt, A. Morse, D. Rascher, and B. Fowler. Coaching salary disparity and team performance: Evidence from the Football Bowl Subdivision. *Journal of Applied Business and Economics*, 22(1), 2020.
28. Y. Zhou, R. Wang, Y.-C. Zhang, A. Zeng, and M. Medo. Improving PageRank using sports results modeling. *Knowledge-Based Systems*, 241:108168, 2022.

Applications of Improvements to the Pythagorean Won-Lost Expectation in Optimizing Rosters

Alexander F. Almeida, Kevin Dayaratna, Steven J. Miller, and Andrew K. Yang

Abstract Bill James' Pythagorean formula has for decades done an excellent job estimating a baseball team's winning percentage from very little data: if the average runs scored and allowed are denoted by RS and RA, respectively, there is some γ such that the winning percentage is approximately $\mathrm{RS}^\gamma/(\mathrm{RS}^\gamma + \mathrm{RA}^\gamma)$. One important consequence is to determine the value of different players to the team, as it allows us to estimate how many more wins we would have given a fixed increase in run production. We summarize earlier work on the subject and extend the earlier theoretical model of Miller (who estimated the run distributions as arising from independent Weibull distributions with the same shape parameter; this has been observed to describe the observed run data well). We now model runs scored and allowed as being drawn from independent Weibull distributions where the shape parameter is not necessarily the same and then use the Method of Moments to solve a system of four equations in four unknowns. Doing so yields a predicted winning percentage that is consistently better than earlier models over the last 30 MLB seasons (1994 to 2023). This comes at a small cost as we no longer have a closed-form expression but must evaluate a two-dimensional integral of two Weibull

---

We thank the editors and referees for comments that improved the analysis and exposition.

A. F. Almeida
McDonough School of Business, Georgetown University, Washington, DC, USA
e-mail: afa66@georgetown.edu

K. Dayaratna
Department of Mathematics and Statistics, Georgetown University, Washington, DC, USA
e-mail: kd871@georgetown.edu

S. J. Miller (✉)
Department of Mathematics and Statistics, Williams College, Williamstown, MA, USA
e-mail: sjm1@williams.edu; Steven.Miller.MC.96@aya.yale.edu

A. K. Yang
Emmanuel College, University of Cambridge, Cambridge, UK
e-mail: aky30@cam.ac.uk

M. J. Blondin et al. (eds.), *Artificial Intelligence, Optimization, and Data Sciences in Sports*, Springer Optimization and Its Applications 218,
https://doi.org/10.1007/978-3-031-76047-1_13

distributions and numerically estimate the solutions to the system of equations; as these are trivial to do with simple computational programs, it is well worth adopting this framework and avoiding the issues of implementing the Method of Least Squares or the Method of Maximum Likelihood.

Keywords James' pythagorean Won-Lost formula · Weibull distribution

2020 Mathematics Subject Classification 62P99

1 Introduction

There are two related problems all baseball teams struggle to solve: win games (and championships) and make money. With finite resources, it is essential for teams to be efficient in determining whom to sign and for how much. It is a two-step process to optimally make these decisions. First, teams must determine the net value of a player to their offense/defense. This can be done through metrics such as the "runs created" statistic [1, 7], or more involved methods such as Monte Carlo simulation, which takes into account that players do not exist in a vacuum and one's contributions depends on the rest of the lineup.

To estimate how valuable a player that improves a team's offensive or defensive output is, we need to convert from runs scored and allowed to wins. In other words, how valuable is a given output? For example, see [2, 6, 5, 13, 10, 18, 22, 23, 24] for a sampling of how teams estimate performance and the impact this has on team construction, as well as recent work [8] which shows that a natural extension of this formula leads to an excellent predictor of the probability one team beats another in the playoffs. This conversion has often been done by use of Bill James' Pythagorean Won-Lost formula [16, 25], which for most teams leads to a simple rule of thumb that roughly every 10 net runs created translates to an additional win. It states that a team's winning percentage is well-estimated by

$$\frac{\#\text{Wins}}{\#\text{Games}} \approx \frac{\text{RS}^2}{\text{RA}^2 + \text{RS}^2}, \tag{1}$$

where RS (resp. RA) is the average number of runs scored (allowed) per game.[1] Though it is simple to compute with a standard calculator (or even with pen and paper) when the exponent is 2, simplicity such as this is not needed in the twenty-first century, and one can explore improvements. As teams are trying to optimize wins, revenue or both, the better they can predict the value of a player, the better they

[1] In ongoing work, Cleary, Jeffries, Miller, Miller, Murray, Pasha and Skiera [8] adjust the Pythagorean formula to take into account the two teams playing by rescaling the runs scored and allowed relative to the league average. Thus if we have for the home team RS_h, RA_h and for theaway team RS_a, RA_a and the league average runs scored per game is R, to calculate

can solve these problems. Thus, rather than have an exponent of 2, sabermetricians explored, both numerically and theoretically, and found values of the exponent that do a better job. These values depend on the era and style of play: is it a pitcher's friendly environment, say the deadball era, or is it from a time when offensive production suddenly exploded?

We begin by summarizing earlier work on the subject. Our starting point is Miller's 2007 paper [21], where he showed that expressions of the form (1) are consequences of reasonable models for run production; this advances the subject from experimental observations to a theoretical justification. The model makes several assumptions which range from the clearly false (runs scored and allowed are drawn from continuous and not discrete random variables) to the perhaps needlessly restrictive, perhaps not (specifically, both are modeled from three parameter Weibulls[2] with the same shape parameter; see Appendix "Appendix A: Moments of the Weibull Distribution" for more details on this family of random variables). These assumptions are deliberately chosen to lead to a tractable mathematical model (see Appendix "Appendix B: Deriving the Pythagorean Formula", where we recall those arguments). While it does a very good job fitting the data, we present below a discussion of some previous improvements, followed by our new results and observations.

There are several earlier works worth noting.

- The reason Miller used the three parameter Weibull distribution is that the needed multivariable integral, namely the probability a team scores more runs than it allows, can be done in closed form (when the shape parameter γ is the same for both), leading to (1). Luo and Miller [19] generalized to modeling the run distributions by linear combinations of Weibulls with the same shape parameter; interestingly, there is no significant improvement in predictive power.
- Just as elections can be confidently called before all the results are in, so too can many games. Also in [19] the authors show that if we call games in late innings when one team is up by a lot, and adjust the runs scored and allowed averages for the team accordingly, there is no significant improvement in predictive power. In other words, the "garbage" runs scored or allowed when a team is winning or losing by a lot makes very little difference. Additionally, there was no noticeable gain in taking into account ballpark effects on run production.
- While a simpler formula than (1) is not needed, it is nice to have one that is easier for the average fan to use and understand, allowing a quick ballpark estimate for the value of decisions. This is similar to how many complicated statistics

the probability the home team wins we use for runs scored $\mathrm{RS}_h(\mathrm{RA_a}/\mathrm{R})$ and for runs allowed $\mathrm{RA}_h(\mathrm{RS_a}/\mathrm{R})$; thus if the home team is playing a team that allows fewer runs than the league average, its run production is decreased. Looking at playoff data from 2001 through 2019, this modification predicted the home team should win 80.18 times and lose 68.82, essentially perfectly matching the observed 80 series victories and 69 series losses by the home team!

[2] A random variable X follows a Weibull distribution with parameters α, β, γ if $\mathrm{Prob}(X \in [a, b]) = \int_a^b f(x; \alpha, \beta, \gamma)dx$, where $f(x; \alpha, \beta, \gamma) = (\gamma/\alpha)u^{\gamma-1}\exp(-u^\gamma)$ with $u = (x - \beta)/\alpha$.

are often normalized and expressed in a way that is relatable to the general public. A linear version exists, originally observed by Jones and Tappin [17] and derived in [9, 11] as a consequence of (1) by doing a multivariate Taylor Series expansion. We reproduce that derivation in Appendix "Appendix C: Linearizing Pythagoras", and compare how easy it is to use and how well it predicts to other methods.

We then turn to our main contribution: exploring the potential improved predictive power when we allow more general distributions for runs scored and allowed. Miller's work led to determining the parameters of the Weibull distributions by approximating the observed distribution of runs scored or allowed by either the Method of Least Squares or the Method of Maximum Likelihood; while these are straightforward computations, it is a bit of a pain to code and use (though quite doable these days; unfortunately due to the intricate relationships, there are no simple closed-form solutions for the values that minimize the difference between predicted and observed run distributions).

Our main contribution here is to explore the consequences of no longer requiring a closed-form expression for the integral of the probability the runs scored exceeds the runs allowed; it was that requirement that restricted earlier work to distributions such as the three-parameter Weibull where both had the same shape parameter. In particular, we concentrate on the case when both runs scored and allowed are drawn from three-parameter Weibulls, but we no longer require the shape parameter γ to be the same for each (we do still take β to be $-1/2$, as this has each bin centered about integer scores; thus the area under the curve from -1/2 to 1/2 corresponds to 0 runs, while 1/2 to 1 corresponds to 1 run and so on).[3]

We now have four free shape parameters: $\alpha_{\mathrm{RS}}, \gamma_{\mathrm{RS}}, \alpha_{\mathrm{RA}}, \gamma_{\mathrm{RA}}$. We can numerically determine these by looking at the observed runs scored and allowed data and choosing the values of these parameters such that our continuous distributions have the same mean and variance as the data. While there are simple expressions for the mean and variance of a Weibull in terms of its parameters, the resulting system of equations cannot be solved in closed form, to say nothing about the subsequent problem of evaluating the multivariable integral which is the probability that the runs scored exceed the runs allowed; however, it is trivial with any reasonable modern computational system to immediately obtain excellent numerical approximations to the systems of equation and resulting integral.

We describe how to do this in Sect. 2 and report on the improvement this has in predictive ability by examining the last 30 years of MLB seasons (1994 to 2023). After this analysis, we use our approach to turn to the motivating question for this research: estimating the value of scoring additional runs given how many runs a team is scoring and allowing. Our improvements in predictive power thus translate to better assessments of the worth of players. We then conclude in Sect. 4 with thoughts on future research.

---

[3] The advantage of this choice of β is that the observed runs are never at the boundary of two bins.

2 Runs Scored and Allowed with Differently Shaped Weibulls

Work by Miller [21], and then extended by others both for baseball and other sports, established a statistical model that explicitly yields the Pythagorean formula as a consequence of the assumptions: runs scored and allowed are independent random variables drawn from Weibulls with the same shape parameter. Our contribution is to remove the assumption that the shape parameter of the Weibull, γ, must be the same for both distributions. By introducing two different shape parameters, which we denote γ_{RS} and γ_{RA}, we are able to obtain a better fit to the data and an improvement in predictive power, though at a cost: we no longer have a closed-form expression for the winning percentage.

While of course runs are not drawn from continuous distributions, doing so leads to a tractable model that is quite close, year after year, to observed data. Further, as remarked earlier, by setting the shift parameter β to be $-1/2$ we remove all edge effects from the discreteness of the observed run distributions, with those values now separated as the centers of our bins. Miller introduced the Method of Least Squares, which fits the values of α_{RS}, α_{RA}, and γ by minimizing the squared error between the Weibull distribution and the bins of the observed, discrete run distribution (see [21], Section 3, footnote 2). Miller showed that the Method of Least Squares gave more accurate win predictions in the 2004 American League season than estimating α_{RS}, α_{RA}, and γ by the Method of Maximum Likelihood. We update the Method of Least Squares to now allow γ_{RS} and γ_{RA} to be different, allowing better fits and improved predictive power.

We now introduce the Method of Moments. Instead of finding the values of the parameters that lead to minimizing errors with the observed run histograms, we find the four values by setting the means and variances equal. This leads to a significantly easier method to implement than the earlier works, which proceeded by varying parameters in applications of the Method of Least Squares or the Method of Maximum Likelihood to find the optimal values; now we just numerically approximate the solution to two different systems of two equations with two unknowns and then estimate the resulting two-dimensional integral. The new predictive value is marginally better than the Pythagorean formula with shape parameter 1.83, which is used by Baseball Reference and believed to be the value with the strongest predictive power [3]. Hereafter we refer to the prediction from James' formula with exponent 1.83 as Pythag(1.83).

The Method of Moments often outperforms even the updated Method of Least Squares, despite requiring considerably less data and computation. It narrowly outperforms Pythag(1.83) and hugely outperforms the old Method of Least Squares in [21]. This leap should be expected as a team likely does not score and allow runs under the same shaped distributions. While forcing the two shape parameters to be equal results in easier integrals which can be done in closed form, it loses important information. While this model improves accuracy, it loses closed-form results.

Fig. 1 The changing probabilities of a family of Weibulls with $\alpha = 1$, $\beta = 0$, and $\gamma \in \{1, 1.25, 1.5, 1.75, 2\}$; $\gamma = 1$ corresponds to the exponential distribution, and increasing γ results in the bump moving rightward

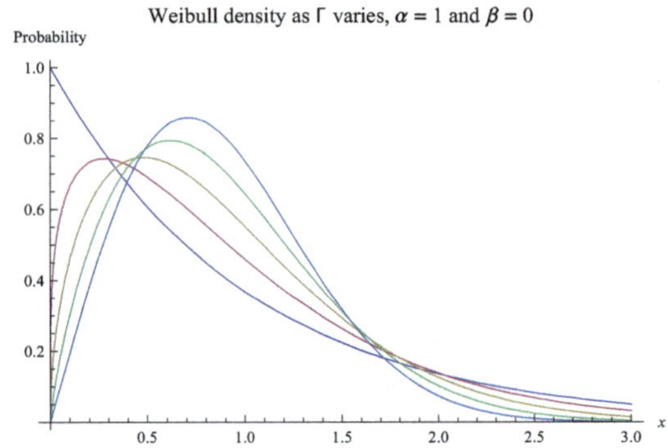

2.1 Method of Moments

As remarked, the probability density function of the Weibull distribution is

$$f(x; \alpha, \beta, \gamma) \ := \ \frac{\gamma}{\alpha}((x - \beta)/\alpha)^{\gamma - 1} e^{-((x-\beta)/\alpha)^{\gamma}} \tag{2}$$

for $x \geq \beta$; α, β, and γ are the three parameters of the distribution. While we are able to model a variety of curves by appropriately choosing values for these parameters, the possibilities are not as extensive as one might think, as α and β just rescale and translate the distribution, respectively; it is only γ that changes the shape.

We illustrate the effect of different choices of γ in Fig. 1 (taken from [9]). As α and β just rescale and translate, without loss of generality we set their values to be 1 and 0, respectively.

We now describe how to use the Method of Moments to estimate the winning percentage. Let the runs scored and runs allowed per game be drawn from independent Weibull distributions with parameters $(\alpha_{\rm RS}, \beta = -1/2, \gamma_{\rm RS})$ and $(\alpha_{\rm RA}, \beta = -1/2, \gamma_{\rm RA})$, respectively. Straightforward integration yields closed-form expressions for the mean $\mu_{\alpha,\beta,\gamma}$ and the variance $\sigma^2_{\alpha,\beta,\gamma}$ of a Weibull distribution in terms of its parameters and the gamma function,[4] $\Gamma(s)$:

$$\Gamma(s) \ := \ \int_0^{\infty} e^{-x} x^{s-1} dx, \quad \text{Re(s)} > 0. \tag{3}$$

After straightforward integration, we find

[4] Though we do not need this result, it is worth noting that the gamma function generalizes the factorial function: $\Gamma(n + 1) = n!$ when n is a nonnegative integer.

$$\begin{aligned} \mu_{\alpha,\beta,\gamma} &= \alpha\Gamma\left(1+\gamma^{-1}\right)+\beta \\ \sigma^2_{\alpha,\beta,\gamma} &= \alpha^2\Gamma\left(1+2\gamma^{-1}\right)-\alpha^2\Gamma\left(1+\gamma^{-1}\right)^2; \end{aligned} \tag{4}$$

for a derivation, see Appendix "Appendix A: Moments of the Weibull Distribution". Note (4) gives us two equations with two unknowns. We thus expect a solution to exist; while we cannot find a closed-form expression for the parameters in terms of the observed mean and variance, we can easily approximate these values.

Let $\widehat{\mu}_{\mathrm{RS}}$ be an estimate for a team's mean runs scored per game, $\widehat{\sigma^2_{\mathrm{RS}}}$ an estimate for the variance in runs scored per game, $\widehat{\mu}_{\mathrm{RA}}$ an estimate for the mean runs allowed per game, and $\widehat{\sigma^2_{\mathrm{RA}}}$ an estimate for the variance in runs allowed per game. In our investigations, these are the sample means and sample variances of a team's runs scored and runs allowed per game over the course of the 2022 season.

We can now solve for α_{RS}, α_{RA}, γ_{RS}, and γ_{RA} in the following system of equations (gradient descent and grid search both work well and efficiently):

$$\begin{aligned} \widehat{\mu}_{\mathrm{RS}} &= \alpha_{\mathrm{RS}}\Gamma\left(1+\gamma_{\mathrm{RS}}^{-1}\right)+\beta \\ \widehat{\sigma^2_{\mathrm{RS}}} &= \alpha^2_{\mathrm{RS}}\Gamma\left(1+2\gamma_{\mathrm{RS}}^{-1}\right)-\alpha^2_{\mathrm{RS}}\Gamma\left(1+\gamma_{\mathrm{RS}}^{-1}\right)^2 \\ \widehat{\mu}_{\mathrm{RA}} &= \alpha_{\mathrm{RA}}\Gamma\left(1+\gamma_{\mathrm{RA}}^{-1}\right)+\beta \\ \widehat{\sigma^2_{\mathrm{RA}}} &= \alpha^2_{\mathrm{RA}}\Gamma\left(1+2\gamma_{\mathrm{RA}}^{-1}\right)-\alpha^2_{\mathrm{RA}}\Gamma\left(1+\gamma_{\mathrm{RA}}^{-1}\right)^2. \end{aligned} \tag{5}$$

Let X and Y be random variables modeling the runs scored and runs allowed per game, respectively, drawn from independent Weibull distributions with parameters $(\alpha_{\mathrm{RS}}, -1/2, \gamma_{\mathrm{RS}})$ and $(\alpha_{\mathrm{RA}}, -1/2, \gamma_{\mathrm{RA}})$. Then the winning percentage is

$$\begin{aligned} \mathrm{Prob}(X>Y) &= \int_{x=\beta}^{\infty}\int_{y=\beta}^{x} f(x;\alpha_{\mathrm{RS}},\beta,\gamma_{\mathrm{RS}})f(y;\alpha_{\mathrm{RA}},\beta,\gamma_{\mathrm{RA}})\mathrm{d}y\,\mathrm{d}x \\ &= \int_{x=0}^{\infty}\frac{\gamma_{\mathrm{RS}}}{\alpha_{\mathrm{RS}}}\left(\frac{x}{\alpha_{\mathrm{RS}}}\right)^{\gamma_{\mathrm{RS}}-1}\exp\left(-\left(\frac{x}{\alpha_{\mathrm{RS}}}\right)^{\gamma_{\mathrm{RS}}}\right) \\ &\quad\cdot\left[\int_{y=0}^{x}\frac{\gamma_{\mathrm{RA}}}{\alpha_{\mathrm{RA}}}\left(\frac{y}{\alpha_{\mathrm{RA}}}\right)^{\gamma_{\mathrm{RA}}-1}\exp\left(-\left(\frac{x}{\alpha_{\mathrm{RA}}}\right)^{\gamma_{\mathrm{RA}}}\right)\mathrm{d}y\right]\mathrm{d}x \\ &= \int_{x=0}^{\infty}\frac{\gamma_{\mathrm{RS}}}{\alpha_{\mathrm{RS}}}\left(\frac{x}{\alpha_{\mathrm{RS}}}\right)^{\gamma_{\mathrm{RS}}-1}\exp\left(-(x/\alpha_{\mathrm{RS}})^{\gamma_{\mathrm{RS}}}\right)\left[1-\exp\left(-(x/\alpha_{\mathrm{RA}})^{\gamma_{\mathrm{RA}}}\right)\right]\mathrm{d}x \\ &= 1-\int_{x=0}^{\infty}\frac{\gamma_{\mathrm{RS}}}{\alpha_{\mathrm{RS}}}\left(\frac{x}{\alpha_{\mathrm{RS}}}\right)^{\gamma_{\mathrm{RS}}-1}\exp\left[-(x/\alpha_{\mathrm{RS}})^{\gamma_{\mathrm{RS}}}-(x/\alpha_{\mathrm{RA}})^{\gamma_{\mathrm{RA}}}\right]\mathrm{d}x. \end{aligned} \tag{6}$$

If the shape exponents are the same, a simple change of variables leads to a closed-form expression for the above (see [9, 21]; for completeness we reproduce this derivation in Appendix "Appendix B: Deriving the Pythagorean Formula"); while we are not so fortunate in this more general setting, the resulting integral can be quickly computed numerically with high accuracy by Riemann sums, or better yet Simpson's rule. Other good numerical methods include quadrature or Monte Carlo methods.

2.2 Analysis

We use the Method of Moments to analyze the 30 teams, which are ordered by the number of overall season wins, from the 2022 season to see how closely our model fits the observed scoring patterns. For each team we compute the sample mean runs scored and allowed per game and the sample variance in runs scored and allowed per game. We solve for the α_{RS}, α_{RA}, γ_{RS}, and γ_{RA} that satisfy (5) and then compute the win percentages by (6).

In Table 1 we find that indeed many teams have large differences between their γ_{RS} and γ_{RA} values. However, every run scored for one team is a run allowed by another, so we expect the league average values of γ_{RS} and γ_{RA} to be similar. Indeed, we find that the league average of γ_{RS} is 1.59, while the league average of γ_{RA} is 1.61.

Across the league in 2022, the mean difference between predicted and observed wins is -0.01, suggesting the Method of Moments is unbiased, but as the difference can take negative values, that alone is no cause for celebration. A statistic more indicative of predictive power to consider is the absolute difference between predicted and observed wins for each team. In Appendix "Appendix C: Linearizing Pythagoras", see Table 2 for a comparison of the absolute differences of different methods in the 2012 and 2022 seasons.

To further justify the merits of the Method of Moments, we repeated this analysis over all 30 seasons since 1994. A common and useful statistic to evaluate estimators is the *mean squared error*, since that accounts for both bias and variance. After finding the mean squared error between predicted and observed wins for each method and team over the last 30 years of data, we find empirical evidence that the Method of Moments outperforms Pythag(1.83) and the Method of Least Squares. We present a boxplot illustrating these results in Fig. 2. The interquartile range of the boxplot clearly exhibits the superiority of the Method of Moments over the last 30 years.

It is clear that the old Method of Least Squares with $\gamma_{\mathrm{RS}} = \gamma_{\mathrm{RA}}$ typically has a larger mean squared error and performs the poorest. Even though it is using the amount of runs observed in *every* game of the season to fit the Weibulls, rather than just the sample first and second moments, we see that forcing $\gamma_{\mathrm{RS}} = \gamma_{\mathrm{RA}}$ leads to worse fits and worse results.

Table 1 Results from the Method of Moments, displaying the observed and predicted number of wins, winning percentage, and difference in games won and predicted for the 2022 season

Team	Obs W	Pred W	Obs %	Pred %	Diff W	γ_{RS}	γ_{RA}
Los Angeles Dodgers	111	113.3	0.685	0.699	−2.3	1.88	1.55
Houston Astros	106	100.4	0.654	0.620	5.6	1.56	1.55
Atlanta Braves	101	98.8	0.623	0.610	2.2	1.80	1.55
New York Mets	101	97.4	0.623	0.601	3.6	1.74	1.51
New York Yankees	99	98.1	0.611	0.605	0.9	1.47	1.70
St. Louis Cardinals	93	91.0	0.574	0.562	2.0	1.51	1.57
Cleveland Guardians	92	85.4	0.568	0.527	6.6	1.60	1.74
Toronto Blue Jays	92	88.5	0.568	0.546	3.5	1.58	1.59
Seattle Mariners	90	87.5	0.556	0.540	2.5	1.70	1.66
San Diego Padres	89	83.6	0.549	0.516	5.4	1.46	1.55
Philadelphia Phillies	87	88.0	0.537	0.543	−1.0	1.58	1.39
Milwaukee Brewers	86	83.9	0.531	0.518	2.1	1.64	1.66
Tampa Bay Rays	86	86.5	0.531	0.534	−0.5	1.70	1.63
Baltimore Orioles	83	79.9	0.512	0.493	3.1	1.59	1.58
Chicago White Sox	81	81.4	0.500	0.503	−0.4	1.64	1.40
San Francisco Giants	81	82.0	0.500	0.506	−1.0	1.58	1.63
Boston Red Sox	78	79.4	0.481	0.490	−1.4	1.60	1.42
Minnesota Twins	78	82.4	0.481	0.509	−4.4	1.64	1.60
Arizona Diamondbacks	74	78.9	0.457	0.487	−4.9	1.68	1.58
Chicago Cubs	74	75.1	0.457	0.464	−1.1	1.45	1.47
Los Angeles Angels	73	78.0	0.451	0.482	−5.0	1.59	1.51
Miami Marlins	69	71.1	0.426	0.439	−2.1	1.47	1.64
Colorado Rockies	68	65.7	0.420	0.406	2.3	1.57	1.76
Texas Rangers	68	76.9	0.420	0.475	−8.9	1.70	1.81
Detroit Tigers	66	63.1	0.407	0.390	2.9	1.57	1.76
Kansas City Royals	65	66.8	0.401	0.413	−1.8	1.57	1.63
Cincinnati Reds	62	65.6	0.383	0.405	−3.6	1.53	1.72
Pittsburgh Pirates	62	63.7	0.383	0.393	−1.7	1.61	1.53
Oakland Athletics	60	61.7	0.370	0.381	−1.7	1.46	1.68
Washington Nationals	55	55.5	0.340	0.342	−0.5	1.44	1.97

We illustrate this in Fig. 3, showing the observed run distribution for the Washington Nationals in the 2022 season—the team with the largest difference between γ_{RS} and γ_{RA}—against the Weibulls produced by the Method of Moments. The runs scored data is heavily packed around 0 to 3 runs, while the runs allowed data is comparatively spread. The flexibility in shape allows the Weibulls to capture this. Compare how well these fit to Fig. 4, with the Weibulls produced by the Method of Least Squares. Even though the Weibulls are fitted as closely as possible to the observed data, the restriction that both distributions have the same γ still results in a slightly weaker fit.

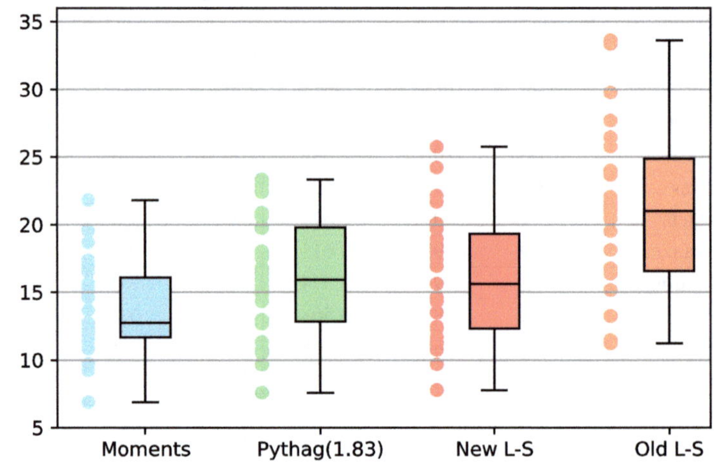

Fig. 2 Scatterplot with boxplot representation for each season of the last 30 years (excl. 1994, 1995, 2020) of the mean squared error in predicted versus observed wins yielded by the four different methods: Moments, Pythag(1.83), "New" Least Squares (γ_{RS}, γ_{RA} free), and "Ol" Least Squares ($\gamma_{\mathrm{RS}} = \gamma_{\mathrm{RA}}$). Tied games were included in the data and counted as 0.5 observed wins for both teams

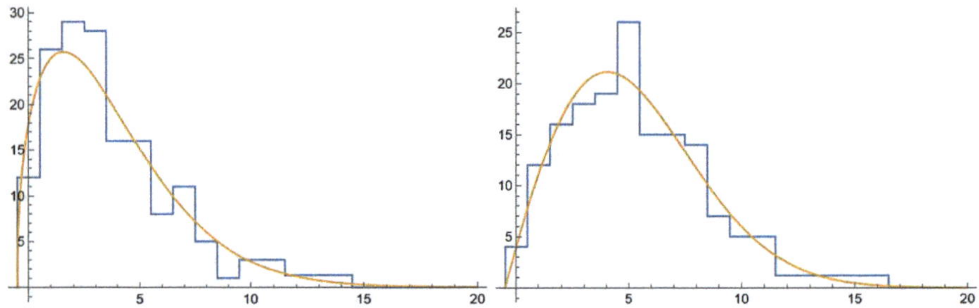

Fig. 3 For the 2022 Washington Nationals, comparison of the Weibulls produced by the *Method of Moments* against the observed distribution of runs scored (left) and runs allowed (right) per game

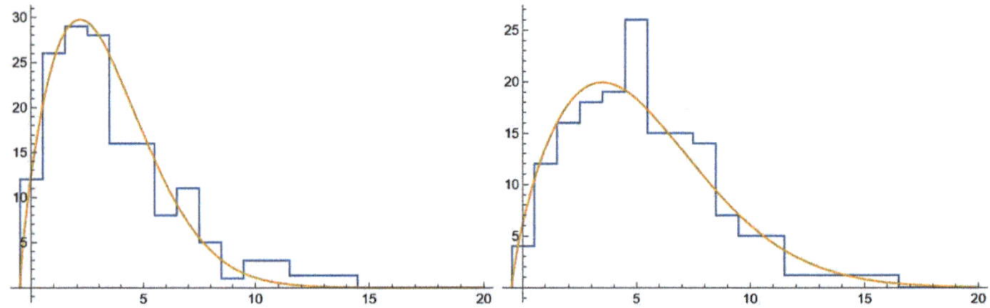

Fig. 4 For the 2022 Washington Nationals, comparison of the Weibulls produced by the *Method of Least Squares* against the observed distribution of runs scored (left) and runs allowed (right) per game

It is interesting that even when we allow γ_{RS}, γ_{RA} to be free, the Method of Least Squares still performs marginally worse than the less computationally intensive Method of Moments and Pythag(1.83). This may due to *overfitting*. By only using the sample means and variances to fit the Weibulls, we have lost information by processing the data, but have enough information to make a reasonable fit and are more robust against data that may be misleading than an approach that uses every bit of information possible.

Finally, observe that Pythag(1.83) is actually using the first moment, since total runs scored/allowed is just the average runs scored/allowed multiplied by a constant number of games played. It does not matter whether we use the average or total runs scored/allowed in the Pythagorean formula. What we have shown is that by introducing just the second moment, our Method of Moments not only outperforms Pythag(1.83), but does so with theoretical backing from our statistical model.

We have made the CSV files of the last 30 years of raw data and the Jupyter Notebook files used to produce these findings publicly available at [26] for easy reproduction. The data was originally sourced from [3].

3 The Pythagorean Formula: Applications

In this section we apply the Pythagorean Formula to a critical economic problem for a baseball team—valuing players. We perform a similar analysis as in Section 5 of [9] to estimate the value a given player brings to their team. Note the answer depends on how many runs the team scores and allows; not surprisingly adding 50 runs to a team that scores few might be significantly more valuable than adding 50 runs to an already productive team.

Specifically, if our team scores x runs and allows y across a season, how much should we pay to sign someone whom we estimate would increase our run production by s? For now we will focus only on how many additional wins they generate, treating all wins equally; this of course is a false assumption, as not all extra wins are created equal. Going from 55 to 65 wins in a season does not alter the fact that the season was a bad one, but going from 85 wins to 95 wins is often the difference between making the playoffs or not!

We proceed with the staple model Pythag(1.83), since this formula is not only the most robust, but also requires very little data—only the total or average runs scored and allowed. Once we estimate the amount of runs a player would contribute to our team, we can immediately compute the change in predicted wins. In Sect. 4, we discuss the possibility of incorporating variance in runs scored and allowed into player analysis.

In Fig. 5, we consider a range of runs scored and allowed per season that a team may currently operate at and plot the additional wins per season that both a player who adds 10 runs a season is expected to give that team and similarly for a player who saves 10 runs a season. We plot in the ballpark of 700 runs scored per season, which is close to the average for most MLB seasons (including 2022, at 693 runs).

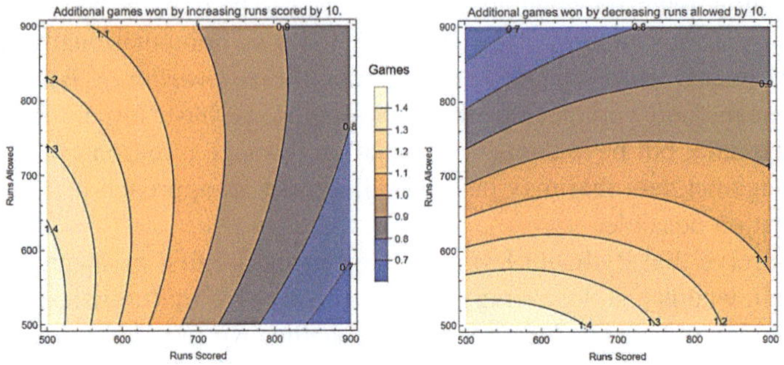

Fig. 5 The predicted number of additional wins under Pythag(1.83) when (left) scoring 10 more per season and (right) preventing 10 more per season

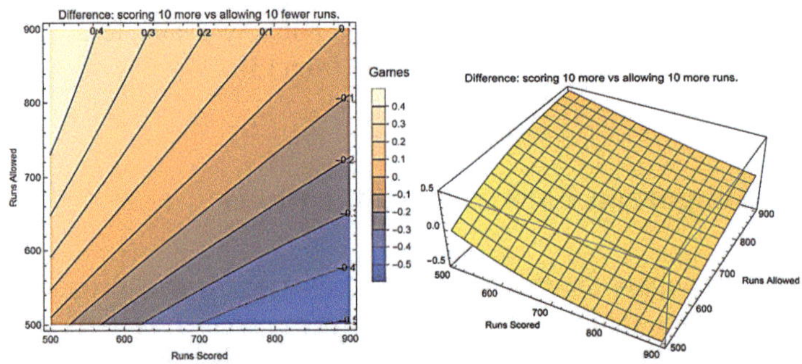

Fig. 6 The difference in the predicted number of additional wins under Pythag(1.83) from scoring 10 more per season versus allowing 10 fewer per season

We deliberately chose $s = 10$, as the common adage goes "every 10 additional runs translates to one more win per season" (see [4]).

The Pythagorean formula allows us to quantify the value of scoring or preventing runs; see Fig. 6, where we plot the difference in wins gained from scoring 10 more runs to wins gained from allowing 10 fewer runs.

While half a win may not sound like much, for the most competitive teams, any edge could be decisive. All teams make hundreds of these decisions every season, and the best teams get them right more often. Across the course of several seasons, both the money saved from better player evaluations, or the wins earned from wiser purchases, could very well provide a winning edge in a sport of extremely fine margins.

3.1 Applications to Other Sports

The Pythagorean formula is flexible, and modelling win percentage from runs/points/goals scored and allowed can be transferred effectively to different sports (see [22]) by choosing a suitable exponent γ. The formula works better for high-scoring sports like basketball than for low-scoring sports like soccer. This is because an average soccer game only has around three goals scored in total [12], meaning very often a single important goal can decide a match, while in basketball a single successful shot holds much less significance over the result of the match. This additional randomness translates to more variance in the performance of Pythagorean expectation.

However, the main issue with soccer in particular is the prevalence of draws, which make up around 20% [12] of results in the English Premier League. A common strategy for mid-table teams is to play very defensively and hardly attempt to score against better opposition, settling for a draw. Taking draws as being worth half a win to each team would work well with the Pythagorean formula, but is not how the soccer league table works—a win is worth *three points*, and a draw is worth just a single point. Thus, a different approach is needed to model the relationship between goals and points, and points are how teams are ranked at the end of the season.

4 Future Possibilities for the Pythagorean Formula

The Pythagorean formula, in particular Pythag(1.83), has proven to be extremely robust. Many attempts have been made to improve upon it, with very little to show for a lot more work. For example, Luo-Miller [19] take into account park effects and see essentially no improvement, while an effort to account for irrelevant runs in blowouts actually led to a worse predictor! Similar adjustments based on pitcher quality and others also do not lead to improvements.

The shape parameter, γ, is equivalent to the exponent of the Pythagorean formula in Miller's work. We have shown that having two different shape parameters, for RS and RA, can be useful for modelling. Future work should explore and see if a Pythagorean formula with flexible exponents for RS and RA could outperform James' original formulation.

Another possibility for further research is to investigate further for which teams and seasons the Method of Moments outperforms Pythag(1.83). We expect an improvement when analyzing teams that score and allow runs with very differently shaped distributions (such as the 2022 Washington Nationals; see Fig. 3).

One can also explore if incorporating the third moment leads to any improvements. Just adding one more equation with the third moment could lead to issues, as we would now have three equations but only two unknowns. A possible resolution would be to let β be a free parameter.

Finally, we discuss the potential of the Method of Moments to be applied to valuing players. In Sect. 3, we assess the value of a player by the runs they add or prevent to a team. However, it is plausible to suggest some players could significantly affect not just the mean runs of a team, but also the variance. For example, a carefree slugger and hardened walker might add a similar number of runs over the course of a season, but certainly one adds more variance than the other. By the Method of Moments, we can now account for that when analyzing how many extra wins such a player might give a team.

Appendix A: Moments of the Weibull Distribution

The Weibull distribution is a continuous, three-parameter distribution, with probability density function:

$$f(x;\alpha,\beta,\gamma) \;=\; \frac{\gamma}{\alpha}((x-\beta)/\alpha)^{\gamma-1}e^{-((x-\beta)/\alpha)^{\gamma}} \tag{7}$$

for $x \geq \beta$. It is a very flexible distribution (see, for example, [20] and the references therein); we saw this in Fig. 1, where it can model many different one-bump distributions by appropriately choosing values of the parameters.

One reason for its popularity is that straightforward integration suffices to obtain closed-form expressions for its moments in terms of its parameters and the gamma function $\Gamma(s)$; for the convenience of the reader we repeat the definition from (3). For $s \in \mathbb{C}$ with the real part of s greater than 0,

$$\Gamma(s) \;:=\; \int_0^{\infty} e^{-u}u^{s-1}\mathrm{d}u \;=\; \int_0^{\infty} e^{-u}u^{s}\frac{\mathrm{d}u}{u}. \tag{8}$$

Denote the kth moment of the Weibull distribution about β by m_k;[5] we can easily find the mean and the variance of our distribution from m_1 and m_2 and thus do the more general calculation below and then specialize. The mean is just

$$\mathbb{E}[X] \;=\; \mathbb{E}[X-\beta+\beta] \;=\; \mathbb{E}[X-\beta]+\mathbb{E}[\beta] \;=\; m_1+\beta,$$

while the variance is

$$\begin{aligned}
\mathrm{Var}(X) &= \mathbb{E}[X^2]-\mathbb{E}[X]^2 \\
&= \mathbb{E}[(X-\beta+\beta)^2]-\mathbb{E}[X-\beta+\beta]^2 \\
&= \mathbb{E}[(X-\beta)^2+2\beta(X-\beta)+\beta^2]-(\mathbb{E}[X-\beta]+\mathbb{E}[\beta])^2 \\
&= \left(\mathbb{E}[(X-\beta)^2]+2\beta\mathbb{E}[X-\beta]+\mathbb{E}[\beta^2]\right)-(m_1+\beta)^2 \\
&= \left(m_2+2\beta m_1+\beta^2\right)-\left(m_1^2+2\beta m_1+\beta^2\right) \;=\; m_2-m_1^2.
\end{aligned}$$

[5] For X a random variable and $\beta \in \mathbb{R}$, the kth moment around β is $\mathbb{E}[(X-\beta)^k]$; thus if X has density p, then $m_k = \int_{-\infty}^{\infty}(x-\beta)^k p(x)\mathrm{d}x$.

We have

$$\begin{aligned} m_k &= \int_{\beta}^{\infty} (x-\beta)^k \cdot \frac{\gamma}{\alpha}\left(\frac{x-\beta}{\alpha}\right)^{\gamma-1} e^{-((x-\beta)/\alpha)^{\gamma}} \mathrm{d}x \\ &= \int_{\beta}^{\infty} \alpha^k \left(\frac{x-\beta}{\alpha}\right)^k \cdot \frac{\gamma}{\alpha}\left(\frac{x-\beta}{\alpha}\right)^{\gamma-1} e^{-((x-\beta)/\alpha)^{\gamma}} \mathrm{d}x. \end{aligned}$$

Substituting $u = \left(\frac{x-\beta}{\alpha}\right)^{\gamma}$, $\mathrm{d}u = \frac{\gamma}{\alpha}\left(\frac{x-\beta}{\alpha}\right)^{\gamma-1} \mathrm{d}x$, we obtain

$$\begin{aligned} m_k &= \int_0^{\infty} \alpha^k u^{k\gamma^{-1}} \cdot e^{-u} \mathrm{d}u \\ &= \alpha^k \int_0^{\infty} e^{-u} u^{1+k\gamma^{-1}} \frac{\mathrm{d}u}{u} \\ &= \alpha^k \, \Gamma(1+k\gamma^{-1}) \end{aligned} \tag{9}$$

by the definition of the gamma function.

Denoting the mean by $\mu_{\alpha,\beta,\gamma}$ and the variance by $\sigma^2_{\alpha,\beta,\gamma}$, we find

$$\begin{aligned} \mu_{\alpha,\beta,\gamma} &= \alpha\Gamma\left(1+\gamma^{-1}\right)+\beta \\ \sigma^2_{\alpha,\beta,\gamma} &= \alpha^2\Gamma\left(1+2\gamma^{-1}\right)-\alpha^2\Gamma\left(1+\gamma^{-1}\right)^2. \end{aligned}$$

Appendix B: Deriving the Pythagorean Formula

For completeness, we reproduce the argument of how James' Pythagorean prediction is a consequence of the assumptions that runs scored and allowed are independently drawn from Weibull distributions with the same parameter; see [9, 21].

Let X and Y be independent random variables with Weibull distributions $(\alpha_{\mathrm{RS}}, \beta, \gamma)$ and $(\alpha_{\mathrm{RA}}, \beta, \gamma)$, respectively, where X is the number of runs scored and Y the number of runs allowed per game. We wish to choose our parameters such that the means of our Weibull match the observed average runs scored and allowed, which we denote by RS and RA, respectively.

We use (9) and find

$$\alpha_{\mathrm{RS}} = \frac{\mathrm{RS}-\beta}{\Gamma(1+\gamma^{-1})}, \qquad \alpha_{\mathrm{RA}} = \frac{\mathrm{RA}-\beta}{\Gamma(1+\gamma^{-1})}. \tag{10}$$

The winning percentage is thus reduced to determining the probability that X exceeds Y:

$$\begin{aligned}
\text{Prob}(X > Y) &= \int_{x=\beta}^{\infty} \int_{y=\beta}^{x} f(x; \alpha_{\text{RS}}, \beta, \gamma) f(y; \alpha_{\text{RA}}, \beta, \gamma) \mathrm{d}y \, \mathrm{d}x \\
&= \int_{x=0}^{\infty} \frac{\gamma}{\alpha_{\text{RS}}} \left(\frac{x}{\alpha_{RS}}\right)^{\gamma-1} e^{-(x/\alpha_{\text{RS}})^{\gamma}} \left[\int_{y=0}^{x} \frac{\gamma}{\alpha_{\text{RA}}} \left(\frac{y}{\alpha_{\text{RA}}}\right)^{\gamma-1} e^{-(y/\alpha_{\text{RA}})^{\gamma}} \mathrm{d}y\right] \mathrm{d}x \\
&= \int_{x=0}^{\infty} \frac{\gamma}{\alpha_{\text{RS}}} \left(\frac{x}{\alpha_{RS}}\right)^{\gamma-1} e^{-(x/\alpha_{\text{RS}})^{\gamma}} \left[1 - e^{-(x/\alpha_{\text{RA}})^{\gamma}}\right] \mathrm{d}x \\
&= 1 - \int_{x=0}^{\infty} \frac{\gamma}{\alpha_{\text{RS}}} \left(\frac{x}{\alpha_{RS}}\right)^{\gamma-1} e^{-(x/\alpha)^{\gamma}} \mathrm{d}x, \qquad (11)
\end{aligned}$$

letting

$$\frac{1}{\alpha^{\gamma}} = \frac{1}{\alpha_{\text{RS}}^{\gamma}} + \frac{1}{\alpha_{\text{RA}}^{\gamma}} = \frac{\alpha_{\text{RS}}^{\gamma} + \alpha_{\text{RA}}^{\gamma}}{\alpha_{\text{RS}}^{\gamma} \alpha_{\text{RA}}^{\gamma}}. \qquad (12)$$

Note we have reduced the problem to integrating a new Weibull with scale parameter[6] α. Continuing, we have

$$\begin{aligned}
\text{Prob}(X > Y) &= 1 - \frac{\alpha^{\gamma}}{\alpha_{\text{RS}}^{\gamma}} \int_{0}^{\infty} \frac{\gamma}{\alpha} \left(\frac{x}{\alpha}\right)^{\gamma-1} e^{(x/\alpha)^{\gamma}} \mathrm{d}x \\
&= 1 - \frac{\alpha^{\gamma}}{\alpha_{\text{RS}}^{\gamma}} \\
&= 1 - \frac{1}{\alpha_{\text{RS}}^{\gamma}} \frac{\alpha_{\text{RS}}^{\gamma} \alpha_{\text{RA}}^{\gamma}}{\alpha_{\text{RS}}^{\gamma} + \alpha_{\text{RA}}^{\gamma}} \\
&= \frac{\alpha_{\text{RS}}^{\gamma}}{\alpha_{\text{RS}}^{\gamma} + \alpha_{\text{RA}}^{\gamma}}. \qquad (13)
\end{aligned}$$

Substituting in the relations for α_{RS} and α_{RA} from (10) gives

$$\text{Prob}(X > Y) = \frac{(\text{RS} - \beta)^{\gamma}}{(\text{RS} - \beta)^{\gamma} + (\text{RA} - \beta)^{\gamma}}, \qquad (14)$$

[6] We often see similar expressions of how items combine; for example, in physics such combinations arise in center of mass calculations, or in adding resistors in parallel.

returning the Pythagorean formula when $\beta = 0$ (which makes sense theoretically, as this is the minimum number of runs a team can score; as remarked above we often take $\beta = -1/2$ for binning purposes).

Appendix C: Linearizing Pythagoras

The Pythagorean won-lost formula (see (1)) was initially suggested by Bill James [16] in the early 1980s. James originally proposed γ to be 2 due to its ease of use, leading to the "Pythagorean" name. As remarked in the introduction, decades later in 2007 Miller [21] offered the first statistical verification of the formula. By presuming that runs scored and runs allowed can be expressed as statistically independent Weibull distributions, he found that the probability of runs scored exceeding runs allowed yields Bill James' formula. Additionally, he found γ to be approximately 1.82 by fitting the Weibull distributions to observed run production. Five years later, Dayaratna and Miller [11] derived a linear predictor for MLB teams' winning percentage by taking a first-order approximation of Bill James' formula. They found the first-order, multivariate Taylor series expansion of James' formula:

$$\frac{\#\text{Wins}}{\#\text{Games}} \approx 0.500 + \frac{\gamma}{4 \cdot \text{R}_{\text{ave}}}(\text{RS} - \text{RA}), \tag{C 1}$$

where R_{ave} is equal to the league-wide average runs scored over the course of a particular season. In doing so, they provided a justification for the simple linear predictor put forth by Jones and Tappin [17], where the winning percentage is $0.500 + \beta(\text{RS} - \text{RA})$, and suggested that β should be approximately $\gamma/(4 \cdot \text{R}_{\text{ave}})$, which is born out from seasonal data.

As this formula is easy to use and allows a quick estimate of the worth of increased run production or run prevention, we summarize its derivation. Our starting point is the second-order Taylor series expansion of a function $f(x, y)$ about the point (a, b):

$$\begin{aligned} f(x,y) &= f(a,b) + \left.\frac{\partial f}{\partial x}\right|_{(a,b)}(x-a) + \left.\frac{\partial f}{\partial y}\right|_{(a,b)}(y-b) + \frac{1}{2}\left.\frac{\partial^2 f}{\partial x^2}\right|_{(a,b)}(x-a)^2 \\ &\quad + \left.\frac{\partial^2 f}{\partial x \partial y}\right|_{(a,b)}(x-a)(y-b) + \frac{1}{2}\left.\frac{\partial^2 f}{\partial y^2}\right|_{(a,b)}(y-b)^2 \\ &\quad + \text{higher-order terms.} \end{aligned}$$

Here, the higher-order terms involve products of $(x - a)$ and $(y - b)$ to the third and higher powers and thus are much smaller than the other terms when x is close to a and y is close to b. A common technique in calculus is to replace a complicated function with a linear approximation, namely, the tangent line in one dimension or the tangent plane in two; for us this means keeping just the constant and linear terms:

$$f(x,y) \approx f(a,b) + \left.\frac{\partial f}{\partial x}\right|_{(a,b)}(x-a) + \left.\frac{\partial f}{\partial y}\right|_{(a,b)}(y-b).$$

Letting R_{ave} denote the average number of runs scored in the league, we apply the above to James' Pythagorean estimate

$$f(x,y) = \frac{x^{\gamma}}{x^{\gamma} + y^{\gamma}}$$

and expand about the point $(a, b) = (R_{ave}, R_{ave})$. Taking $x = RS$ and $y = RA$ yields

$$\begin{aligned}
f(R_{ave}, R_{ave}) &= 0.500\\
\frac{\partial f}{\partial x} = \frac{\gamma x^{\gamma-1} y^{\gamma}}{(x^{\gamma}+y^{\gamma})^2} &\Rightarrow \left.\frac{\partial f}{\partial x}\right|_{(R_{ave},R_{ave})} = \frac{\gamma}{4 \cdot R_{ave}}\\
\frac{\partial f}{\partial y} = -\frac{\gamma x^{\gamma} y^{\gamma-1}}{(x^{\gamma}+y^{\gamma})^2} &\Rightarrow \left.\frac{\partial f}{\partial y}\right|_{(R_{ave},R_{ave})} = -\frac{\gamma}{4 \cdot R_{ave}}.
\end{aligned}$$

Noting that the predicted winning percentage is $f(RS, RA)$, we see that the first-order, multivariate Taylor series expansion about (RS, RA) implies

$$\begin{aligned}
\text{Winning percentage} &\approx 0.500 + \frac{\gamma}{4 \cdot R_{ave}}(RS - R_{ave}) - \frac{\gamma}{4 \cdot R_{ave}}(RA - R_{ave})\\
&= 0.500 + \frac{\gamma}{4 \cdot R_{ave}}(RS - RA).
\end{aligned}$$

The slope coefficient $\gamma/(4 \cdot R_{ave})$ can be easily computed using standard linear regression techniques. Thus, the value of γ can be directly estimated by multiplying the slope coefficient by $4 \cdot R_{ave}$. Note of course that this analysis crucially depends on the shape of James' Pythagorean predictor; a different function would have different partial derivatives, leading to another estimator. For other candidates, see the work of Hammond, Johnson, and Miller [14].

With Bill James's original formula, the use of squared powers renders expected won-lost percentages easy to compute on a calculator. Although improvements to statistical computing in the decades since have certainly made dealing with powers of γ (including estimates around 1.8 as estimated in Miller [21]) even easier, it is nevertheless useful to have good approximations that are quick and easy to use and give a "ballpark" sense of what is going on. The linear approximation presented above does precisely this by offering a much simpler method suitable for a nontechnical audience and can be easily implemented in commonly used programs such as Microsoft Excel or Google Sheets.

Table 2 Comparing the mean, standard deviation, and median of the absolute difference between the observed wins and the predicted wins by different methods in the 2022 and 2012 MLB seasons

Method	'22 avg	'12 avg	'22 standard deviation	'12 standard deviation	'22 median	'12 median
Moments	2.84	3.10	2.02	2.28	2.23	2.56
Pythag(1.83)	2.63	2.94	1.99	2.43	2.18	2.19
Least Squares with $\gamma_{RS} = \gamma_{RA}$	2.47	3.66	2.30	2.79	1.63	2.80
Least Squares with γ_{RS}, γ_{RA} free	2.72	3.42	2.01	2.32	2.08	3.36
Linear Predictor	2.56	2.76	3.21	3.76	2.30	1.70

Using data from the 2012 and 2022 MLB seasons, we estimate win percentage (WP) by the linear regression model: $\text{WP} = 0.500 + \beta(\text{RS} - \text{RA})$. We find the slope β, for 2012 and 2022 to be 0.11236 and 0.09835, respectively. From here, we can simply use Dayaratna and Miller's [11] first-order approximation to estimate γ. We find γ to be approximately 1.94 for the 2012 season and 1.69 for the 2022 season by multiplying each season's slope by $4 \cdot \text{R}_{\text{ave}}$.

Assuming a significance level of $\alpha = 0.05$, the slope β for both the 2012 and 2022 seasons are highly significant with $p < 0.01$. Even after adjusting for multiple comparisons via Bonferroni corrections to a threshold of $\alpha = 0.025$, both slopes are still significant. Additionally, the high R-squared statistics indicate the model's strong fit to the data given; in 2012, 89.81% and, in 2022, 94.82% of the variability in winning percentage can be explained by the average run differential per game (RS − RA). Finding a 95% confidence interval for the slope β is useful because we can then calculate an interval for γ by multiplying by $4 \cdot \text{R}_{\text{ave}}$. Using a t-distribution with 28 degrees of freedom, we get a confidence interval of (0.08944, 0.10725) for the 2012 season's slope and (0.09770, 0.12701) for the 2022 season's slope. Translating these intervals into γ yields confidence intervals of approximately (1.69, 2.20) for the 2012 season's γ and (1.53, 1.84) for the 2022 season's γ.

Using our linear estimate of γ for each respective season as the exponent in the Pythagorean won-lost formula, $\text{RS}^{\gamma}/(\text{RS}^{\gamma} + \text{RA}^{\gamma})$, we can compare our linear predictor's accuracy to other methods. Table 2 above shows the mean, standard deviation, and median of the absolute difference between the observed wins and predicted wins. The linear model's low mean and median absolute difference are on par, if not marginally better than other methods. However, it is important to note the higher variability in the linear model's predictions, as shown by its notably larger standard deviation of absolute differences. As a result, the linear approximation has less predictive power than other methods, but remains a reasonable option for the average fan due to its ease of use and understanding.

References

1. J. Albert, *A Breakdown of a Batter's Plate Appearance—Four Hitting Rates*, By the Numbers **16** (2006), no. 1, 23–29.
2. J. Adler, *Baseball Hacks: Tips & Tools for Analyzing and Winning with Statistics*, O'Reilly Media, 2006
3. S. Foreman, https://www.baseball-reference.com.
4. P. Birnbaum, *Sabermetric Research: Saturday, April 24, 2010*, see http://blog.philbirnbaum.com/2010/04/marginal-value-of-win-in-baseball.html.
5. J. C. Bradbury, *The Baseball Economist: The Real Game Exposed*, Dutton, 2007. (page 181: Step 1: What is a win worth?)
6. B. Baumer and A. ZImbalist, *The Sabermetric Revolution: Assessing the Growth of Analytics in Baseball*, University of Pennsylvania Press, 2014. (pages 39, 57, 72, 76, 168: Pythagorean formula)
7. F. M. Chimkin, *Another Look at Runs Created*, Baseball Research Journal (2003), https://sabr.org/journal/article/another-look-at-runs-created.
8. R. Cleary, J. Jeffries, C. Miller, S. J. Miller, J. Murray, S. Palma and N. Skiera, *Adjusting James' Pythagorean Theorem for Head-to-Head Matchups*, preprint.
9. T. Corcoran, J. Gossels, V. Luo, S. J. Miller and J. Porfilio, *Pythagoras at the Bat*, in Social Networks and the Economics of Sports (edited by Panos M. Pardalos and Victor Zamaraev), Springer-Verlag, 2014, pages 89–114.
10. D. Decatur, *Behind-the-Scenes Baseball*, ACTA Sports, 2006.
11. K. Dayaratna and S. J. Miller, *First Order Approximations of the Pythagorean Won-Loss Formula for Predicting MLB Teams Winning Percentages*, By The Numbers—The Newsletter of the SABR Statistical Analysis Committee **22** (2012), no 1, 15–19.
12. J. Tatsubana, https://footystats.org/england/premier-league/draws, https://footystats.org/england/premier-league/average-total-goals-table.
13. S. Goldman (editor), *Extra Innings: More Baseball Between the Numbers from the Team at Baseball Prospectus*, Basic Books, 2012. (pages 278, 337, 359: are relievers being used properly, hitting vs fielding)
14. C. N. B. Hammond, W. P. Johnson and S. J. Miller, *The James Function*, Mathematics Magazine **88** (2015) 54–71.
15. H. Hundel, *Derivation of James' Pythagorean Formula*, 2003; see https://groups.google.com/forum/#!topic/rec.puzzles/O-DmrUljHds.
16. B. James, *1981 Baseball Abstract*, self-published, Lawrence, KS, 1981.
17. M. Jones and L. Tappin, *The Pythagorean Theorem of Baseball and Alternative Models*, The UMAP Journal **26** (2005), no. 2, 12 pages. https://www.comap.com/membership/member-resources/item/the-pythagorean-theorem-of-baseball-and-alternative-models-umap.
18. J. Keri (editor), *Baseball Between the Numbers: Why Everything You Know About the Game Is Wrong*, Basic Books, 2006. (page 145, 279, 318, 362, 380, 402, 409: Pythagorean formula and record)
19. V. Luo and S. J. Miller, *Relieving and Readjusting Pythagoras*, By The Numbers—The Newsletter of the SABR Statistical Analysis Committee **25** (2015), no. 1, 5–14.
20. B. McShane, M. Adrian, E. T. Bradlow, P. S. Fader, *Count Models Based on Weibull Interarrival Times*, Journal of Business & Economic Statistics **26** (2008), no 3, 369–378.
21. S. J. Miller, *A derivation of the Pythagorean Won-Loss Formula in baseball*, Chance Magazine **20** (2007), no. 1, 40–48 (an abridged version appeared in The Newsletter of the SABR Statistical Analysis Committee **16** (February 2006), no. 1, 17–22, and an expanded version is online at http://arxiv.org/pdf/math/0509698).
22. R. B. Minton, *Sports Math: An Introductory Course in the Mathematics of Sports Science and Sports Analytics*, CRC Press, 2017 (page 122: The Pythagorean Cult)
23. D. Perry, *Winners: How Good Baseball Teams Become Great Ones*, Wiley, 2006. (page 224: A matter of luck, or how teams create their luck)

24. R. Vollman (with T. Awad and I Fyffe), *Stat Shot: The Ultimate Guide to Hockey Analytics*, ECW Press, 2016. (page 243: Predicting Future Success)
25. Wikipedia, *Pythagorean Expectation*, http://en.wikipedia.org/wiki/Pythagorean_expectation.
26. A. K. Yang and S. J. Miller, *Code and Data for Applications of Improvements to the Pythagorean Won-Lost Expectation in Optimizing Rosters*, https://web.williams.edu/Mathematics/sjmiller/public_html/math/papers/PythagCodeAndFiles.zip.

MIX
Papier aus verantwortungsvollen Quellen
Paper from responsible sources
FSC® C105338

If you have any concerns about our products,
you can contact us on
ProductSafety@springernature.com

In case Publisher is established outside the EU,
the EU authorized representative is:
Springer Nature Customer Service Center GmbH
Europaplatz 3, 69115 Heidelberg, Germany

Printed by Libri Plureos GmbH
in Hamburg, Germany